Acid-Base Chemistry

Acid-Base Chemistry

Clifford W. Hand

Harry Lyon Blewitt

Department of Chemistry
University of Alabama

Macmillan Publishing Company
New York
Collier Macmillan Publishers
London

Printed in the United States of America

Macmillan Publishing Company
866 Third Avenue, New York, New York 10022

Collier Macmillan Canada, Inc.

Library of Congress Cataloging in Publication Data

Hand, Clifford W.
Acid–base chemistry.
Includes index.
1. Acid–base equilibrium. I. Blewitt, Harry Lyon.
II. Title.
QD477.H36 1986 547'.037 84–20195
ISBN 0-02-349910-9

Printing: 1 2 3 4 5 6 7 8 Year: 6 7 8 9 0 1 2 3 4

ISBN 0-02-349910-9

Preface

First and second year college chemistry courses are necessarily limited in depth of treatment; one such limited area is acid-base chemistry. Typically, first year students are rushed through the Brφnsted-Lowry treatment and given a definition, but little more, of Lewis acids and bases. Yet these topics are of central importance as students progress in their study of chemistry. At the second year level the beginning student of quantitative analysis is presumed to have mastered subjects such as ionic equilibra, the effect of pH on solubility, and buffers. In our experience the mastery is far from complete. The student of organic chemistry is required to memorize a vast array of reactions, which may appear to be unrelated to one another. The task is much easier for those students with a firm grasp of the unifying principles of charge distribution in molecules, but the first year treatment of charge distribution is usually cursory at best. Our text, *Acid-Base Chemistry*, was written in an attempt to meet these needs. Specifically, the book:

1. is a mix of qualitative and quantitative approaches, well separated in the chapters. In the quantitative sections, no details are omitted from the worked examples; even good students do not like to be told "it can easily be shown that ... ."
2. includes a treatment of errors in pH calculations, as well as methods for exact solutions; these are rarely seen in other textbooks. Rules are developed for deciding when the exact method must be used. Programs written in BASIC are appended for student use.
3. provides many examples of numerical problems, completely worked, as well as end-of-chapter problems suitable for homework or exams.
4. can be used to ease the transition between first and second year chemistry. One example: introductory texts rarely use the concept of formal charge, yet organic texts use formal charge extensively, sometimes without good explanations. Our text includes a section explaining the concept and the rules, together with their significance and usefulness.
5. provides many examples of reaction mechanisms, taken primarily from organic chemistry. In all of these the importance of charge distribution is stressed. It is only a slight exaggeration to say that electron distribution within molecules is what determines the course of chemical reactions. That's what this book is about.
6. provides helpful supplementary material for three groups of students. The concepts and examples are simple enough for the first year student; there is a good review as

well as additional mathematical sophistication for the quantitative analysis student; there is, as has been pointed out by our reviewers, excellent material for the organic student.

Most projects have unsung heroes, and this is no exception. As a live-in editor, Dr. Elli O. Hand was a constant source of detailed critical analysis, suggestions for improvement, and ideas for additional topics. She deserves a great deal of credit. We are grateful also to those people who reviewed the manuscript in detail: Wayne Dunbar, I. C. Hisatsune, Alan Pribula, and especially Charles U. Pittman, Jr., and Joe Murphy. Finally, we would like to thank Nancie, Laura, Larry, Lance, Linda, Lisa, Alex, and Peggy—and each other for having a friendship strong enough to endure the rigors of coauthorship.

C. W. H.
H. L. B.

Contents

1. INTRODUCTION TO THE STUDY OF ACID-BASE CHEMISTRY **1**

1.1 What Is Acid-Base Chemistry? 1
1.2 Chemical Reactions 1
1.3 Bond Cleavage Processes 2
1.4 Energy and Stability 3
1.5 Importance of Acid-Base Chemistry 5

2. GENERAL CONCEPTS OF ACIDS AND BASES **6**

2.1 Historical Background 6
2.2 Properties of Acids and Bases 7
2.3 Arrhenius Concept 8
2.4 Brønsted-Lowry Theory 9
2.5 Strength of Acids and Bases 11
2.6 Nature of the Aqueous Proton 12
2.7 Amphiprotic Substances 13
2.8 Autoinoization 17
2.9 The Leveling Effect of Solvent 17
2.10 Preparation of Inorganic Acids 18
2.11 Preparation of Some Common Bases 22
Problems 26

3. QUANTITATIVE ASPECTS OF ACIDITY **29**

3.1 Introduction and Review of Elementary Concepts 29
3.2 Strong Acids 36
3.3 Weak Acids 37
3.4 Calculations of Acid Ionization 41
3.5 Weak Bases 43
3.6 Basicity 45

3.7 Autoionization 45
3.8 An Approximate Method 48
3.9 Very Dilute Solutions and Very Weak Acids 50
3.10 Summary 53
Problems 54

4. LEWIS ACIDS AND BASES 58
4.1 Introduction 58
4.2 What Is a Lewis Acid? 58
4.3 Strength of Lewis Acids 60
4.4 Strength of Lewis Bases 60
4.5 Types of Lewis Acids 61
4.6 Types of Lewis Bases 63
4.7 Displacement Reactions 64
4.8 Comparison of Lewis and Brønsted-Lowry Views 68
4.9 Usanovich Acid-Base Theory 68
4.10 Some Generalizations 69
Problems 70

5. THE EFFECTS OF STRUCTURE ON ACID-BASE STRENGTH 71
5.1 Introduction 71
5.2 Electronegativity 71
5.3 Resonance 72
5.4 Inductive Effect 78
5.5 Size of the Ion 83
5.6 Magnitude of the Charge 84
5.7 Applications: Hydrogen Halides and Boron Trihalides 84
Problems 87

6. APPLICATIONS AND EXAMPLES IN ORGANIC CHEMISTRY 89
6.1 Introduction 89
6.2 Hydrocarbons: The Effect of Hybridization 89
6.3 Organic Oxygen Acids 92
6.4 Basic Functional Groups 99
6.5 Amides and Imides 104
6.6 Sulfonamides and the Hinsberg Test 105
6.7 Carbanions: Their Use in Organic Synthesis 107
6.8 Alkenes: Their Formation and Basic Character 111
6.9 Benzene as a Base 113
6.10 Acid-Catalyzed Ester Formation 115
6.11 Conclusion 116
Problems 117

7. SALT HYDROLYSIS AND BUFFER ACTION 124

7.1 Hydrolysis 124
7.2 The Relationship Between K_h and K_a 127
7.3 Buffer Solutions 130
7.4 Buffer Calculations 132
7.5 Buffer Capacity 136
7.6 Preparation of Buffers 138
7.7 Exact Treatment 139
7.8 Validity of the Henderson-Hasselbalch Equation 141
7.9 Errors and Validity Ranges 142
Problems 144

8. TITRATIONS 148

8.1 Introduction and Definitions 148
8.2 Calculations 152
8.3 Weak Acid Titration Curves 159
8.4 Indicators 163
8.5 Very Weak Acids 167
Problems 167

9. POLYPROTIC ACIDS AND AMPHIPROTIC SUBSTANCES 169

9.1 Polyprotic Acids 169
9.2 Polyhydric Bases 172
9.3 Exact Calculations 172
9.4 Concentration Versus pH Graphs 176
9.5 Approximate Calculations 180
9.6 Polyprotic Buffer Solutions 183
9.7 Titration and Titration Curves 184
9.8 Amphiprotic Substances 187
9.9 Amphiprotic Salts 188
9.10 Inner Salts 193
Problems 195

10. RELATION EQUILIBRIA 197

10.1 Complexation 197
10.2 Solubility 201
10.3 Simultaneous Equilibria 204
10.4 Successive Approximations 207
10.5 Precipitation of Hydroxides by pH Control 209
10.6 Precipitation of Sulfides by pH Control 211
10.7 Other pH-Dependent Precipitations 212
Problems 213

11. EXPERIMENTAL DETERMINATION OF pK 215

11.1 Weak Acids 215
11.2 Potentiometric Titrations 216
11.3 Spectrophotometric Methods 219
11.4 Other Methods 221
11.5 Strong Acids 222
11.6 Nonaqueous Solvents 222
11.7 Very Weak Acids: Carbon Acids and Carbanions 225
11.8 Gas-Phase Acidities 226
11.9 Cinclusion 226

APPENDIX I Autoionization Constant of Water 227
APPENDIX II Computer Programs in Basic 228
APPENDIX III Ionization Constants (25°C) of Selected Acids and Bases 246
APPENDIX IV Activity Coefficients 255
APPENDIX V Absorbance 258
APPENDIX VI Approximations, Errors, and the Exact Solution 260

INDEX 267

Acid-Base Chemistry

Chapter 1

Introduction to the Study of Acid-Base Chemistry

1.1 WHAT IS ACID-BASE CHEMISTRY?

The area of acid-base chemistry is probably not a completely unfamiliar one. Most people, for instance, have heard that acids "turn litmus red" and know in general that acids and bases neutralize each other. However, there is no single, commonly used definition that adequately describes all the phenomena included under the heading "acid-base chemistry." The three best known approaches are the Arrhenius, the Brønsted-Lowry, and the Lewis theories. These define a *base* as a source of hydroxide ions, a proton acceptor, or an electron-pair donor, respectively, and an *acid* as a source of protons, a proton donor, or an electron-pair acceptor. Each of these theories has some limitations, and each emphasizes a different aspect of acid-base behavior. Of the three, the Lewis concept is the most widely applicable. These concepts are discussed in detail in Chapters 2 and 4.

In the broadest sense, *acid-base chemistry* includes any interaction that takes place between electron-rich species (bases) and electron-poor species (acids). In this context *electron-rich* means a capability for giving up or sharing electrons with another species, and *electron-poor* means a capability for accepting or sharing electrons from another species. Many bases are negative ions or neutral molecules, and many acids are positive ions or neutral species. Free radicals and radical ions[1] are included in this definition, although radical reactions are often discussed separately from acid-base reactions. Most chemical reactions, in fact, occur between electron-rich and electron-poor species, and therefore may be categorized as acid-base reactions. The attraction and interaction between such species are discussed in detail in this text; the principles derived in the discussion give insight into *how* chemical reactions occur and aid in understanding *why* they occur.

1.2 CHEMICAL REACTIONS

In general, when a chemical reaction occurs, one or more chemical bonds are broken and/or one or more new chemical bonds are formed. Both bond breaking and bond formation are important aspects of reactions, although in most reactions these processes

[1] A *free radical* is any atom or molecule containing one or more unpaired electron(s). Examples include highly reactive molecular fragments such as Cl or CH_3, as well as more stable species such as nitric oxide (NO) and molecular oxygen (O_2). A positive or negative ion having unpaired electrons is called a *radical ion*.

do not occur as separate sequential steps. There is instead a gradual, continuous transition between reactants and products; the bond-making and bond-breaking processes occur simultaneously, or nearly so. Such processes may be analyzed in terms of acid-base principles. For example, in reaction (1-1) methyl bromide (an acid) reacts with hydroxide ion (a base) to form methanol and bromide ion.

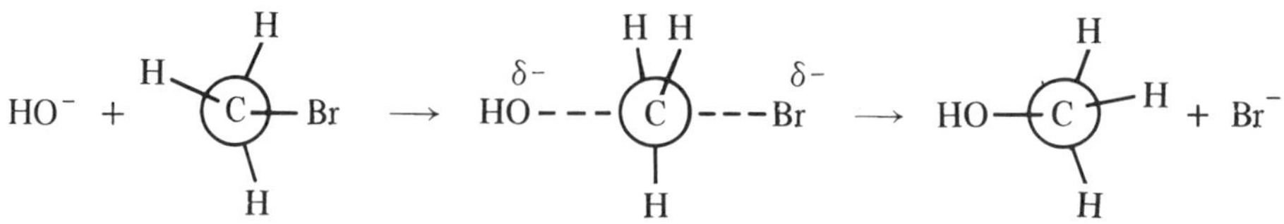

(1-1)

reactants   transition state   products

The dashed lines represent partial bonds: the Br–C bond is breaking while simultaneously the C–O bond is forming. The symbols δ– represent a partial negative charge and imply a continuous transfer of charge from oxygen to bromine. A significant portion of the driving force for this reaction can be attributed to the attraction between an electron-rich species (the oxygen atom in OH^-) and an electron- deficient species (the carbon atom in CH_3Br). If we know the relative base strength of Br^- and OH^-, we can predict the direction in which the reaction proceeds. The reaction goes to the right and not to the left because a strong (OH^-) displaces a weaker base (Br^-) from the carbon atom in CH_3Br.

If a reaction does not involve bond breaking or bond making, but rather the transfer of an electron from one species to another, it is called an *oxidation–reduction* or *redox reaction*. Since electron transfer from one species to another necessarily involves electron-rich and electron-poor species, the principles of acid-base interactions are applicable to redox reactions as well. Many reactions involving catalysts can be discussed with acid-base principles in mind. In short, acid-base concepts can be applied to most reactions; they provide a valuable tool for understanding chemical mechanisms.

1.3 BOND CLEAVAGE PROCESSES

A covalent chemical bond consists of a pair of electrons, located between and shared by two atoms. Such a bond can be broken in three ways, shown in reactions (1-2), (1-3), and (1-4).

$$A{:}B \longrightarrow A^+ + {:}B^- \qquad (1\text{-}2)$$

$$A{:}B \longrightarrow A\bar{:} + B^+ \qquad (1\text{-}3)$$

$$A{:}B \longrightarrow A\cdot + B\cdot \qquad (1\text{-}4)$$

The electrons of the bond remain paired in the first two of these cleavage modes and reside on one or the other of the products. This process is called *heterolytic cleavage*. In reaction (1-4), on the other hand, the bonding electrons become unpaired. One electron goes with each of the product fragments, in a process called *homolytic cleavage*. Homolytic cleavage in a neutral molecule leads to uncharged free radicals, whereas heterolytic cleavage leads to two oppositely charged ionic species.

1.4 ENERGY AND STABILITY

It is usually helpful to think about energy relationships, because the energy absorbed or released during a process is an important factor determining whether or not the process takes place. Most spontaneous processes, including most spontaneous chemical reactions, move in the direction of decreasing energy. In such a reaction, illustrated in Figure 1.1, the products are at a lower energy than the reactants. As the reaction proceeds, the arrangement of the atoms in space changes continuously from the configuration of the reactants to the configuration of the products. The energy of the configuration rises, reaches a maximum, and then decreases again. At the energy maximum, the atoms are

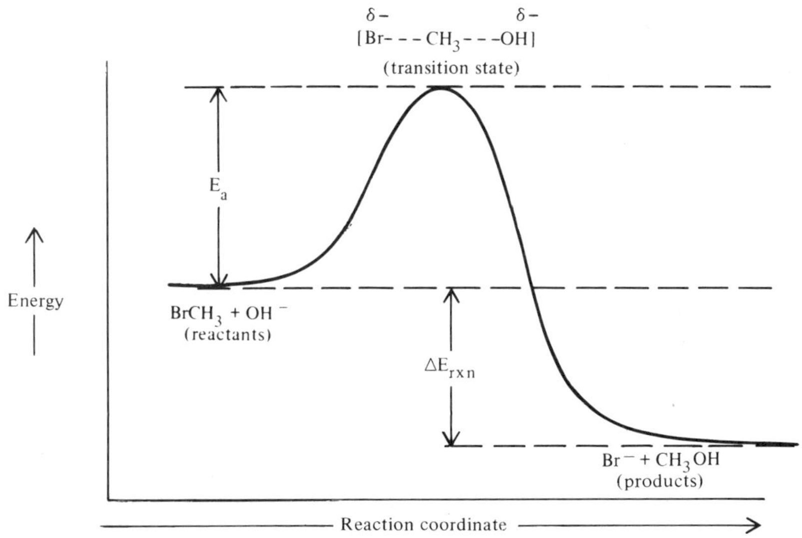

Figure 1.1 Potential energy changes during the basic hydrolysis of methyl bromide. The energy change of the reaction is indicated by ΔE_{rxn}; E_a is the energy of activation. The term "reaction coordinate" refers to the change in geometrical configuration as the seven atoms involved are transformed from the separate $BrCH_3$ and OH^- species to the separate CH_3OH and Br^- species. It does not refer to fraction of reactant species consumed.

clustered in a configuration known as the *transition state.* The energy increase from reactants to transition state, called the *activation energy* (E_a), exerts a controlling influence on the *rate* of the reaction. The energy change from reactants to products, the *reaction energy* (ΔE_{rxn}), helps to determine the *extent* of reaction. When ΔE_{rxn} is a negative quantity, as in Figure 1.1, the reaction is *exothermic* and gives off heat to the environment. In *endothermic* reactions, ΔE_{rxn} is positive, and the reaction absorbs heat from the environment. Both reaction types are common. In an equilibrium reaction ΔE_{rxn} is usually small, although it may be either positive or negative.

The *stability* of a species refers to its energy. A stable species is low in energy, and a species that is more stable than another has lower energy. The stability of a species is closely related to the energy of the process in which the species was formed. A species formed in a highly exothermic reaction is in general more stable than those formed in less exothermic, thermoneutral, or endothermic reactions. The energy of the formation reaction can be divided into two components—the energy inherent in the process, and the energy resulting from interaction of the species with solvent. For example, the energies of the three possible cleavage reactions of hydrogen chloride are given in reactions (1-5), (1-6), and (1-7), in order of increasing energy. Note that the energies given are for

$$HCl(g) \longrightarrow H\cdot(g) + Cl\cdot(g) \qquad \Delta E_{rxn} = 431 \text{ kJ/mol} \tag{1-5}$$

$$HCl(g) \longrightarrow H^+(g) + Cl^-(g) \qquad \Delta E_{rxn} = 1393 \text{ kJ/mol} \tag{1-6}$$

$$HCl(g) \longrightarrow H^-(g) + Cl^+(g) \qquad \Delta E_{rxn} = 1610 \text{ kJ/mol} \tag{1-7}$$

processes taking place in the gas phase. Since there are no solvent interactions in the gas phase, these are the inherent energies of the cleavage reactions. Not one of these reactions takes place spontaneously; all involve an increase in energy. That is, H^- and Cl^+, H^+ and Cl^-, and H• and Cl• are less stable than HCl in the gas phase. Furthermore, the heterolytic reactions require more energy than the homolytic reaction, by a considerable margin: the ionic cleavage products are inherently less stable than the free radicals. When comparisons are made between gas-phase species, it is in general true that free radicals are more stable than the corresponding ions. Finally, we note that the pair (H^+, Cl^-) is more stable than the pair (H^-, Cl^+).

The relationships are different when the species are in solution. The solvent plays a crucial role, and the solvent interaction energy or *solvation energy* is usually the dominant factor. This is especially true for ions dissolved in polar solvents such as water. Solvation energies of ions in water are large, so that aqueous ionic species are quite stable. Thus although HCl shows no tendency to separate into fragments in the gas phase, it does so when dissolved in water. The dissociation reaction (1-8) is spontaneous and exothermic. The species $H^+(aq)$ and $Cl^-(aq)$ are considerably more stable than the undissociated

$$HCl(g) \longrightarrow H^+(aq) + Cl^-(aq) \qquad \Delta E_{rxn} \approx -70 \text{ kJ/mol} \tag{1-8}$$

molecule HCl. In general, a solvated species has lower energy and greater stability than it has when in the pure state.

In reactions that take place in solution, the reactant molecules as well as the product species are stabilized by interaction with the solvent. Whether the reaction is helped or hindered by the solvent hinges on whether products or reactants are stabilized more. If the products are stabilized more than the reactants, the reaction proceeds to a greater extent in solution than it would if there were no solvent present. In reaction (1-5) neither product nor reactant is charged, and therefore neither interacts particularly strongly with solvent. The reaction proceeds (or in this case fails to proceed) to about the same extent in solution as in the gas phase. Since in reaction (1-6) the products are charged, they can be stabilized considerably by solvation. When HCl is dissolved in water, reaction (1-8) proceeds essentially to completion. [Reaction (1-7) does not take place in solution, because H^- and Cl^+ are inherently much less stable than H^+ and Cl^-]. In contrast to the situation in the gas phase, ions are more stable than the corresponding free radicals when dissolved in a polar solvent such as water.

These considerations lead to a useful general rule. Processes involving ions are facilitated in polar solvents, even though the *inherent* stability of ions is less than that of free radicals. In nonpolar solvents no special stabilization is available for charged species, and the greater inherent stability of free radicals facilitates free-radical processes compared to ionic processes.

A cleavage reaction is usually not the final step, however, as the product fragments may undergo further reaction with other species present in solution. Consider the heterolytic cleavage of a molecule A–B, which forms ions according to reaction (1-9):

$$A{-}B \rightleftharpoons A^+ + B^- \tag{1-9}$$

The subsequent reactions of A^+ and B^- to form the final products [reactions (1-10) and (1-11)], should be considered according to acid-base principles. The interaction of A^+ with reagent$_1$ depends on the electron-rich nature, that is, the basicity, of reagent$_1$.

$$A^+ + \text{reagent}_1 \longrightarrow \text{product}_1 \tag{1-10}$$

$$B^- + \text{reagent}_2 \longrightarrow \text{product}_2 \tag{1-11}$$

Similarly, the reaction of B^- depends on the electron-poor nature of reagent$_2$. Acid-base principles are useful tools for understanding both of these processes.

1.5 IMPORTANCE OF ACID-BASE CHEMISTRY

Perhaps the most significant feature of this introduction is the idea that acid-base principles have broad applicability in chemical thinking. Most chemical interactions occur between electron-rich and electron-poor species, and can be discussed in terms of the tendency to acquire or release electrons. It is both possible and useful to extend the reach of acid-base thinking to such topics as inorganic, organic, and biochemical (enzymatic) catalysis; oxidation-reduction reactions; and free-radical reactions.

Chapter 2

General Concepts of Acids and Bases

2.1 HISTORICAL BACKGROUND

Many different kinds of acids and bases have been known for centuries. The name *acid* is derived from the Latin *acidus,* which means sour, and describes the characteristic taste of acids. The terms *alkali* and *alkaline*, which refer to inorganic bases, are derived from the Arabic word *al-qili* (meaning "ashes of a plant"; a related English word is "potash"[1]).

Perhaps the earliest known acidic substance was vinegar, which the ancient Egyptians produced by exposing wine to air. When treated in this way, wine can contain as much as 3 to 6% acid. We now know that acetic acid is formed by the air oxidation of ethyl alcohol, one of the constituents of wine, and that bacteria present in fermenting fruit juices produce an enzyme that catalyzes the oxidation. During the Middle Ages, the alchemists first prepared the strongly acidic substances nitric acid, hydrochloric acid, and sulfuric acid. The acid *aqua regia* was prepared by mixing hydrochloric acid and nitric acid. The Latin name means "royal water," so called because of its ability to dissolve the precious metals platinum, silver, and gold.

The base potash, obtained by slowly leaching wood ashes with water, has been used for centuries in the manufacture of soap. The base calcium oxide (CaO) can be prepared by roasting ground seashells, which consist largely of calcium carbonate, $CaCO_3$ [reaction (2-1)]. Calcium oxide is sometimes referred to as "quicklime" or "unslaked lime" and can be converted to "slaked lime" by the addition of water [reaction (2-2)].

$$CaCO_3 \xrightarrow{\text{heat}} CaO + CO_2 \tag{2-1}$$

$$CaO + H_2O \longrightarrow Ca(OH)_2 \tag{2-2}$$

As the scope of activity in chemistry grew, so did the understanding of these compounds and their properties. It was known even in medieval times that if acidic and basic substances were mixed, a solution was produced which had neither the characteristic

[1] The term *potash* is not precisely defined in the chemical sense; it can refer to potassium oxide, potassium hydroxide, or potassium carbonate. All three of these substances are bases.

properties of acids nor those of bases. Acids and bases were said to *neutralize* each other. This was an important discovery, and attempts were made to classify *all* chemicals as either acids or bases. It was hoped that all of chemistry could be understood in terms of acid-base interactions. Robert Boyle, who lived in the seventeenth century, was among the first to realize that all materials were not either acids or bases; his experiments proved that many materials do not fit into either category.

The fact that all acids exhibit certain similar physical and chemical properties prompted a search for the *acidifying principle,* believed to be a substance common to all acids. An important early theory of acids was proposed by Antoine Lavoisier (1743-1794). He thought that the acidifying principle was oxygen, that is, that all acids must contain oxygen. Hydrochloric acid (HCl), which was known at the time, contains no oxygen and should therefore have disproved the theory. However, the element chlorine had not yet been discovered and Lavoisier thought that the chlorine part of the molecule was really a mixture of elements, one of which was oxygen. Chemists of the time generally supported the oxygen theory; not until much later was it realized that Lavoisier's idea was incorrect.

The successor to Lavoisier's theory was the *hydrogen theory of acids*, in which the acidifying principle was identified as hydrogen rather than oxygen. The work of Humphry Davy in the early nineteenth century presented convincing evidence against Lavoisier's theory. Davy established that hydrochloric acid did not contain oxygen, but only hydrogen and chlorine. The work of Justus von Liebig and Jon Jakob Berzelius also provided support for the hydrogen theory. In the late nineteenth century Svante Arrhenius developed his theory (Section 2.3), which is accepted today as essentially correct, although limited in scope.

2.2 PROPERTIES OF ACIDS AND BASES

Early chemists classified substances as either acids or bases according to the chemical and physical properties of their aqueous solutions.

Acids are substances which:

1. Have a sour taste.
2. React with many metals to liberate a flammable gas (hydrogen), and with limestone and some other minerals to produce a nonflammable gas (carbon dioxide).
3. Change the color of litmus (a vegetable dye) from blue to red.
4. Cause inflammation of the skin, or severe burns if the acid is concentrated.
5. Conduct electricity.
6. React with bases, causing the disappearance of acidic as well as basic properties.

Bases are substances which:

1. Have a slippery or soapy feel.
2. Have a bitter taste.
3. Change the color of litmus from red to blue.
4. Cause inflammation of the skin, or severe burns if the base is concentrated.
5. Conduct electricity.
6. React with acids, causing the disappearance of basic as well as acidic properties.

2.3 ARRHENIUS'S CONCEPT

The first modern approach to acid-base chemistry was offered in 1887 by Svante Arrhenius. His ideas evolved from his doctoral work on the electrical conductance of solutions, particularly of acid solutions. The prevailing belief at the time was that the passage of an electric current creates charged species, called *ions,* in solutions of certain substances. In contrast, Arrhenius suggested that ions exist in solutions even in the absence of an electric current. Specifically, he proposed that certain substances, as they dissolve in water, *dissociate* spontaneously into positive and negative ions. Arrhenius defined an acid as a substance capable of producing hydrogen ions (protons) by dissociation in water solution, according to reaction (2-3). In this reaction A refers to the remainder of the molecule (i.e., to all parts other than hydrogen). We use this notation throughout the text.

$$HA \xrightarrow{H_2O\ (\text{solvent})} H^+ + A^- \tag{2-3}$$

The theory provides a satisfactory explanation of the fact that all acid solutions have similar properties. Acidic properties, according to Arrhenius, are due entirely to H^+, which is present in all acids. He defined a base as a substance capable of producing hydroxide ions by dissociation in water solution. This is illustrated in reaction (2-4), in

$$MOH \xrightarrow{H_2O\ (\text{solvent})} M^+ + OH^- \tag{2-4}$$

which M refers to all parts of the base except OH. Basic properties are attributed to hydroxide ion, which is present in all solutions of bases. *Neutralization,* which causes the disappearance of both acidic and basic properties, results from the reaction of a proton and a hydroxide ion to produce a molecule of water:

$$H^+ + OH^- \longrightarrow H_2O \tag{2-5}$$

Although the Arrhenius acid-base theory is generally useful, there are two features that restrict its use. One limitation arises from the fact that the theory requires dissociation into ions prior to undergoing any characteristic reaction of acids or bases. Since, for example, the gas-phase reaction (2-6) between hydrochloric acid and ammonia takes

$$HCl(g) + NH_3(g) \longrightarrow NH_4Cl(s) \tag{2-6}$$

place without dissociation into ions, it cannot be regarded as a neutralization reaction in the framework of Arrhenius theory. This is in spite of the fact that HCl has all the properties of acids, NH_3 has all the properties of bases, and NH_4Cl has the properties of neither. Furthermore, there are many commonly used solvents in which dissociation does not occur. Arrhenius theory does not permit us to regard as an acid-base reaction any interaction taking place in such a solvent.

A second limitation is that the theory does not explain why some substances, such as $AlCl_3$, have acidic properties and yet lack hydrogen, and some substances, such as NH_3, lack hydroxide and yet have basic properties. As examples, consider the compounds ammonia, sodium amide ($NaNH_2$), and sodium carbonate (Na_2CO_3). All these materials are bases, because their aqueous solutions show all the properties of bases. Yet they do not contain the hydroxide ion. Arrhenius theory circumvents this embarrassment by saying that ammonia reacts with water to produce ammonium hydroxide, which then ionizes to yield OH^- ions [reaction (2-7)]. We are thus encouraged to believe that

$$NH_3 + H_2O \longrightarrow NH_4OH \longrightarrow NH_4^+ + OH^- \qquad (2\text{-}7)$$

NH_4OH is a base and NH_3 is not. The Arrhenius explanation of the basicity of sodium carbonate or sodium amide solutions is even more cumbersome.

Even with these restrictions, however, Arrhenius theory is useful. It is simple and easy to understand, and as long as water is the solvent, it explains a wide variety of acid-base behavior.

2.4 BRØNSTED-LOWRY THEORY

A very important theory of acid-base behavior was introduced independently by Johannes Brønsted (Denmark) and Martin Lowry (England) in 1923. In the Brønsted-Lowry theory, an acid is defined as any species that can *donate* or transfer a proton to another species. This is similar to the Arrhenius definition, but differs in that it is not limited to aqueous solutions and does not require dissociation into ions. A base is defined as any species that can *accept* a proton from another species: this is considerably different from the Arrhenius definition. These definitions are illustrated in reaction (2-8). In this reaction the acid (hydrofluoric) donates or transfers a proton to the base (ammonia).

$$NH_3 + HF \longrightarrow NH_4^+ + F^- \qquad (2\text{-}8)$$

The Brønsted-Lowry concept is broader and more inclusive than the Arrhenius theory. Arrhenius acids and Brønsted-Lowry acids are essentially the same; that is, each is a source of protons. But whereas all Arrhenius bases, such as NaOH, are also Brønsted-Lowry bases [reaction (2-9)], there are Brønsted-Lowry bases that contain no hydroxide

$$\underset{\substack{H^+\\ \text{donor}}}{HCl} + \underset{\substack{H^+\\ \text{acceptor}}}{NaOH} \longrightarrow H_2O + NaCl \qquad (2\text{-}9)$$

and hence cannot be Arrhenius bases. Furthermore, the Brønsted-Lowry definitions do not refer to a solvent. A reaction such as (2-8) has the same Brønsted-Lowry description whether it takes place in water, another solvent, or no solvent at all.

The concept of *conjugate acid-base pairs* is an important aspect of Brønsted-Lowry acid-base theory. When an acid such as HCl donates its proton, the species that is left behind (Cl^-) is called the *conjugate base* of the acid. This relationship is indicated by a diagonal line: HCl/Cl^-. When the base NH_3 accepts a proton, the result (NH_4^+) is called the *conjugate acid* of the base. The same notation, a diagonal line with the acid on the left and the base on the right, is used to denote this relationship: NH_4^+/NH_3. Table 2.1 lists some common conjugate pairs. Note that the difference between the two paired species is a single proton.

The concept of reversible acid-base interactions is also of central importance in Brønsted-Lowry theory. *Reversibility* implies that a species can acquire a proton, then lose it in the reverse of the acquisition process. Reaction (2-8) may be rewritten as follows:

$$\underset{\text{acid}_1}{HF} + \underset{\text{base}_2}{NH_3} \rightleftharpoons \underset{\text{base}_1}{F^-} + \underset{\text{acid}_2}{NH_4^+} \tag{2-10}$$

As this reaction proceeds from left to right it is an ordinary acid-base reaction between the proton donor HF (acid_1) and the proton acceptor NH_3 (base_2). The products of this reaction are ammonium ion, the conjugate acid of ammonia, and fluoride ion, the conjugate base of hydrofluoric acid. The labeling of NH_4^+ as acid_2 and of F^- as base_1 is appropriate, since when the reaction proceeds from right to left, the ammonium ion acts as an acid by donating a proton to the fluoride ion. Likewise the fluoride ion accepts a proton and functions as a base. The reaction is reversible; it is an acid-base reaction in both directions. The species on the left of reaction (2-10) bear a conjugate-pair relationship to those on the right; the conjugate pairs are HF/F^- ($\text{acid}_1/\text{base}_1$) and NH_4^+/NH_3 ($\text{acid}_2/\text{base}_2$). A Brønsted-Lowry acid-base reaction can always be written in the general form

$$\text{acid}_1 + \text{base}_2 \rightleftharpoons \text{acid}_2 + \text{base}_1 \tag{2-11}$$

In a mixture containing all four substances, whether the reaction proceeds from left to

Table 2.1 Common Acid-Base Conjugate Pairs

Conjugate Acid		Conjugate Base	
Name	*Formula*	*Formula*	*Name*
Hydrogen cyanide	HCN	CN^-	Cyanide
Methane	CH_4	CH_3^-	Methyl carbanion
Nitric acid	HNO_3	NO_3^-	Nitrate
Perchloric acid	$HClO_4$	ClO_4^-	Perchlorate
Phenol	C_6H_5OH	$C_6H_5O^-$	Phenoxide
Phosphoric acid (orthophosphoric acid)	H_3PO_4	$H_2PO_4^-$	Dihydrogen phosphate
Dihydrogen phosphate	$H_2PO_4^-$	HPO_4^{2-}	Hydrogen phosphate
Pyridinium		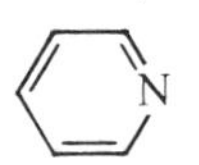	Pyridine
Sulfuric acid	H_2SO_4	HSO_4^-	Hydrogen sulfate (bisulfate)
Water	HOH	OH^-	Hydroxide
Hydroxide	OH^-	O^{2-}	Oxide
Protonated sulfuric acid	$H_3SO_4^+$	H_2SO_4	Sulfuric acid

right or from right to left depends on the relative strengths and concentrations of the acids and bases involved.

2.5 STRENGTH OF ACIDS AND BASES

The Brønsted-Lowry theory defines acid strength in terms of proton-donating tendency. The greater the tendency of a substance to donate a proton, the greater is its acid strength. Consider the reaction between acetic acid and the solvent, water [reaction (2-12)]. This is an acid-base reaction in which the solvent acts as a base by accepting a

$$\underset{\text{acid}_1}{CH_3C\overset{\displaystyle O}{\underset{\displaystyle OH}{\lessgtr}}} + \underset{\text{base}_2}{H_2O} \rightleftharpoons \underset{\text{base}_1}{CH_3C\overset{\displaystyle O}{\underset{\displaystyle O^-}{\lessgtr}}} + \underset{\text{acid}_2}{H_3O^+} \qquad (2\text{-}12)$$

proton from acetic acid. In the process the solvent is converted to its *protonated* form, that is, to its conjugate acid. In the Brønsted-Lowry system acid-base interactions can take place in nonaqueous solvents, so that reaction (2-12) can be written in the more general form (2-13). The strength of acetic acid depends not only on its inherent tendency to donate a proton but also on the ability of the solvent to accept a proton.

$$\underset{\text{acid}_1}{CH_3C\overset{\displaystyle O}{\underset{\displaystyle OH}{\lessgtr}}} + \underset{\text{base}_2}{\text{S (solvent)}} \rightleftharpoons \underset{\text{base}_1}{CH_3C\overset{\displaystyle O}{\underset{\displaystyle O^-}{\lessgtr}}} + \underset{\text{acid}_2}{\text{SH}^+\text{ (protonated solvent)}} \qquad (2\text{-}13)$$

If we wish to discuss relative strengths of different acids, we must compare their proton-donating tendencies in the same solvent. The solvent used for such comparisons, almost always, is water. Thus relative acid strength is most commonly defined as the strength of the tendency of an acid to donate a proton to water. If all or almost all of the acid molecules originally present transfer a proton to water, the resulting concentration of H_3O^+ is large and the acid is said to be *strong*. If only a small number of the acid molecules undergo a proton-donation reaction, while the rest remain unchanged, the resulting concentration of H_3O^+ is small, and the acid is called *weak*. Acetic acid is an example of a weak acid: in ordinary concentrations only about 3% of the acetic acid molecules donate a proton to water.

The fact that acetic acid is a weak acid means that reaction (2-12) has less of a tendency to proceed from left to right than it has to proceed from right to left. The opposing tendencies reach a condition of equilibrium, and since the acid ionizes only slightly, the equilibrium is said to lie to the left. Reactions such as (2-12) are called *ionization reactions* or *ionization equilibria.* The *equilibrium constant* is a quantitative measure of acid strength. For reaction (2-12), and others like it, the equilibrium constant is defined as the molar concentrations of the products divided by the molar concentrations of the reactants.[2] If the acid is weak and the equilibrium lies to the left, the product of terms in the numerator of (2-14) is smaller than the product of terms in the

$$\text{equilibrium constant} = K_{eq} = \frac{[CH_3CO_2^-][H_3O^+]}{[H_2O][CH_3CO_2H]} \qquad (2\text{-}14)$$

denominator, and the equilibrium constant is a small number, much less than 1. The weaker the acid, the smaller its equilibrium constant. Strong acids have large equilibrium constants, usually greater than several hundred.

The same reasoning applies to the relative strength of Brønsted–Lowry bases. As an example consider (2-15), the reaction of methylamine and water. In this reaction the

$$CH_3NH_2 + H_2O \rightleftharpoons CH_3NH_3^+ + OH^- \qquad (2\text{-}15)$$

tendency of methylamine to accept a proton from the solvent (water) is slight, and the equilibrium lies to the left. The equilibrium constant K_{eq}, defined by equation (2-16),

$$K_{eq} = \frac{[CH_3NH_3^+][OH^-]}{[CH_3NH_2][H_2O]} \qquad (2\text{-}16)$$

is small and most of the amino compound is present as unprotonated methylamine (CH_3NH_2) rather than the protonated form, methylammonium ion ($CH_3NH_3^+$).

2.6 NATURE OF THE AQUEOUS PROTON

In the Arrhenius view an acid HA simply dissociates when dissolved, giving a dissolved proton and a dissolved anion:

$$HA \overset{H_2O}{\rightleftharpoons} H^+(aq) + A^-(aq) \qquad (2\text{-}17)$$

However, we now know that a "bare" proton is unstable in aqueous solution; it exists in a form known as the *hydronium ion*, H_3O^+, in which it is covalently bonded to a water molecule.[3]

$$H^+ + H_2O \longrightarrow H_3O^+ \qquad (2\text{-}18)$$

In the Brønsted–Lowry view reactions (2-17) and (2-18) are considered to be a single process [reaction (2-19)] rather than a two-step sequence.

$$HA + H_2O \longrightarrow H_3O^+ + A^- \qquad (2\text{-}19)$$

[2] In Section 3.3 we introduce a modification of this definition, which is more widely used for aqueous acids.

[3] C. A. VanderWerf has estimated the fraction of "bare" or unbonded protons to be 10^{-190}, a number chemically indistinguishable from zero (*Acids, Bases and the Chemistry of the Covalent Bond*, Reinhold, New York, 1961).

The hydronium ion, like other ions that contain hydrogen and oxygen, is hydrogen-bonded to additional water molecules. This association results in a species of molecular formula $H_9O_4^+$.

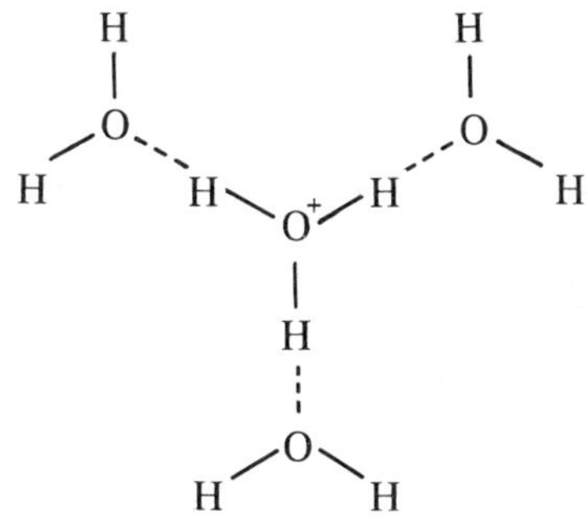

This species in turn, like all ions, interacts with other water molecules via ion-dipole forces. It is loosely associated with several additional neighboring water molecules; these become ordered or oriented in the process, as illustrated below.

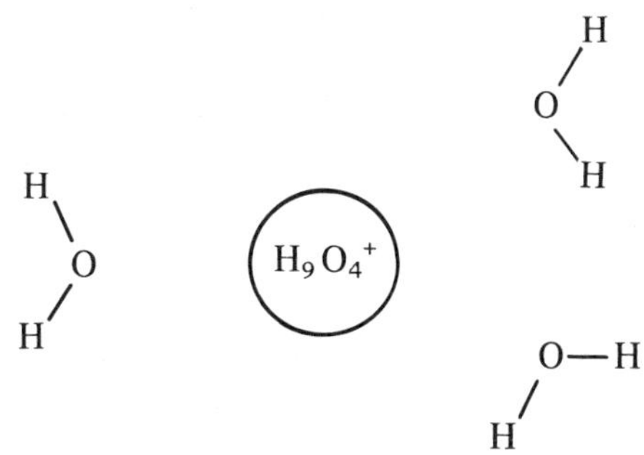

2.7 AMPHIPROTIC SUBSTANCES

There are many substances that can act as either an acid or a base, depending on the conditions. An amphiprotic substance is defined as one that may either donate or accept a proton. For example, in reaction (2-20) water acts as a base by accepting a proton from the acid HCl, whereas in reaction (2-21), water acts as an acid by donating a proton to ammonia.

$$H_2O + HCl \rightleftharpoons H_3O^+ + Cl^- \quad (2\text{-}20)$$

$$H_2O + NH_3 \rightleftharpoons OH^- + NH_4^+ \quad (2\text{-}21)$$

Another common amphiprotic substance is the hydrogen carbonate (bicarbonate) ion, HCO_3^-. In reaction (2-22) HCO_3^- donates a proton and in reaction (2-23) it accepts one.

$$HCO_3^- + NH_3 \rightleftharpoons CO_3^{2-} + NH_4^+ \quad (2\text{-}22)$$

$$HCO_3^- + HCl \rightleftharpoons H_2CO_3 + Cl^- \quad (2\text{-}23)$$

Table 2.2 Some Common Amphiprotic Substances and Their Acid–Base Reactions[a]

Name	*Formula*	*Acid and Base Reactions*
Water	H_2O	$H_2O + NH_3 \rightleftharpoons OH^- + NH_4^+$ $H_2O + HCl \rightleftharpoons H_3O^+ + Cl^-$
Ammonia	NH_3	$NH_3 + CH_3MgBr \rightleftharpoons CH_4 + MgBrNH_2$ $NH_3 + H_2O \rightleftharpoons NH_4^+ + OH^-$
Hydrogen carbonate	HCO_3^-	$HCO_3^- + OH^- \rightleftharpoons CO_3^{2-} + H_2O$ $HCO_3^- + H_2SO_4 \rightleftharpoons H_2CO_3 + HSO_4^-$
Hydrogen phthalate	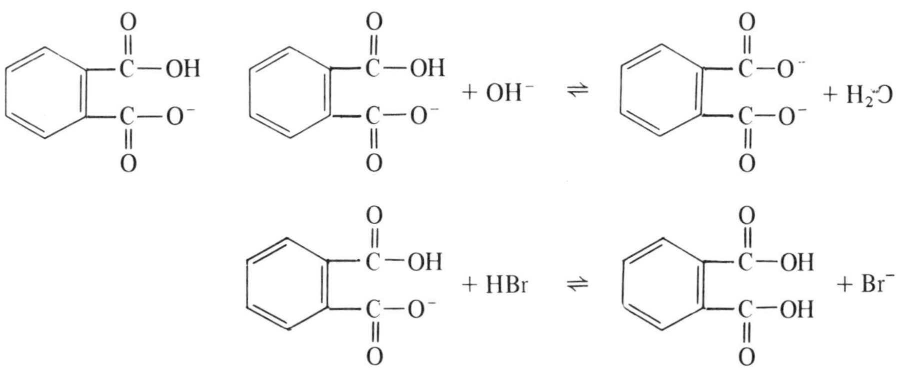	
Glycine	$NH_2CH_2-\overset{O}{\overset{\|}{C}}-OH$	$NH_2CH_2-\overset{O}{\overset{\|}{C}}-OH + NH_3 \rightleftharpoons NH_2CH_2-\overset{O}{\overset{\|}{C}}-O^- + NH_4^+$ $NH_2CH_2-\overset{O}{\overset{\|}{C}}-OH + HNO_3 \rightleftharpoons .\overset{+}{N}H_3CH_2-\overset{O}{\overset{\|}{C}}-OH + NO_3^-$
Acetic acid	$CH_3-\overset{O}{\overset{\|}{C}}-OH$	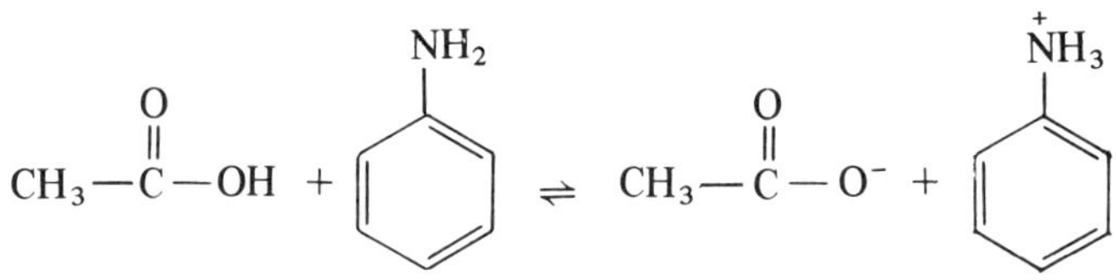 $CH_3-\overset{O}{\overset{\|}{C}}-OH + H_2SO_4 \rightleftharpoons CH_3-\overset{H-O^+}{\overset{\|}{C}}-OH + HSO_4^-$

[a]In this table the first reaction given for each substance illustrates an acid reaction; this is followed by a reaction showing the substance acting as a base.

Table 2.2 (cont'd.)

Name	*Formula*	*Acid and Base Reactions*
Hydroxide	OH^-	$OH^- + NH_2^- \rightleftharpoons O^{2-} + NH_3$
		$OH^- + HClO_4 \rightleftharpoons H_2O + ClO_4^-$
Sulfuric acid	H_2SO_4	$H_2SO_4 + H_2O \rightleftharpoons HSO_4^- + H_3O^+$
		$H_2SO_4 + HClO_4 \rightleftharpoons H_3SO_4^+ + ClO_4^-$
Phenol	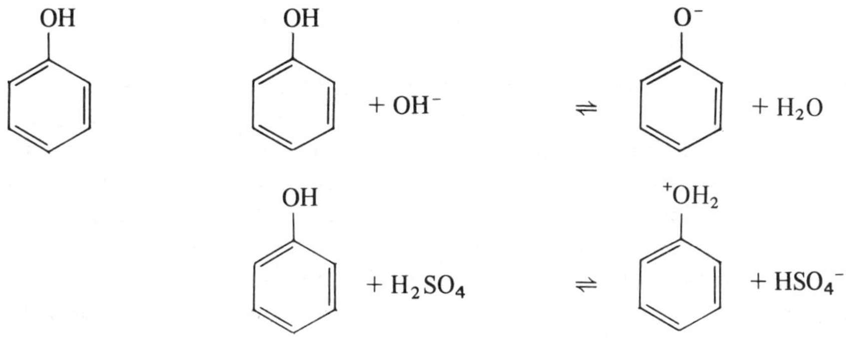	
Ethyl acetate	$C_2H_5O-\overset{O}{\overset{\Vert}{C}}-CH_3$	$C_2H_5O-\overset{O}{\overset{\Vert}{C}}-CH_3 + CH_3O^- \rightleftharpoons C_2H_5O-\overset{O}{\overset{\Vert}{C}}-CH_2^- + CH_3OH$
		$C_2H_5O-\overset{O}{\overset{\Vert}{C}}-CH_3 + H_3O^+ \rightleftharpoons C_2H_5O-\overset{HO^+}{\overset{\Vert}{C}}-CH_3 + H_2O$
Benzoic acid	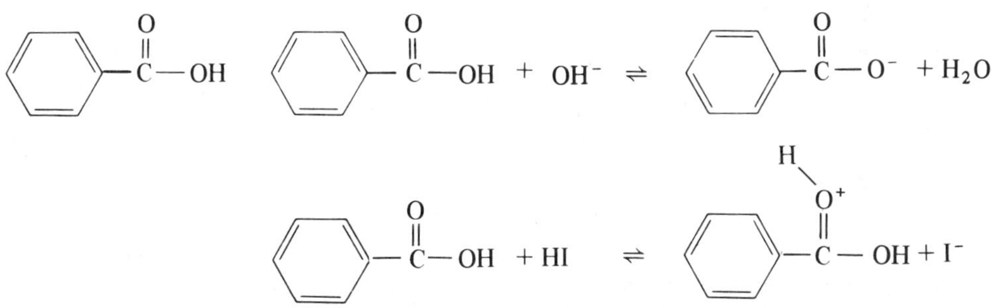	

A word closely related to "amphiprotic" is "amphoteric." Although many people use these words synonymously, there is a distinction. An *amphiprotic* substance is one that can either donate or accept a proton; the emphasis in amphiprotism is on proton transfer. An *amphoteric* substance is one that is capable of reacting with either acids or with bases. The emphasis here is on the reaction partner rather than on proton transfer.

An example of an amphoteric substance is zinc oxide, illustrated in reactions (2-24) and (2-25). It is clear that ZnO can react with either acid or base. It is not correct to

$$ZnO + 2H^+ \rightleftharpoons Zn^{2+} + H_2O \qquad (2\text{-}24)$$

$$ZnO + 2OH^- + H_2O \rightleftharpoons Zn(OH)_4{}^{2-} \qquad (2\text{-}25)$$

describe ZnO as amphiprotic since reactions (2-24) and (2-25) are not simple proton-transfer reactions. Water, on the other hand, as illustrated by reactions (2-20) and (2-21), may be described as either amphiprotic or amphoteric because it reacts with both acids and bases (amphoterism) by either donating or accepting protons (amphiprotism). Some additional examples of amphiprotic substances are shown in Table 2.2.

Tables 2.1 and 2.2 introduce a shorthand method for representing the structures of organic molecules such as pyridine (Table 2.1) and substituted benzenes such as phenol (Table 2.2). The complete structures, which include all atoms, bonds, and nonbonded electron pairs, are shown below.

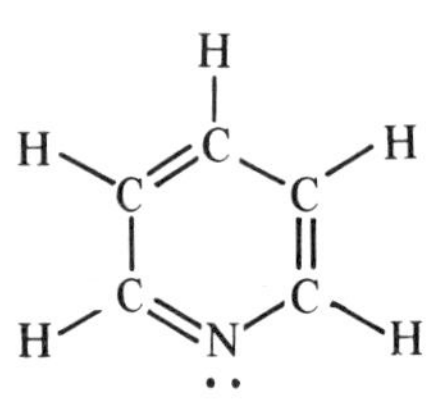
pyridine

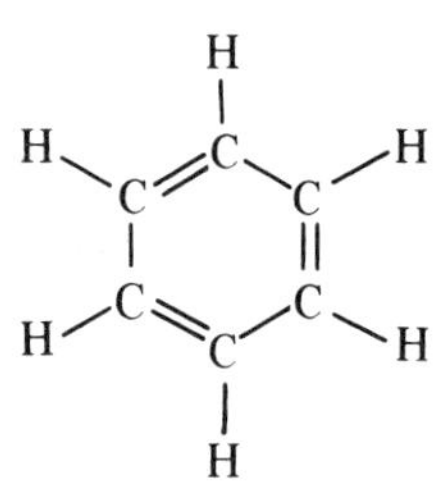
benzene

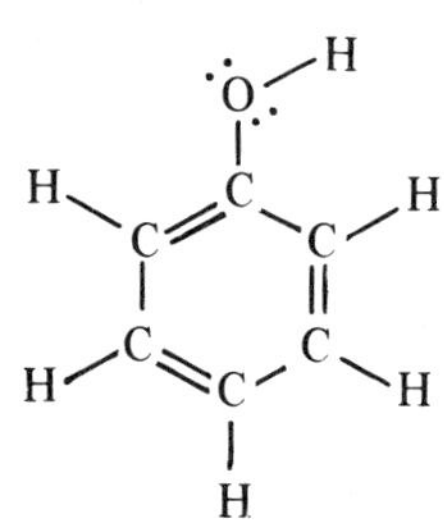
phenol
(hydroxybenzene)

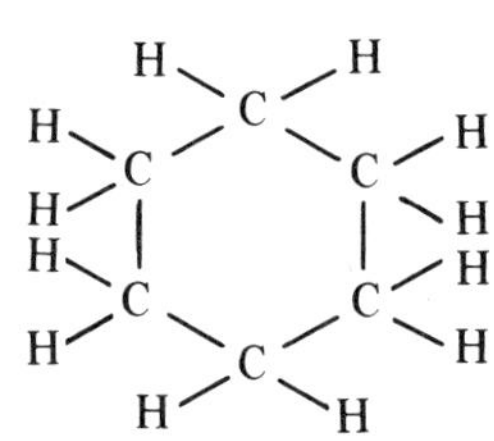
cyclohexane

The shorthand representations of the same structures are:

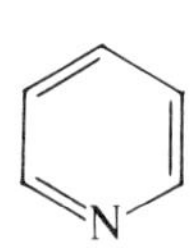
pyridine

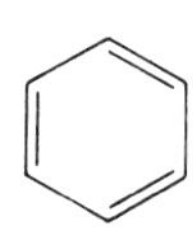
benzene

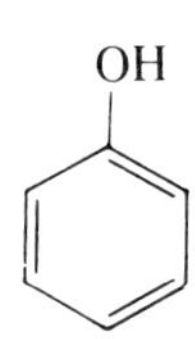
phenol

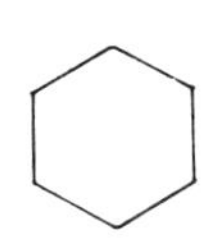
cyclohexane

Carbon atoms and hydrogen atoms are not explicitly represented in shorthand. Instead, each angle vertex represents either (1) a carbon atom with two bonds and two hydrogen atoms, as in cyclohexane; (2) a carbon atom with three bonds and one hydrogen atom, as in benzene; or (3) a carbon atom with four bonds and no hydrogen atoms, as in phenol.

If a vertex is occupied by an atom other than carbon, it is indicated explicitly, as in pyridine. Thus

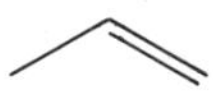 represents 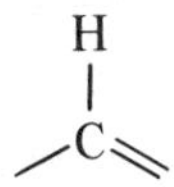 and 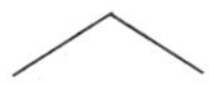 represents 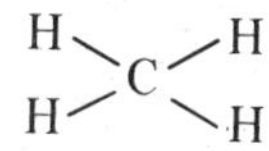

in the shorthand notation. Explicit recognition of the presence of nonbonded electron pairs is optional; pyridine may be represented as

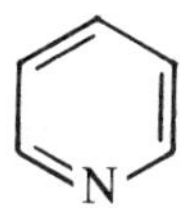 or 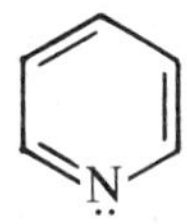

In this text we usually omit nonbonded pairs from structural representations, unless they are important to the discussion.

2.8 AUTOIONIZATION

Certain compounds are amphiprotic even in the absence of any other materials; that is, they can protonate themselves. This is known as *autoionization* and is illustrated for water in reaction (2-26). The autoionization of water is discussed in more detail in Chapter 3.

$$H_2O + H_2O \rightleftharpoons H_3O^+ + OH^- \tag{2-26}$$

2.9 THE LEVELING EFFECT OF SOLVENT

When a Brϕnsted–Lowry acid is placed in water, an equilibrium reaction (2-27) occurs.

$$HA + H_2O \rightleftharpoons H_3O^+ + A^- \tag{2-27}$$

The acid is called *strong* if the equilibrium lies far to the right, that is, if essentially 100% of the HA has been converted to H_3O^+ and A^-. Two different acids can both be strong in the sense of undergoing 100% ionization in water, yet one of them can have a greater inherent tendency to donate a proton. However, *there is no way to measure this difference in aqueous solution*, since both are ionized to the maximum extent possible. Even though the acids may differ greatly in their inherent abilities to donate protons, they give rise to aqueous solutions which have the same acidic properties and which are essentially indistinguishable from one another as far as acid-base reactions are concerned. This effect is known as the *leveling effect of solvent*. All acids stronger than hydronium ion are leveled to its strength when dissolved in water. The most important of these substances, known as the mineral acids, are HCl, HBr, HI, HNO_3, H_2SO_4 and $HClO_4$. Although our discussion has been in terms of water and its conjugate acid H_3O^+, it is in general true that *the strongest Brønsted-Lowry acid that can exist in any solvent is the conjugate acid of the solvent.*

The leveling effect is also observed when strong bases are dissolved in water. A strong base is one in which the ionization reaction [e.g., (2-28)] proceeds completely to the right. When bases stronger than hydroxide ion are dissolved in water, they are leveled to the strength of the hydroxide ion. Any stronger base is completely converted to OH^- in a reaction similar to (2-28). All strong bases form aqueous solutions having the maxi-

$$\underset{\text{base}_1}{NH_2^-} + \underset{\text{acid}_2}{H_2O} \rightleftharpoons \underset{\text{acid}_1}{NH_3} + \underset{\text{base}_2}{OH^-} \qquad (2\text{-}28)$$

mum possible OH^- concentration, and a group of such bases will form solutions that are indistinguishable as far as basic strength (as measured by OH^- concentration) and properties are concerned. Some common strong bases are given in Table 2.3. In Table 2.3, and elsewhere throughout the text, the symbol R denotes a group of atoms, usually containing carbon. Examples of R groups are methyl (CH_3), ethyl (C_2H_5), and *n*-butyl ($CH_3CH_2CH_2CH_2$). The symbol M is used to denote a metal atom.

Table 2.3 Some Common Bases Stronger Than Hydroxide Ion

Name	*Formula*	*Comments*
Metal hydrides	MH_X	Two good examples are NaH and MgH_2. The H^- is the strongly basic portion of the molecule, and reacts (violently) with water according to: $H^- + H_2O \longrightarrow H_2 + OH^-$
Sodamide (sodium amide)	$NaNH_2$	Sodamide is the product of the reaction of metallic sodium with liquid NH_3.
Alkoxides	RO^-	Sodium ethoxide ($CH_3CH_2O^-Na^+$) is formed by reacting sodium with ethyl alcohol.
Metal alkyls	RM	In *n*-butyllithium ($CH_3CH_2CH_2CH_2Li$) the carbon bonded to lithium carries a significant negative charge and behaves as a strong base.

2.10 PREPARATION OF INORGANIC ACIDS

This section lists the processes by which the major mineral acids may be prepared. Some physical properties of these acids and their solutions are shown in Table 2.4.

Table 2.4 Physical Properties of the Common Mineral Acids[a]

Name	*Formula*	*m.p.(°C)*	*b.p.(°C)*	*Comments*
Hydrofluoric acid	HF	−92.3	19.4	Colorless liquid or gas
Hydrochloric acid	HCl	−112	−83.7	Colorless liquid or gas
Hydrobromic acid	HBr	−88.5	−67.0	Colorless gas or pale yellow liquid
Hydriodic acid	HI	−50.8	−35.4	Colorless gas or pale yellow liquid
Nitric acid	HNO_3	−42	86	Colorless liquid
Nitrous acid	HNO_2	—	—	Liquid is unknown. Has been prepared and studied in solution and gas phase only
Sulfuric acid	H_2SO_4	10.5	330	Colorless, oily liquid
Phosphoric acid	H_3PO_4	42.4	213	Colorless liquid. Loses water when boiled, forming $H_4P_2O_7$
Perchloric acid	$HClO_4$	−112	39 (at 6 mm Hg)	Colorless gas or liquid, with a very offensive odor

[a]*CRC Handbook of Chemistry & Physics*, 1978 ed. All data, unless stated otherwise, are given for a pressure of one atmosphere.

A. Hydrofluoric Acid

Calcium fluoride is reacted with concentrated sulfuric acid:

$$CaF_2 + H_2SO_4 \longrightarrow CaSO_4 + 2HF \qquad (2\text{-}29)$$

Reaction (2-29) gives the reagents used, and reactions (2-30) show the two steps of the mechanism. This preparation is an application of Brϕnsted-Lowry principles. The strong

$$H_2SO_4 + F^- \longrightarrow HSO_4^- + HF \qquad (2\text{-}30a)$$

$$HSO_4^- + F^- \longrightarrow SO_4^{2-} + HF \qquad (2\text{-}30b)$$

acids, H_2SO_4 and HSO_4^-, provide protons to the fluoride ions, producing the weaker acid, HF. If the reaction is carried out at moderate temperature, HF (bp 19.4°) is removed as a gas. Although hydrofluoric acid may be obtained commercially in purity greater than 99%, it is normally shipped as a 49% aqueous solution. Hydrofluoric acid is stored in special plastic, metal, or wax-lined containers because HF reacts with glass, a mixture of SiO_2 and silicates such as $CaSiO_3$. Indeed, HF(aq) is used to etch glass, as shown in reactions (2-31) and (2-32). Hydrofluoric acid should be handled with care;

$$SiO_2(s) + 4HF(aq) \longrightarrow SiF_4(g) + 2H_2O \tag{2-31}$$

$$CaSiO_3(s) + 6HF(aq) \longrightarrow CaF_2(s) + SiF_4(g) + 3H_2O \tag{2-32}$$

formation and deposition of CaF_2 in the flesh causes HF burns to heal very slowly.

B. Hydrochloric Acid

Also known as muriatic acid, HCl may be prepared by direct union of its elements [reaction (2-33)]. Reaction of a strong acid with a chloride salt such as NaCl [reaction (2-34)] liberates gaseous HCl and can be used as a laboratory method of generating small quantities of water-free HCl. Although hydrochloric acid is a gas at room temperature, it is highly soluble in water. It is sold either as a compressed gas or as a 37% aqueous solution (12 molar).

$$H_2 + Cl_2 \longrightarrow 2HCl \tag{2-33}$$

$$NaCl + H_2SO_4 \longrightarrow NaHSO_4 + HCl \tag{2-34}$$

C. Hydrobromic and Hydriodic Acids

These strong acids are prepared commercially via direct union of their elements [reactions (2-35) and (2-36)]. Both are gaseous substances at room temperature, and are normally handled and sold as 48% aqueous solutions.

$$H_2 + Br_2 \longrightarrow 2HBr \tag{2-35}$$

$$H_2 + I_2 \longrightarrow 2HI \tag{2-36}$$

D. Nitric Acid

A nitrate salt (e.g., $NaNO_3$) can be treated with sulfuric acid to produce nitric acid [reaction (2-37)]. The *Ostwald process*, used commercially for the preparation of large

$$NaNO_3 + H_2SO_4 \longrightarrow HNO_3 + NaHSO_4 \tag{2-37}$$

quantities of HNO_3, is carried out in three stages [reactions (2-38), (2-39), and (2-40)].

$$4NH_3 + 5O_2 \xrightarrow{Pt} 4NO + 6H_2O \tag{2-38}$$

$$2NO + O_2 \longrightarrow 2NO_2 \tag{2-39}$$

$$3NO_2 + H_2O \longrightarrow 2HNO_3(aq) + NO \tag{2-40}$$

The first is the air oxidation of NH_3 in the presence of a metal catalyst to produce nitric oxide. The second step is the further oxidation to nitrogen dioxide. The final step is the reaction of nitrogen dioxide with water, producing nitric acid; the other product, NO, is recycled in step (2-39). The acid is sold commercially as a water solution, either 71% or 90 to 92%.

E. Sulfuric Acid

Sulfuric acid is the most important industrial chemical produced in the United States. Over 80 billion pounds is prepared annually. It is produced commercially by the *contact process*, beginning with sulfur, which is mined and used without further purification. Sulfur is burned in oxygen to produce SO_2, which is then further oxidized in air to SO_3 in a process catalyzed by V_2O_5. In the final step [reaction (2-41)], which is highly

$$H_2O + SO_3 \longrightarrow H_2SO_4 \qquad (2\text{-}41)$$

exothermic and very hazardous, the anhydride SO_3 is added to water to produce sulfuric acid. Sulfuric acid is usually sold as a 96 to 98% water solution. If additional SO_3 is added, a solution known as *fuming sulfuric acid* is produced. Sulfuric acid combines the properties of strong acidity and dehydrating power. One example of the dehydrating power of H_2SO_4 is its reaction (2-42) with sucrose (table sugar), converting it to carbon.

$$C_{12}H_{22}O_{11} \xrightarrow{H_2SO_4} 12C + 11H_2O \qquad (2\text{-}42)$$

F. Phosphoric Acid

Also known as orthophosphoric acid, phosphoric acid is prepared by the reaction of powdered phosphate rocks with concentrated sulfuric acid. Small quantities may be prepared in the laboratory by the highly exothermic reaction of P_4O_{10} with water. Concentrated phosphoric acid is usually sold as a syrupy 85% water solution.

$$PO_4^{3-} + 3H_2SO_4 \longrightarrow H_3PO_4 + 3HSO_4^- \qquad (2\text{-}43)$$

$$P_4O_{10} + 6H_2O \longrightarrow 4H_3PO_4 \qquad (2\text{-}44)$$

G. Perchloric Acid

Perchloric acid is one of the strongest mineral acids known. It is usually used as an approximately 70% solution in water. The concentrated acid is a very good oxidizing agent and reacts violently (even explosively in some cases) with a wide variety of organic materials. Great caution should be taken in its use. Perchloric acid is prepared commercially by reacting hydrochloric acid with a concentrated solution of sodium perchlorate [reaction (2-45)]. The partially soluble sodium chloride is removed by filtration and the

$$HCl + NaClO_4 \longrightarrow NaCl + HClO_4 \qquad (2\text{-}45)$$

filtrate is distilled. The distillate contains H_2O, HCl, and $HClO_4$, and is distilled again to remove HCl and to produce a concentrated solution (70%) of perchloric acid. Pure perchloric acid can be obtained by distilling a 70% solution of $HClO_4$ from concentrated sulfuric acid, which acts as a dehydrating agent.

Pure perchloric acid is not explosive; however, it is usually contaminated with Cl_2O_7, and in this form is sensitive to both mechanical shock and heat and can explode violently under a variety of conditions.

H. General Preparations of Acids

Several of the preparative methods that have already been discussed involve the reaction of water with an oxide of a nonmetal. This reaction provides a general method of preparing a variety of inorganic acids, as illustrated by reactions (2-46) through (2-50). A compound derived from an acid by the removal of one or more H_2O units is called an acid anhydride.

$$SO_3 + H_2O \longrightarrow \underset{\text{sulfuric acid}}{H_2SO_4} \quad (2\text{-}46)$$

$$SeO_2 + H_2O \longrightarrow \underset{\text{selenous acid}}{H_2SeO_3} \quad (2\text{-}47)$$

$$3NO_2 + H_2O \longrightarrow \underset{\text{nitric acid}}{2HNO_3} + NO \quad (2\text{-}48)$$

$$CO_2 + H_2O \longrightarrow \underset{\text{carbonic acid}}{H_2CO_3} \quad (2\text{-}49)$$

$$P_4O_{10} + 6H_2O \longrightarrow \underset{\text{phosphoric acid}}{4H_3PO_4} \quad (2\text{-}50)$$

2.11 PREPARATION OF SOME COMMON BASES

Methods of preparation of some of the more common bases are discussed in this section. A partial list of strong bases is shown in Table 2.3.

A. Sodium Hydroxide

The *lime-soda process* for the production of NaOH has been in commercial use for many years. In the first step calcium carbonate (limestone) is heated to convert it to calcium oxide and drive off gaseous carbon dioxide. Addition of water to the residue produces calcium hydroxide (slaked lime), which is then reacted with sodium carbonate (soda). The insoluble $CaCO_3$ is removed by filtration and the residual water is evaporated to yield essentially pure NaOH.

$$CaCO_3 \longrightarrow CaO + CO_2 + CO_2 \quad (2\text{-}51)$$

$$CaO + H_2O \longrightarrow Ca(OH)_2 \quad (2\text{-}52)$$

$$Ca(OH)_2 + Na_2CO_3 \longrightarrow CaCO_3 + 2NaOH \quad (2\text{-}53)$$

A more modern method for preparing NaOH is the electrolysis of aqueous NaCl [reaction (2-54)]. This method is more desirable than the lime-soda process because the

$$2NaCl + 2H_2O \xrightarrow{\text{electrolysis}} 2NaOH + H_2 + Cl_2 \qquad (2\text{-}54)$$

by-products H_2 and Cl_2 are also commercially valuable. Sodium hydroxide prepared in this fashion is contaminated with sodium chloride and is further purified by fractional crystallization.

B. Other Metal Hydroxides

An excellent general method for preparing hydroxide bases is the reaction of a metal oxide, peroxide, or superoxide with water. Examples of some of these reactions are shown in reactions (2-55) through (2-58).

$$\underset{\text{sodium oxide}}{Na_2O} + H_2O \longrightarrow 2NaOH \qquad (2\text{-}55)$$

$$\underset{\text{sodium peroxide}}{2Na_2O_2} + 2H_2O \longrightarrow 4NaOH + O_2 \qquad (2\text{-}56)$$

$$\underset{\text{potassium superoxide}}{4KO_2} + 2H_2O \longrightarrow 4KOH + 3O_2 \qquad (2\text{-}57)$$

$$\underset{\text{magnesium oxide}}{MgO} + H_2O \longrightarrow Mg(OH)_2 \qquad (2\text{-}58)$$

Hydroxide bases may also be prepared by the reaction of the group I or group II metals with water, as illustrated by reactions (2-59) through (2-61). One difficulty with

$$2Na + 2H_2O \longrightarrow 2NaOH + H_2 \qquad (2\text{-}59)$$

$$2K + 2H_2O \longrightarrow 2KOH + H_2 \qquad (2\text{-}60)$$

$$Mg + 2H_2O \longrightarrow Mg(OH)_2 + H_2 \qquad (2\text{-}61)$$

this method is the violent nature of the reactions. If the metal is sufficiently reactive, as both sodium and potassium are, the reaction quickly generates enough heat to ignite the hydrogen produced. This method of preparation is not in wide use. The base-forming tendencies of the group I and group II metals are quite general, and reagents other than water may be used to produce bases other than hydroxide bases. These reactions are illustrated in reactions (2-62), (2-63), and (2-64), in which the strong bases sodamide, sodium ethylamine, and potassium isopropoxide are produced from ammonia, ethylamine, and isopropanol, respectively.

$$2NH_3 + 2Na \longrightarrow H_2 + 2NaNH_2 \quad (2\text{-}62)$$

$$2CH_3CH_2NH_2 + 2Na \longrightarrow H_2 + 2CH_3CH_2NHNa \quad (2\text{-}63)$$

$$2(CH_3)_2CHOH + 2K \longrightarrow H_2 + 2(CH_3)_2CHOK \quad (2\text{-}64)$$

To understand these reactions it may be helpful to regard ethylamine as *substituted* ammonia and isopropanol as *substituted* water. The term "substituted ammonia" refers to compounds having the same general structure as ammonia, but in which one or more of the hydrogen atoms have been replaced by a different atom or group of atoms. Examples are chloramine ($ClNH_2$), dimethylamine [$(CH_3)_2NH$], and propylamine ($CH_3CH_2CH_2NH_2$). Most substituted ammonias are called amines (Section 6.4); the chemical behavior of amines is similar to that of ammonia, as illustrated by reactions (2-62) and (2-63).

Similarly, a "substituted water" has the same general structure as water, but one or both of the hydrogen atoms have been replaced by a different atom or group. In *alcohols*, for example, methanol (CH_3OH), ethanol (CH_3CH_2OH), and *n*-butanol ($CH_3CH_2CH_2CH_2OH$), one of the hydrogen atoms of water has been replaced by a larger group. The chemical behavior of alcohols is similar to that of water, a fact that is useful in predicting the outcome of reactions such as (2-64).

C. Ammonia

Some of the early methods for the preparation of ammonia are quite interesting. One method was the destructive distillation (heating in the absence of air) of waste animal products such as hides, hoofs, and horns. A term once used by pharmacists, "spirits of hartshorn," reflects this source of ammonia. Another early method was the production of ammonia by the destructive distillation of coal. During this process a gas containing a mixture of hydrocarbons and ammonia is produced. The ammonia-rich gas was bubbled through sulfuric acid to produce ammonium sulfate, which was then crystallized and isolated. Ammonium sulfate prepared in this manner was once widely used as a fertilizer. Gaseous ammonia can be obtained by treating ammonium sulfate with a strong base.

Another method for the preparation of ammonia is illustrated in reactions (2-65) and (2-66). Calcium cyanamide can be used as a fertilizer; it is added to moist soil, where it slowly hydrolyzes to produce ammonia.

$$\underset{\text{calcium carbide}}{CaC_2} + N_2 \longrightarrow \underset{\text{calcium cyanamide}}{CaNCN} + C \quad (2\text{-}65)$$

$$CaNCN + 3H_2O \longrightarrow 2NH_3 + CaCO_3 \quad (2\text{-}66)$$

These early preparations of ammonia have been entirely replaced by the more economical *Haber process* [reaction (2-67)], which uses a catalyst to effect the direct combination of N_2 with H_2 at elevated temperature and pressure.[4]

$$N_2 + 3H_2 \xrightarrow{\text{catalyst}} 2NH_3 \qquad (2\text{-}67)$$

D. Sodium Carbonate and Sodium Hydrogen Carbonate

Sodium carbonate and sodium hydrogen carbonate are produced commercially by the *Solvay process,* which uses sodium chloride and calcium carbonate as raw materials. In the first step, reaction (2-51), calcium carbonate is heated to produce calcium oxide and

$$CaCO_3 \longrightarrow CaO + CO_2 \qquad (2\text{-}51)$$

carbon dioxide. The carbon dioxide is mixed with ammonia and added to an aqueous solution of sodium chloride, producing sodium hydrogen carbonate and ammonium chloride [reaction (2-68)].

$$H_2O + CO_2 + NH_3 + NaCl \longrightarrow NaHCO_3 + NH_4Cl \qquad (2\text{-}68)$$

Sodium hydrogen carbonate is insoluble and is isolated by filtration. Sodium hydrogen carbonate may be heated to produce sodium carbonate [reaction (2-69)].

$$2NaHCO_3 \longrightarrow Na_2CO_3 + CO_2 + H_2O \qquad (2\text{-}69)$$

The by-products from (2-51) and (2-68), NH_4Cl and CaO, are used (2-70) to regenerate ammonia, which is used over and over again in the process. The overall reaction of the Solvay process is shown in reaction (2-71).

$$2NH_4Cl + CaO \longrightarrow 2NH_3 + H_2O + CaCl_2 \qquad (2\text{-}70)$$

$$CaCO_3 + 2NaCl \longrightarrow Na_2CO_3 + CaCl_2 \qquad (2\text{-}71)$$

E. Sodamide

The basic compound sodamide, also called sodium amide, is frequently used in chemistry when a base stronger than OH^- is required. The amide ion is strong enough to remove a proton from a carbon-hydrogen bond in many compounds, forming a carbanion. Sodamide may be generated by adding metallic sodium to liquid ammonia.

$$2Na + 2NH_3 \longrightarrow 2NaNH_2 + H_2 \qquad (2\text{-}72)$$

F. Metal Hydrides

During the past 40 years the chemistry of the metal hydrides has revolutionized many areas of organic synthesis. Although extended discussion of these substances is outside

[4] The Haber process had important consequences for European history, for its development freed warring countries from dependence on imported nitrates to produce munitions. Nitric acid, produced from ammonia in the Ostwald process, is used to produce explosives such a TNT.

the scope of this text, it should be mentioned that metal hydrides function in most reactions as a source of the strongly basic hydride ion (H^-).[5] Two of the most commonly used metal hydrides are lithium aluminum hydride ($LiAlH_4$) and sodium borohydride ($NaBH_4$). These materials are prepared by the following reactions:

$$4LiH + AlCl_3 \xrightarrow[\text{ethyl ether (solvent)}]{(C_2H_5)_2O} LiAlH_4 + 3LiCl \qquad (2\text{-}73)$$

$$4NaH + \underset{\text{methyl borate}}{B(OCH_3)_3} \xrightarrow{250^\circ C} NaBH_4 + \underset{\text{sodium methoxide}}{3NaOCH_3} \qquad (2\text{-}74)$$

Alkali-metal hydrides are produced by direct combination of the elements in high-temperature processes [reactions (2-75) and (2-76)].

$$2Li + H_2 \xrightarrow{750^\circ C} 2LiH \qquad (2\text{-}75)$$

$$2Na + H_2 \xrightarrow{200^\circ C} 2NaH \qquad (2\text{-}76)$$

G. Miscellaneous Strong Bases

Two other types of strong bases frequently encountered in organic syntheses are carbanions and alkoxides. Methods for the preparation of the alkali-metal salts of these species are shown in reactions (2-77) and (2-78). Their formation and chemistry are discussed in greater detail in Chapter 6.

$$\underset{n\text{-butyl chloride}}{CH_3CH_2CH_2CH_2Cl} + 2Li \longrightarrow \underset{n\text{-butyllithium}}{CH_3CH_2CH_2CH_2Li} + LiCl \qquad (2\text{-}77)$$

$$\underset{\text{methyl alcohol}}{2CH_3OH} + 2Na \longrightarrow \underset{\text{sodium methoxide}}{2CH_3ONa} + H_2 \qquad (2\text{-}78)$$

PROBLEMS

2.1 Complete and balance each of the following equations.

(a) $CH_3CH_2CH_2OH + Na$

(b) $H_2SO_4 + NaOH$

(c) $(CH_3CH_2)_2NH + K$

(d) Ethyl alcohol + Mg

(e) $Mg_3N_2 + H_2O$

[5] Do not be confused by the fact that H^- is a base while H^+ is an acid. Hydride ion is a proton acceptor and reacts with proton donors to produce gaseous hydrogen; $H^- + HA \rightarrow H_2 + A^-$.

(f) 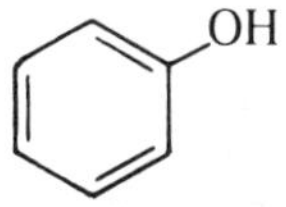 $+ Na_2O$

(g) 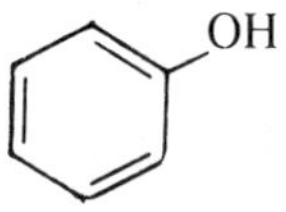 $+ NaOH$

(h) $CH_3CH_2NH_2 + HCl$

(i) Product from part (h) reacted with NaOH

(j) $CH_3-\overset{\overset{\displaystyle O}{\|}}{C}-OH + Na$

(k) $CH_3-\overset{\overset{\displaystyle O}{\|}}{C}-OH + Ca(OH)_2$

(l) 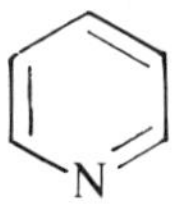 $+ HCl$

(m) $NaHCO_3 + HBr$

(n) $CH_3CH_2CH_2Li + H_2O$

2.2 Define or explain each of the following terms. Write out a complete answer first, then check in the text or other sources.

(a) Neutralization
(b) Aqua regia
(c) Oil of vitriol
(d) Muriatic acid
(e) Brønsted-Lowry acid
(f) Brønsted-Lowry base
(g) Arrhenius acid

2.3 Show examples (include the balanced equations) of three acids that may be prepared by reacting nonmetal oxides (acid anhydrides) with water.

2.4 Suggest how the following acid anhydrides react with water to form acids.

(a) $CH_3-\overset{\overset{\displaystyle O}{\|}}{C}-O-\overset{\overset{\displaystyle O}{\|}}{C}-CH_3$

acetic anhydride

(b) 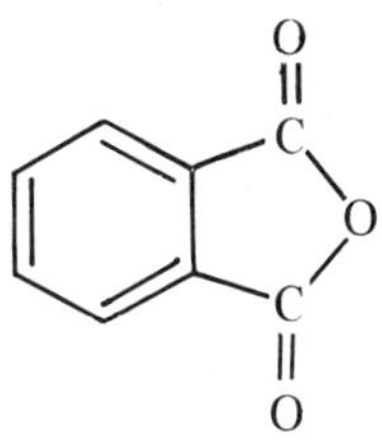

phthalic anhydride

2.5 Show examples (include the balanced equations) of three bases that may be prepared by the reaction of metal oxides with water.

2.6 Explain the difference between amphiprotic and amphoteric. Illustrate with examples.

2.7 Write equations illustrating the autoionization of the following solvents.

(a) H_2O
(b) CH_3OH
(c) NH_3
(d) CH_3CO_2H

2.8 Show all the steps involved in:

(a) The Ostwald process
(b) The Haber process

2.9 Write the conjugate base of each of the following acids.

(a) H_2SO_4
(b) H_2O
(c) CH_3OH
(d) H_2F^+
(e) H_3O^+
(f) OH^-
(g) NH_4^+
(h) CH_3CH_3
(i) acetylene

2.10 Write the conjugate acid of each of the following.

(a) H_2SO_4
(b) H_2O
(c) CH_3OH
(d) HF
(e) OH^-
(f) CH_3^-
(g) $CH_3CH_2NH_2$
(h) 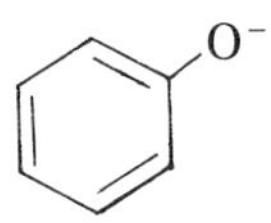
(i) HSO_4^-

Chapter 3

Quantitative Aspects of Acidity

3.1 INTRODUCTION AND REVIEW OF FUNDAMENTAL CONCEPTS

Those chemical and physical properties that are common to all aqueous acid solutions are due to the presence of hydrogen ions. Dissolved hydrogen ions, or *aqueous protons*, are commonly represented by $H^+(aq)$, H^+, or H_3O^+. In this chapter we use the term *acidity* to mean "concentration of aqueous protons"[1] and represent it by the symbol $[H_3O^+]$ (i.e., the molarity of hydronium ions).

The concentration of dissolved substances can be expressed in a variety of ways: weight percent, mole fraction, and others. A common measure of concentration is the *molarity*, which is defined as the number of moles of solute per liter of solution. The molarity is calculated from the number of moles of solute, and the solution volume, as shown in equation (3-1).

$$\text{molarity} = \frac{\text{moles of solute}}{\text{liters of solution}} \tag{3-1}$$

Example 3.1

Calculate the molarity of a solution, 500 mL of which contains 2.55 g of a solute whose molecular weight is 60.0 g/mol.

Answer: We begin with the definition of molarity [equation (3-1)] and note that the problem contains part of the necessary information explicitly; the volume of the solution is given as 500 mL or 0.500 liter. We must then calculate the number of moles of solute, using equation (3-2):

$$\text{moles} = \frac{\text{grams of solute}}{\text{molecular weight of solute}} \tag{3-2}$$

Insertion of the given values yields

$$\text{moles} = \frac{2.55\ \text{g}}{60.0\ \text{g/mol}} = 0.0425\ \text{mol}$$

[1] Acid behavior is found in nonaqueous solutions as well, so that a more general definition might refer to *dissolved* rather than *aqueous* protons. The symbol to represent acidity might then be $[SH^+]$, where S is the formula of the solvent. Some aspects of nonaqueous acidity are treated in Sections 11.6 and 11.7.

Substitution of this into equation (3-1) gives

$$\text{molarity} = \frac{\text{moles of solute}}{\text{liters of solution}} = \frac{0.0425 \text{ mol}}{0.500 \text{ liter}} = 0.0850\,M$$

Note that the volume must be expressed in liters, not milliliters, and that the units of molarity can be written as *moles/liter, molar*, or *M*.

Example 3.2

How many grams of solute, whose molecular weight is 60.0 g/mol, is required to prepare 250 mL of a 0.100 *M* solution?

Answer: We begin with the defining equation (3-1):

$$\text{molarity} = \frac{\text{moles of solute}}{\text{liters of solution}}$$

We insert the values of the given quantities, molarity and solution volume, and solve for the number of moles of solute.

$$0.100 \frac{\text{mol}}{\text{liter}} = \frac{\text{moles of solute}}{0.250 \text{ liter}}$$

$$(0.250 \cancel{\text{liter}})\ 0.100 \frac{\text{mol}}{\cancel{\text{liter}}} = \frac{\text{moles of solute}}{\cancel{0.250 \text{ liter}}} \quad (\cancel{0.250 \text{ liter}})$$

$$0.0250 \text{ mol} = \text{moles of solute}$$

(Note the conversion of 250 mL to 0.250 liter.)

It is essential to include the units in such operations. An experimental number, in science, has two parts: the numerical part and the dimensional part, usually called the units. Both parts are important, and it is obvious that omission of the numerical part makes problem solving impossible. It is less obvious but no less true that failure to write down and manipulate the units often leads to confusion and error.

Since the problem asked for weight rather than moles of solute, we use equation (3-2) to convert:

$$\text{moles} = \frac{\text{grams}}{\text{molecular weight}}$$

We insert the number of moles just calculated and the given molecular weight (using x for the unknown weight)

$$0.0250 \text{ mol} = \frac{x \text{ grams}}{60.0 \text{ g/mole}}$$

and solve:

$$60.0 \frac{\text{g}}{\cancel{\text{mol}}}(0.0250 \cancel{\text{mol}}) = \frac{(x \text{ grams})(\cancel{60.0 \text{ g/mol}})}{\cancel{60.0 \text{ g/mol}}}$$

$$1.50 \text{ g} = x \text{ grams}$$

This problem could have been solved in one step rather than two if we had substituted the relationship moles = grams/(molecular weight) into equation (3-1) to begin with:

$$\text{molarity} = \frac{\text{moles of solute}}{\text{liters of solution}} = \frac{\text{grams of solute/molecular weight}}{\text{liters of solution}}$$

Let x equal the unknown number of grams, and insert the given numerical values and units.

$$0.100\,\frac{\text{mol}}{\text{liter}} = \frac{(x\text{ grams})/(60.0\text{ g/mol})}{0.250\text{ liter}}$$

Clear fractions and cancel.

$$(0.250\,\cancel{\text{liter}})(60.0\,\frac{\text{g}}{\cancel{\text{mol}}})(0.100\,\frac{\cancel{\text{mol}}}{\cancel{\text{liter}}}) = \frac{(x\text{ grams})\cancel{(0.250\text{ liter})}\cancel{(60.0\text{ g/mol})}}{\cancel{(0.250\text{ liter})}\cancel{(60.0\text{ g/mol})}}$$

$$1.50\text{ g} = x\text{ grams}$$

Another problem, which frequently occurs in laboratory operations, involves the preparation of a solution not from a weighed quantity of solute, but by dilution of a stock solution (a more concentrated solution of known molarity).

Example 3.3

An experimental procedure calls for the use of 500 mL of 0.10 *M* sodium hydroxide. How much stock solution (3.0 *M*) and how much water must be used to prepare the desired solution?

Answer: This problem is not difficult once it is realized that the *amount* of solute in the final solution is the same as the amount in the required volume of stock solution. That is, addition of water to a beaker containing a few milliliters of stock solution does not change the amount of solute in the beaker. This problem is solved more easily if it is broken down into two steps:

Step 1. Calculate the amount of solute, x moles, which must be present in the final solution. Begin with equation (3-1), insert the numerical values given in the problem, and solve.

$$\text{molarity} = \frac{\text{moles of solute}}{\text{liters of solution}}$$

$$0.10\text{ mol/liter} = \frac{x\text{ moles}}{0.500\text{ liter}}$$

$$x\text{ moles} = (0.10\text{ mol/liter})(0.500\text{ liter})$$

$$= 0.050\text{ mol}$$

This is the amount of solute that must be supplied by the stock solution.

Step 2. Calculate the amount of stock solution, y liters, that contains 0.050 mol of solute. We again insert known values into equation (3-1) and solve. (Remember that the unit M is identical to mol/liter.)

$$\text{molarity} = \frac{\text{moles of solute}}{\text{liters of solution}}$$

$$3.0\ M = \frac{0.050\ \text{mol}}{y\ \text{liter}}$$

$$y\ \text{liter} = \frac{0.050\ \text{mol}}{3.0\ \text{mol/liter}}$$

$$= 0.017\ \text{liter} = 17\ \text{mL}$$

The procedure, then, is to measure out 17 mL of stock solution and add to it sufficient pure water to bring the final volume to 500 mL. The amount of water added will be very close to 500 − 17 = 483 mL.

This problem can also be solved in a single step; equation (3-1) is first rearranged to give

$$\text{moles of solute} = (\text{molarity})(\text{liters of solution})$$

We designate the stock solution with the subscript 1 and the diluted solution we wish to prepare with the subscript 2, so that

$$\text{moles of solute}_1 = (\text{molarity}_1)(\text{liters}_1)$$

$$\text{moles of solute}_2 = (\text{molarity}_2)(\text{liters}_2)$$

The number of moles in the dilute solution is of course the same as the number of moles in the stock solution before we add water to it, so that

$$(\text{molarity}_1)(\text{liters}_1) = (\text{molarity}_2)(\text{liters}_2)$$

or

$$M_1 V_1 = M_2 V_2 \qquad \text{where } V = \text{volume}$$

Rearranging, inserting the value given in the problem, and solving, we find:

$$V_1 = \frac{M_2 V_2}{M_1}$$

$$= \frac{(0.10\ \cancel{\text{mol/liter}})(0.500\ \text{liter})}{3.0\ \cancel{\text{mol/liter}}}$$

$$= 0.017\ \text{liter} = 17\ \text{mL}$$

The ability to solve problems such as these in a quick and reliable way is an essential prerequisite to working with the quantitative aspects of acid-base chemistry; those who are confident about concentration calculations will not be distracted and will find the main task all the easier.

A concept closely related to molarity is *normality*, which in the context of the chemical reactions of acids may be thought of as the "effective" molarity of an acid. Normality is frequently used as a measure of concentration in acid–base chemistry. For many acids, such as HCl, normality and molarity have the same value. The number of moles of H^+ available for reaction in a given volume of water is numerically equal to the number of moles of HCl that was used when the solution was prepared. For other acids, however, this is not the case. H_2SO_4, for example, can provide 2 moles of H^+ for each mole of H_2SO_4 dissolved. In this sense ½ mole of H_2SO_4 is equivalent to 1 mole of a single-proton acid, and the term *equivalent* is defined as "that quantity of acid which can provide 1 mole of H^+ ions in acid-base reactions." Thus 50 mL of a 0.1 *M* H_2SO_4 solution can provide as many H^+ ions as an equal volume of 0.2 *M* HCl because on a mole-for-mole basis H_2SO_4 can provide twice as many protons as HCl. The molarity of the H_2SO_4 solution is 0.1, but its "effective" molarity, that is, its *normality*, is 0.2. The units of normality are equivalents per liter.

Acids such as H_2SO_4 are called *diprotic*, whereas those such as HCl, yielding a single proton when dissolved, are called *monoprotic*. Diprotic acids are not at all uncommon. There are also a few acids, for example, phosphoric acid (H_3PO_4), which are triprotic and capable of yielding 3 moles of H^+ ions. For triprotic acids 1 mole equals 3 equivalents, and the normality of solutions of triprotic acids is equal to three times the molarity. Phosphoric acid and its derivatives are of considerable importance to us, industrially, environmentally, and as biochemical compounds fundamental to life itself. *Tetra-* and higher *polyprotic* acids are not common in inorganic chemistry. However, their solutions may be discussed in terms analogous to the methods used for di- and triprotic acids. Most examples of polyprotic acids fall in the realm of organic, biological, and polymer chemistry.

A parallel situation exists in the case of bases, for which the normality is defined in terms of the capability of supplying OH^-. A *monohydric* base such as KOH yields 1 mole of OH^- per mole of base dissolved, and the normality of its solution is equal to the molarity. For *dihydric* bases such as $Ba(OH)_2$ the normality is two times the molarity, and for *trihydric* bases like $Al(OH)_3$ the normality is three times the molarity. An *equivalent* of base is that quantity which can provide 1 mole of OH^-, or, as it is more properly defined, in the Brønsted–Lowry sense, that quantity which can accept 1 mole of protons.

Example 3.4

Calculate the normality and molarity of a solution prepared by dissolving 10.0 g of H_2SO_4 in water to give 500 mL of solution.

Answer: Calculate the molarity first, using the same approach as in the preceding examples.

$$\begin{aligned}\text{molarity} &= \frac{\text{moles of solute}}{\text{liters of solution}} \\ &= \frac{(10.0\text{ g})/(98.0\text{ g/mol})}{0.500\text{ liter}} \\ &= 0.204\text{ mol/liter}\end{aligned}$$

Since sulfuric acid is diprotic, it can provide *2* moles of H^+ for each mole of acid, and the normality is thus twice the molarity:

$$\text{normality} = \text{molarity} \times 2$$
$$= 0.408\ N$$

Note that the units of normality may be written as *equivalents/liter, normal*, or simply *N*.

Many introductory chemistry textbooks define normality in a more general (and more complicated) way than we have done. However, provided that we restrict ourselves to acid-base chemistry, our definition is both correct and simple. Our method for calculating normality will be to calculate the molarity first, then multiply it by a factor of 1, 2, or 3, depending on whether the acid is mono-, di-, or triprotic (mono-, di-, or trihydric for base normality).

The number of ionizable hydrogens (H atoms that produce H_3O^+ in aqueous solution) in a compound is not necessarily the same as the number of H atoms in its molecular formula. Sulfuric acid, H_2SO_4, is diprotic and has two ionizable hydrogens, but formic acid, H_2CO_2, is monoprotic and only one of its hydrogen atoms is ionizable. A more useful way of writing the molecular formula of formic acid is HCO_2H, which explicitly indicates the presence of two different kinds of hydrogen atoms. Benzoic acid, $C_6H_5CO_2H$, is monoprotic because five of its six hydrogen atoms are not ionizable and thus do not contribute to its acidic properties; only the hydrogen atom of the $-CO_2H$ group is acidic. In formic acid as well, the acidic hydrogen atom is the one in the $-CO_2H$ group.

Example 3.5

A stock solution of oxalic acid is 0.10 *N*. How much water must be added to 50 mL of the stock solution to dilute it to 0.020 *N*?

Answer: As in Example 3.3, the amount of solute in 50 mL of stock solution is the same as that in the final solution.

Step 1. Starting from equation (3-3), which defines normality, rearrange it and insert the values given in the problem to calculate the number of equivalents of acid in 50 mL of stock.

$$\text{normality} = \frac{\text{equivalents of acid}}{\text{liters of solution}} \tag{3-3}$$

$$\text{equivalents} = (\text{normality})(\text{liters of solution})$$
$$= 0.10\ \frac{\text{equivalent}}{\cancel{\text{liter}}}\ (0.050\ \cancel{\text{liter}})$$
$$= 0.0050 \text{ equivalent of acid}$$

Step 2. Calculate the final volume of the diluted solution, which contains the same number of equivalents of acid as that calculated in step 1. First rearrange equation (3-3):

$$\text{liters of solution} = \frac{\text{equivalents of acid}}{\text{normality}}$$

$$\text{volume} = \frac{0.0050 \cancel{\text{ equivalent of acid}}}{0.020 \frac{\cancel{\text{equivalent}}}{\text{liter}}}$$

$$= 0.25 \text{ liter} = 250 \text{ mL}$$

Step 3. Calculate the amount of water required. We began with 50 mL of stock solution and finished with 250 mL of diluted solution, so $250 - 50 = 200$ mL of water is required.

Example 3.5 illustrates an exceedingly useful point. Nowhere in the problem or its solution is it stated whether the acid is mono-, di-, or triprotic, because this information is simply not needed for solving the problem. This is the case in almost all calculations; once the normality of a solution is known, it is usually unnecessary to know the acid type. (In fact, oxalic acid, HO_2CCO_2H, is diprotic. What is the molarity of the original oxalic acid solution in Example 3.5?)

The range of proton concentration in aqueous solutions is enormous. Chemists routinely deal with solutions in which $[H_3O^+]$ is very large, sometimes as high as 10 to 20 *M*. In solutions that are strongly basic, on the other hand, $[H_3O^+]$ can be as low as 10^{-15} *M*. This large range of common acidity values can be dealt with more conveniently by referring to the logarithm of the H_3O^+ concentration rather than to the concentration itself. Thus as $[H_3O^+]$ ranges from 10^{-15} to 10 *M*, the (base 10) logarithm goes from −15 to +1. If the sign of the logarithm is changed so that the range is +15 to −1, most acidities are described by positive rather than negative numbers. This approach was first used in 1909 by the Danish chemist S. P. L. Sørenson, who defined pH as the negative logarithm of the molar concentration of the hydronium ion:

$$\text{pH} = -\log_{10}\,[H_3O^+] \tag{3-4}$$

Writing in the German language, Sørenson called it *Wasserstoffionenexponent* (hydrogen-ion exponent) and gave it the symbol p_H. The "p" is the initial letter of *Potenz*, the German noun meaning "power"; it refers not to acid strength of the solution but to the power of 10 if $[H_3O^+]$ is expressed in exponential form. Note the effect of the negative sign in equation (3-4): the *lower* the pH, the *greater* the acidity of the solution. For example, if $[H_3O^+]$ is 1.0×10^{-3} *M*, the pH is 3.0; if $[H_3O^+]$ is increased to 5.0×10^{-3}, the pH decreases to 2.3. Examples 3.6 and 3.7 provide some practice with pH calculations.

From its original use in pH, the "p" notation has been extended to take on the general meaning of "the negative logarithm of." Thus pA means "the negative logarithm of A," and so on. Particular usages of importance in acid-base chemistry are pOH, the negative logarithm of hydroxide-ion concentration $[OH^-]$, and p*K*, the negative logarithm of the equilibrium constant; both of these are used frequently throughout this text.

3.2 STRONG ACIDS

Some acids ionize completely when placed into water. Nitric acid, for example, ionizes according to the reaction

$$HNO_3 \xrightarrow{H_2O} H_3O^+ + NO_3^- \tag{3-5}$$

By "complete ionization" we mean that all of the original HNO_3 ionizes, and the species $HNO_3(aq)$ (nonionized HNO_3) does not exist. Such acids contribute the maximum possible concentration of H_3O^+ to the solution and are called *strong acids*. In such solutions $[H_3O^+]$ is equal to the original molarity of the acid. For diprotic strong acids (which are rare) $[H_3O^+]$ is equal to twice the original molarity, and so on.

Example 3.6

Calculate the pH of a 0.010 M solution of nitric acid, a strong acid.

Answer: If the molarity of the strong (monoprotic) acid is known, $[H_3O^+]$ is equal to it and the pH is found by taking the log of the molarity and changing the sign. Almost all pocket calculators are equipped with a "log x" key, so that it is no longer necessary to use a numerical table of logarithms.

$$[H_3O^+] = \text{molarity} = 0.010\,M$$

$$\text{pH} = -\log[H_3O^+] = -\log(0.010) = -(-2)$$

The pH of the nitric acid solution is 2.

Example 3.7

Calculate the pH of a 3% solution of the strong acid HCl.

Answer: First we must convert the composition expressed as percent by weight to composition expressed as molarity; 100 g of a 3% solution contains 3 g of HCl and 97 g of H_2O, and occupies a volume of (approximately) 100 mL. The acid molarity is therefore

$$M = \frac{\text{moles of solute}}{\text{liters of solution}} = \frac{\dfrac{3\text{ g}}{36.5\text{ g/mol}}}{0.1\text{ liter}} = 0.8\text{ mol/liter}$$

Since the acid is strong, $[H_3O^+]$ is numerically equal to acid molarity, so that

$$\text{pH} = -\log\,[H_3O^+] = -\log M = -(-0.09) = +0.09$$

Note that no units are attached to pH; it is a *dimensionless* number.

It is of equal importance to be able to calculate $[H_3O^+]$ from the known pH of a solution. Since pH is defined [equation (3-4)] in terms of $[H_3O^+]$ and not in terms of acid molarity, when making conversions between pH and $[H_3O^+]$ we do not have to take into account whether the acid is mono- or polyprotic. The calculation is the same in either case.

Example 3.8

What is $[H_3O^+]$ in a solution of pH 2.6?

Answer: Since most calculators have a "10^x" key (or its mathematical equivalent, the "antilog x" key), this problem is more of an exercise in calculator use than in chemistry. Starting from the definition of pH,

$$pH = -\log [H_3O^+]$$

we multiply by -1 and take the antilogarithm of both sides:

$$\text{antilog}(-pH) = \text{antilog}(\log [H_3O^+])$$

The definition of a logarithm is that $\log(10^x) = x$, so that $\text{antilog}(x) = 10^x$. Also, $\text{antilog}(\log x) = x$; thus

$$[H_3O^+] = 10^{-pH}$$

In this example, $[H_3O^+] = 0.0025$ mol/liter.

Do not forget to change sign before evaluating the antilog: antilog(−2.6) = 0.0025, whereas antilog(+2.6) = 400.

3.3 WEAK ACIDS

The discussion and examples of the foregoing section have been limited to those acids that are 100% ionized in aqueous solution. Such acids are in the minority, and for most acids the extent of ionization is less than 100%. Both the nonionized acid and its ionized form (the conjugate base) are present in solution. Such acids are called *weak acids*. Calculation of the pH of a solution of a weak acid involves the prior step of calculating the extent of ionization. If, for example, a 0.01 *M* solution of a (monoprotic) weak acid is 10% ionized, $[H_3O^+]$ is 10% of the molarity, or 0.001, and the pH is 3. In contrast, the pH of a 0.01 *M* solution of a strong acid such as HBr is 2, since $[H_3O^+]$ is the same as the molarity of the acid.

The ionization of a weak acid is an equilibrium reaction in which the concentrations of all species are related to one another through the equilibrium expression. The strength of an acid is indicated by the value of its ionization equilibrium constant; the smaller its numerical value, the weaker is the acid. Acid strength is determined by the fundamental properties of the acid molecule itself; the relationship between acidity and structure is discussed in Chapter 5.

In all matters involving the extent of ionization, the question of whether the ionization of, say, benzoic acid[2] is regarded as ionization by Arrhenius dissociation,

$$HBz \rightleftharpoons H^+ + Bz^-$$

or as a Brϕnsted-Lowry ionization by proton transfer,

$$HBz + H_2O \rightleftharpoons H_3O^+ + Bz^-$$

is irrelevant; as we will see, both lead to the same value of $[H_3O^+]$ and pH.

In the Brϕnsted-Lowry view the general expression K = [products]/[reactants] reduces in the case of benzoic acid to[3]

$$K = \frac{[H_3O^+][Bz^-]}{[HBz][H_2O]} \tag{3-6}$$

Even though this equilibrium expression is not very complicated as it stands, it is possible to simplify it further by noting that the concentration of H_2O has very nearly the same numerical value in all solutions of acids or bases. This is true if the solutions are not too concentrated and the temperatures are all about the same. As a means of driving this point home, let us calculate the concentration of water in several different acid solutions. This calculation requires the experimentally determined density of the solutions and the molecular weights of the components of the solutions. The density of pure water at 20°C is 0.9982 g/mL or 998.2 g/liter. Dividing this by the molecular weight gives $[H_2O]$ = (998.2 g/liter)/(18.02 g/mole) = 55.39 mol/liter for the molarity of water when pure. Calculation of the water concentration in an acid solution involves the additional step of subtracting the contribution of the acid. For example, a 2% solution (by weight) of hydrochloric acid (about 0.6 N) has a density at 20°C of 1.0082 g/mL or 1008.2 g/liter. Since 1 liter of this solution by definition contains (1008.2 g) × 2% = 20.164 g of HCl, the mass of water is 1008.2 − 20.2 = 988.0.[4] The molarity of water is found by converting this amount to moles:

$$[H_2O] = \frac{988.0 \text{ g/liter}}{18.02 \text{ g/mol}} = 54.83\,M$$

Proceeding in this way we can calculate the molarity of water in a variety of solutions and support our assertion that the water concentration is nearly the same in all. Some of these results are collected in Table 3.1, from which it may be seen that to two-place accuracy, at least, and provided that we do not consider concentrated solutions, water concentration is approximately 55 M. Even at a concentration as high as 0.5 N, the maximum

[2] There are several common abbreviations for benzoic acid; in this chapter we use HBz for benzoic acid, Bz^- for benzoate ion, and NaBz for the sodium salt. See Table 2.2 for the structural formula of benzoic acid.

[3] Equilibrium constants are sometimes written in terms of reactant and product activities rather than concentrations. This approach is not used in this text but is developed briefly in Appendix IV.

[4] Note that in doing this subtraction we have rounded off the mass of HCl to the nearest tenth of a gram. This is done because the solution mass, 1008.2 g, is known only to the nearest 0.1 g and there is no use pretending that it is accurate in the 0.001 place. The notion of significant figures is central to our argument in this section and should be reviewed if necessary.

Table 3.1 Water Molarity in Dilute Solutions[a]

Solute	*Solute Normality*				
	1.0 N	*0.5 N*	*0.10 N*	*0.05 N*	*0.01 N*
Formic acid	53.5	54.5	55.2	55.3	55.3
Acetic acid	52.5	54.0	55.1	55.2	55.3
Hydrochloric acid	54.4	54.9	55.3	55.3	55.3
Sulfuric acid	54.6	54.9	55.3	55.3	55.3
Ammonia	54.0	54.7	55.3	55.3	55.3
Sodium hydroxide	55.5	55.5	55.4	55.4	55.3
Ammonium nitrate	52.8	54.1	55.1	55.2	55.3
Potassium carbonate	54.3	54.8	55.3	55.3	55.3

[a]All data refer to 20°C, at which temperature the molarity of pure water is 55.39 *M*.
Source: CRC Handbook of Chemistry and Physics, CRC Press, Boca Raton, Fla., 1978 ed.

deviation from the value of pure water is only 2.5%. As a matter of interest, the usual definition of "concentrated solution" is one of greater than 0.01 *N*.

Since the density of solutions depends on temperature, a quite different set of numbers would have been obtained if we had chosen to do the calculation at 10 or 50°C. However, for an acid ionization the equilibrium constant itself changes if the temperature is changed. For this reason it is necessary to carry out calculations (and laboratory experiments) without variations in temperature when making comparisons of acid and base strength.

In an equilibrium expression involving dilute acids, such as equation (3-6), we may substitute the number 55 for $[H_2O]$:

$$K = \frac{[H_3O^+]\,[Bz^-]}{55\,[HBz]} \tag{3-7}$$

Now, the equilibrium constant K is just that, a constant, whose value at 20°C in the case of benzoic acid is 1.23×10^{-6}. There is little point, in working problems or at any other time, in continually writing "$55K$" or "$55 \times 1.23 \times 10^{-6}$". One could, much more easily and with no loss of meaning or accuracy, write the value of this product, 6.8×10^{-5}. What we have done here is to define a new constant, the *ionization constant*, denoted by the symbol K_a:

$$K_a = 55K$$

or, more rigorously,

$$K_a = [H_2O]\,K \tag{3-8}$$

It is always the ionization constant K_a, not the equilibrium constant, which is listed in tables both in this book and elsewhere. The equilibrium expression (3-6) is now written

$$K_a = \frac{[H_3O^+]\,[Bz^-]}{[HBz]} \tag{3-9}$$

We can now return to the statement (page 38) that the question of Arrhenius versus Brφnsted-Lowry is irrelevant to this discussion. The expression (3-9) was developed starting from a Brφnsted-Lowry point of view. Suppose, however, that we had started with the notion that the chemical process was one of dissociation into ions:

$$HBz \rightleftharpoons H^+ + Bz^-$$

The equilibrium constant K for this reaction is

$$K = \frac{[H^+]\,[Bz^-]}{[HBz]} \tag{3-10}$$

Bearing in mind that H^+ and H_3O^+ (see Sec. 2.6) are two different names for the aqueous proton, we see that expressions (3-9) and (3-10) are identical. Henceforth we will use the ionization constant K_a rather than the equilibrium constant K, and we will feel free to use either of the chemical equations since they lead to the same numerical result.

Brφnsted-Lowry:

$$HBz + H_2O \rightleftharpoons H_3O^+ + Bz^-$$

$$K_a = 6.8 \times 10^{-5} = \frac{[H_3O^+]\,[Bz^-]}{[HBz]} \qquad pH = -\log\,[H_3O^+]$$

Arrhenius:

$$HBz \rightleftharpoons H^+ + Br^-$$

$$K_a = 6.8 \times 10^{-5} = \frac{[H^+]\,[Bz^-]}{[HBz]} \qquad pH = -\log\,[H^+]$$

The *extent of ionization*, which is a much used concept in acid-base chemistry, is defined as the ratio of the concentration of the ionized form of the acid to the initial concentration of the nonionized acid. As an example let us consider the ionization of benzoic acid in a solution prepared by dissolving C moles of acid in enough water so that the total volume of the resulting solution is 1 liter. Because the acid must be in either the protonated form HBz or the nonprotonated form Bz^-, the *sum* of the concentrations of these two forms is the molarity C, regardless of the extent to which the ionization proceeds. Such solutions are referred to as "C molar benzoic acid," in spite of the fact that it is the sum $[HBz] + [Bz^-]$, and not $[HBz]$ alone, which is equal to C.

Example 3.9

In a 2.50×10^{-4} M benzoic acid solution the concentration of benzoate anion is 1.05×10^{-4} M. What is the extent of ionization?

Answer:

$$\text{extent of ionization} = \frac{[\text{ionized form}]}{[\text{total acid}]}$$

$$= \frac{1.05 \times 10^{-4}}{2.50 \times 10^{-4}} = 0.420$$

Because of the manner in which it is defined, the extent of ionization is a dimensionless number that cannot be less than zero nor greater than 1. It is a fraction, often called the *fractional ionization* or the *degree of dissociation*. *Percent* ionization is defined as 100 times the fractional ionization:

$$\%\ \text{ionization} = \frac{[\text{ionized form}]}{[\text{total acid}]} \times 100$$

The extent (fractional or percent) of ionization depends on the magnitude of the ionization constant; the smaller the value of K_a, the smaller is the extent of ionization. Less obvious is the dependence on concentration; a given weak acid or base is ionized to a lesser extent at high concentrations than at low concentrations. This concentration dependence is discussed in the following section.

3.4 CALCULATION OF ACID IONIZATION

In order to discuss the ionization of weak acids from a quantitative point of view we will adopt a convenient fiction about the nature of the process: namely, that it occurs in two distinct, sequential steps. These steps are (1) the mixing of acid and water to form a homogeneous solution of *nonionized* acid and (2) the ionization of the dissolved acid. It makes no difference whether we regard step 2 as an Arrhenius dissociation or as a Brønsted-Lowry proton transfer. Our two-step model will allow us to calculate the correct concentrations of all species present in solution even though intuition tells us that in reality the processes of mixing, dissolution, and ionization are more likely to be simultaneous than sequential.[5]

With these considerations in mind we can develop a systematic approach to the quantitative solving of problems involving the ionization of weak acids or bases. As an example let us calculate the concentrations of the species HCO_2H, HCO_2^-, and H_3O^+ present at equilibrium in a 0.0150 *M* solution of formic acid, for which $K_a = 1.77 \times 10^{-4}$ at 25°C. In step 1 the HCO_2H is considered to dissolve and mix with the solvent, but not to react. The concentration of HCO_2H is thus 0.0150 *M*, while the concentrations of H_3O^+ and HCO_2^-, the products of the ionization step, are considered to be zero. ($[HCO_2^-]$ is zero because none has yet been produced; $[H_3O^+]$, at 10^{-7} *M*, is negligibly small in comparison to the amount that will be produced in step 2.) In step 2, the ionization reaction is "turned

[5]The reason that a model, which is known to differ markedly from reality (as our two-step model does), can nonetheless lead us to correct numerical values is bound up with the fact that we are discussing an *equilibrium* process. It is a fundamental fact that a system in equilibrium is stable and unchanging, and moreover *has no "memory" of how the equilibrium condition was achieved*. In a solution of a weak acid, the equilibrium concentrations depend not at all on the path by which the equilibrium was achieved, and thus we may make our calculations on the basis of the most convenient route, free from worry that the actual path may have been different.

on" and allowed to come to equilibrium. In this step part of the HCO_2H ionizes; since the extent of ionization is as yet undetermined, let us denote by x the amount of HCO_2H that ionizes. The HCO_2H concentration is reduced by this amount, so that at equilibrium $[HCO_2H] = 0.0150 - x$. Since each molecule of HCO_2H that ionizes results in the formation of one H_3O^+ and one HCO_2^-, the concentration of both of these species is increased (from the initial value of zero) by the same amount. This information may be conveniently summarized as follows:

	$HCO_2H + H_2O \rightleftharpoons$	H_3O^+ +	HCO_2^-
initial molarity:	0.0150	0	0
equilibrium molarity:	$0.0150 - x$	x	x

$$x = [H_3O^+] = [HCO_2^-] \quad \text{(at equilibrium)}$$

It is very important to write down the preceding summary, abbreviated by whatever personal system you choose to devise, *every time* you begin an acid-base problem. The time it takes is minimal, and the help it provides is great. In our summary we have not written anything for the concentration of water, whether initially or at equilibrium; its concentration, which is a constant 55 M, is incorporated in the value of K_a, and does not enter into the numerical calculations.

It is a simple matter now to substitute the equilibrium concentrations into the ionization constant expression.

$$K_a = \frac{[H_3O^+][HCO_2^-]}{[HCO_2H]} \tag{3-11}$$

$$1.77 \times 10^{-4} = \frac{x^2}{0.0150 - x} \tag{3-12}$$

We multiply both sides by $(0.0150 - x)$ and move all terms to the same side of the equation, to get

$$0 = x^2 + (1.77 \times 10^{-4})x - 1.77 \times 10^{-4}(0.0150) \tag{3-13}$$

In this form the equation is recognized as a quadratic equation whose general form is

$$ax^2 + bx + c = 0 \tag{3-14}$$

and whose general solution is

$$x = \frac{-b \pm \sqrt{b^2 - 4ac}}{2a} \tag{3-15}$$

In our example, therefore, we have

$$\begin{aligned} x &= \frac{-1.77 \times 10^{-4} \pm \sqrt{(1.77 \times 10^{-4})^2 + (4)(1)(1.77 \times 10^{-4})(0.0150)}}{(2)(1)} \\ &= \frac{-1.77 \times 10^{-4}}{2} \pm \frac{3.26 \times 10^{-3}}{2} \\ &= \frac{-3.44 \times 10^{-3}}{2} \quad or \quad \frac{3.08 \times 10^{-3}}{2} \end{aligned} \tag{3-16}$$

Since x represents $[H_3O^+]$ and $[HCO_2^-]$ and cannot be a negative number, there is no difficulty in selecting the correct value of x.

$$[H_3O^+] = [HCO_2^-] = \frac{3.08 \times 10^{-3}}{2} = 1.54 \times 10^{-3}\ M$$

This is generally true of numerical problems in science; in cases where, as with quadratic equations, the *mathematical* problem has more than one correct solution, one finds that only one of them corresponds to a physically meaningful situation. All other answers are physically impossible, as, for example, a negative concentration.

The concentration of nonionized HCO_2H is readily determined: it is just the initial value less the amount that has disappeared via the ionization reaction, or $[HCO_2H] = 0.0150 - 0.00154 = 0.0135\,M$. The extent of ionization is $[HCO_2^-]/([HCO_2H] + [HCO_2^-])$ or $1.54 \times 10^{-3}/0.0150 = 0.103$; the percent ionization is 100 times that, or 10.3%.

We can easily generalize this result to any acid, with ionization constant K_a and initial concentration C molar. The ionization-constant expression is

$$K_a = \frac{x^2}{C - x} \tag{3-17}$$

the quadratic equation is

$$x^2 + K_a x - K_a C = 0 \tag{3-18}$$

and the solution is

$$x = \frac{-K_a + \sqrt{K_a^2 + 4K_a C}}{2} \tag{3-19}$$

As before, the percent ionization is $100x/C$. Once again we stress that these results do not depend on whether a Brønsted-Lowry or an Arrhenius frame of reference is used.

3.5 WEAK BASES

The quantitative description of the ionization of weak bases is very similar to that of acids. A monohydric base such as methylamine, CH_3NH_2, ionizes according to

$$CH_3NH_2 + H_2O \rightleftharpoons CH_3NH_3^+ + OH^-$$

and at equilibrium obeys the equilibrium expression

$$K = \frac{[CH_3NH_3^+]\,[OH^-]}{[H_2O]\,[CH_3NH_2]} \tag{3-20}$$

As before and for precisely the same reasons, the water concentration in dilute solutions is essentially constant (55 M) at varying base concentrations and can be incorporated into the equilibrium constant. We define the new constant, the *ionization constant*, and give it the symbol K_b:

$$K_b = [H_2O]\,K \tag{3-21}$$

so that the equilibrium expression is

$$K_b = \frac{[CH_3NH_3^+][OH^-]}{[CH_3NH_2]} \tag{3-22}$$

It is at this point that we notice a difference from our treatment of acids, and, although it will have no effect on the numerical calculations, it can present conceptual problems to students. The difference is this: whereas the expression for *acid* ionization (in the case of benzoic acid)

$$K_a = \frac{[H^+][Bz^-]}{[HBz]} = \frac{[H_3O^+][Bz^-]}{[HBz]} \tag{3-23}$$

corresponds exactly to an Arrhenius-type dissociation,

$$HBz \rightleftharpoons H^+ + Bz^-$$

the expression (3-22) for the *base* ionization does not. If we follow the usual rules for writing equilibrium-constant expressions, the (hypothetical) reaction that corresponds to the base equilibrium expression (3-22) is

$$CH_3NH_2 \rightleftharpoons CH_3NH_3^+ + OH^-$$

This is obviously *not* a balanced equation for base dissociation; indeed it cannot be, and Arrhenius theory cannot explain why CH_3NH_2, which contains no hydroxide, is a base. The hydroxide ion, which gives such solutions their basic properties, comes not from the base but from the water. This is consistent with the Brønsted-Lowry definition of a base as a H^+ acceptor rather than a OH^- donor.

In spite of this discrepancy, people continue to speak and write of "base dissociation," although, happily, the practice is not as common as it once was. The habit persists because the treatment of acids and bases is identical in most other respects; in particular, the numerical calculations and the algebraic form of the equations are the same. As an example we will calculate the hydroxide ion concentration of a 0.300 *M* solution of methylamine, for which $K_b = 4.17 \times 10^{-4}$ at 25°C. Underneath the stoichiometric equation we write the initial concentrations of all species except water. We define the amount of ionization as x and write the equilibrium concentrations for all species except water as follows:

	CH_3NH_2	$+ H_2O \rightleftharpoons$	$CH_3NH_3^+$	$+ OH^-$
initial molarity:	0.300		0	0
equilibrium molarity:	$0.300 - x$		x	x

x is the amount of CH_3NH_2 ionized, and also $x = [OH^-] = [CH_3NH_3^+]$ formed at equilibrium. Again we have used the two-step model to reach equilibrium. Note that the hydroxide-ion and methylammonium-ion concentrations must be equal, since the ionization reaction is their only source. The algebraic expression for the equilibrium is thus

$$K_b = \frac{[CH_3NH_3^+][OH^-]}{[CH_3NH_2]}$$

$$4.17 \times 10^{-4} = \frac{x^2}{0.300 - x} \tag{3-24}$$

We rearrange this to get the quadratic equation

$$0 = x^2 + (4.17 \times 10^{-4})x - (4.17 \times 10^{-4})(0.300) \tag{3-25}$$

which has the two solutions

$$x = \frac{-4.17 \times 10^{-4} \pm \sqrt{(4.17 \times 10^{-4})^2 + 4(4.17 \times 10^{-4})(0.300)}}{2} \tag{3-26}$$

$$= -1.14 \times 10^{-2}\ M \quad \text{or} \quad +1.10 \times 10^{-2}\ M$$

Once again we reject the negative root because a molarity cannot be a negative number.

$$[OH^-] = [CH_3NH_3^+] = x = 1.10 \times 10^{-2}\ M$$

$$[CH_3NH_2] = 0.300 - x = 0.289\ M$$

The percent ionization is

$$\frac{1.10 \times 10^{-2}}{0.300} \times 100 = 3.67\%$$

3.6 BASICITY

Just as the term *acidity* is defined as the concentration of the aqueous protons that give the solution its acid properties, so we may define *basicity* as the concentration of the hydroxide ions that give solutions their basic properties. Thus in our example we have calculated the basicity of a 0.300 *M* methylamine solution. Further, it is common to define, by analogy to pH, the parameter pOH:

$$\text{pOH} = -\log_{10}[OH^-] \tag{3-27}$$

In our example, then, pOH is $-\log_{10}(1.10 \times 10^{-2})$ or $-(-1.96) = 1.96$. Solution pOH commonly falls within the range of 0 to 14, and the lower the pOH, the greater the basicity. It should be noted, however (and we will discuss this in the following section), that solutions of bases are often referred to by pH rather than by pOH.

3.7 AUTOIONIZATION

As mentioned in Section 2.7, water is a member of that class of substances called *amphiprotic*; it can accept as well as donate a proton and thus act as *either an acid or a base*. This behavior is implicit in the example reactions we have used:

$$\underset{\text{}}{HBz} + \underset{\text{acceptor}}{H_2O} \rightleftharpoons H_3O^+ + Bz^-$$

$$CH_3NH_2 + \underset{\text{donor}}{H_2O} \rightleftharpoons CH_3NH_3^+ + OH^-$$

and occurs to a lesser extent in pure water as well.

$$\underset{\text{acceptor}}{H_2O} + \underset{\text{donor}}{H_2O} \rightleftharpoons H_3O^+ + OH^-$$

A small number of water molecules undergo this *autoionization* or *autoprotolysis* reaction, with half of them acting as acids and half of them acting as bases. The ionization reaction is an equilibrium in which

$$K = \frac{[H_3O^+]\,[OH^-]}{[H_2O]\,[H_2O]} \tag{3-28}$$

As before, we recognize the constancy of water concentration and define

$$K_w = K[H_2O]^2$$

so that the equilibrium expression becomes

$$K_w = [H_3O^+]\,[OH^-] \tag{3-29}$$

(Note that this relationship no longer even looks like an equilibrium-constant expression, since, lacking a denominator, it would have to correspond to a stoichiometric equation lacking reactants.)

K_w is often referred to as the *ion-product constant* of water. K_w, like almost all other equilibrium constants, varies with temperature. At 20°C, $K_w = 6.81 \times 10^{-15}$, at 25°C it equals 1.01×10^{-14}, and at 30°C it equals 1.47×10^{-14}. A more complete set of values is given in Appendix I.

The existence of water autoionization means that the equilibrium expression

$$K_w = [H_3O^+]\,[OH^-]$$

or its equivalent form,

$$[H_3O^+] = \frac{K_w}{[OH^-]}$$

may be used to determine $[H_3O^+]$ when $[OH^-]$ is known. Thus in the 0.3 *M* methylamine example the aqueous proton concentration at 25°C is calculated to be $[H_3O^+] = 1.00 \times 10^{-14}/1.10 \times 10^{-2} = 9.09 \times 10^{-13}$ *M*, from which the pH is found as $-\log(9.09 \times 10^{-13}) = 12.04$. It is possible, then, to express the acidity-basicity of a solution by *either* its pH or its pOH, regardless of whether the solution is acidic or basic, and this is commonly done. The pH is the accepted measure of the strengths of both acids and bases.

Conversions between proton and hydroxide concentrations are somewhat easier if the transaction is handled in pH units rather than in moles per liter. If we take the common log of both sides of equation (3-29),

$$\log K_w = \log [H_3O^+] + \log [OH^-]$$

and multiply both sides by −1,

$$-\log K_w = -\log [H_3O^+] - \log [OH^-]$$

$$pK_w = pH + pOH \tag{3-30}$$

The sum of pH and pOH in *any* aqueous solution, regardless of other species which may also be present, is pK_w.

We defined neutral solutions previously as those which have neither acidic nor basic properties; a more precise and useful definition can be put in terms of pH and pOH. In a neutral solution the concentrations of H_3O^+ and OH^- equal one another, and pH = pOH in neutral solutions. At 25°C, pK_w = 14, so that pH = pOH = 7.00. This leads to the casual statement that "the pH of a neutral solution is 7.00." The statement is careless because it is true only at 25°C.

Example 3.10

Calculate the pH of water at 30°C.

Answer: From Appendix I we see that at 30°C K_w has a value of 1.47×10^{-14}. Let us solve this problem using the same method developed previously for acid or base solutions; that is, we will treat water as an acid (and base) that ionizes according to the proton-transfer reaction

$$H_2O + H_2O \rightleftharpoons H_3O^+ + OH^-$$

We define x as the concentration of H_3O^+ produced in the ionization, and from the stoichiometry the concentration of OH^- produced is also equal to x, so that

	$H_2O + H_2O \rightleftharpoons$	H_3O^+	$+ OH^-$
initial molarity:		0	0
equilibrium molarity:		x	x

As before, the value of H_2O concentration is incorporated in K_w and is not used explicitly in problems. At equilibrium

$$K_w = [H_3O^+][OH^-] = x^2$$

$$x = \sqrt{K_w} = \sqrt{1.47 \times 10^{-14}} = 1.21 \times 10^{-7}$$

$$pH = pOH = -\log_{10}(x) = 6.92$$

We have solved this problem in a roundabout fashion to emphasize the *essential similarity of all calculations involving ionic equilibria*, a point we will continue to stress. A shorter method of working this problem begins with equation (3-30):

$$\text{pH} + \text{pOH} = -\log_{10}(K_w)$$

Since in neutral solutions pH = pOH, we have

$$\text{pH} = \text{pOH} = -\frac{1}{2}\log_{10}(K_w)$$

and at 30°C,

$$\begin{aligned}\text{pH} &= (-\tfrac{1}{2})\log(1.47 \times 10^{-14})\\ &= -\frac{1}{2}(-13.833)\\ &= 6.92\end{aligned}$$

3.8 AN APPROXIMATE METHOD

We have seen that the calculation of pH in both acids and bases involves solving the same quadratic equation,

$$K = \frac{x^2}{C - x} \tag{3-17}$$

or

$$x^2 + Kx - KC = 0 \tag{3-18}$$

where K = either K_a or K_b
x = either $[H_3O^+]$ or $[OH^-]$
C = initial molarity of acid or base

the solution is

$$x = \frac{-K + \sqrt{K^2 + 4KC}}{2} \tag{3-31}$$

Note that we have explicitly rejected the negative root as being physically meaningless. This expression, although correct, is somewhat difficult to use (even with a calculator). It is certainly not an equation that permits quick estimates to be made in one's head. As it happens, under most conditions it is possible to use a much simpler expression which, although not exact, nonetheless yields results of acceptable accuracy.

Suppose that we have an acid or base solution in which the percent ionization is quite small, say 1% or less. This means that the fraction x/C has a value of 0.01 or less, and $C - x$, the denominator of equation (3-17), becomes approximately equal to C. This result is obtained algebraically by factoring $C - x$ and substituting $x/C = 0.01$:

$$C - x = C - x\left(\frac{C}{C}\right) = C - C\left(\frac{x}{C}\right)$$
$$= C\left(1 - \frac{x}{C}\right) = C(1 - 0.01) = 0.99C$$

Now if we are willing to accept as correct an answer that is within 1% of the true value, we may say that 0.99 is approximately equal to 1, and

$$C - x = 0.99C \approx C$$

Equation (3-17) now becomes

$$K = \frac{x^2}{C - x} \approx \frac{x^2}{C} \qquad (3\text{-}32)$$

and the solution is

$$x \approx \sqrt{KC} \qquad (3\text{-}33)$$

The extent of ionization is

$$\frac{x}{C} \approx \frac{\sqrt{KC}}{C} = \sqrt{K/C} \qquad (3\text{-}34)$$

Example 3.11

Calculate $[OH^-]$ and pH for 0.150 *M* methylamine, using both the exact equation and the approximation. $K_b = 4.17 \times 10^{-4}$.

Answer: Exact equation:

$$[OH^-] = \frac{-K_b + \sqrt{K_b^2 + 4K_bC}}{(2)(1)}$$
$$= \frac{-4.17 \times 10^{-4} \times \sqrt{(4.17 \times 10^{-4})^2 + 4(4.17 \times 10^{-4})(0.150)}}{2}$$
$$= 0.00770\ M$$
$$\text{pOH} = -\log\,[OH^-] = 2.11$$
$$\text{pH} = 14 - \text{pOH} = 11.89$$

Approximate form:

$$[OH^-] = \sqrt{K_bC}$$
$$= \sqrt{4.17 \times 10^{-4}(0.150)} = 0.00791\ M$$
$$\text{pOH} = 2.10$$
$$\text{pH} = 14 - \text{pOH} = 11.90$$

Error:

$$\% \text{ error} = \frac{[OH^-]_{approx} - [OH^-]_{true}}{[OH^-]_{true}}$$

$$= \frac{0.00791 - 0.00770}{0.00770} \times 100 = 2.7\%$$

The question now is: When is it permissible to calculate pH with this approximate method? The answer depends to a great extent on the problem to be solved, or rather on the requirements for accuracy that are inherent in the problem. That is, if for some reason the problem demands an answer accurate to better than 0.1%, then one cannot approximate when the percent ionization is even as great as 1%. However, if the extent of ionization $\sqrt{K/C}$ is less than .01 (equation 3-34), that is, if the quotient K/C is no larger than 10^{-4}, the approximate method will yield answers that are within 1% of the correct value. An accuracy of ±1% is adequate for many calculations. Further information about accuracy of calculations is developed in Appendix VI.

3.9 VERY DILUTE SOLUTIONS AND VERY WEAK ACIDS

Although the Brønsted-Lowry theory is quite explicit as to the nature and extent of the participation of water in acid-base ionization, the fact that water is subject to autoionization leads to the participation of water in quite another sense. This second mode of participation becomes important only in very dilute solutions and/or with very weak acids or bases. Simply put, if an acid is so dilute or so weak that the amount of H_3O^+ produced by the ionization of the acid is comparable to or less than that initially present as a result of the autoionization of water, the mathematical procedures developed thus far cannot give accurate answers. Correct results should not, in fact, be expected of these procedures since autoionization was ignored in their development. This point is illustrated by the following examples.

Example 3.12

Calculate the pH of a 3.00×10^{-8} M solution of the weak base aniline ($K_b = 1.10 \times 10^{-6}$), using the approximate method.

Answer: The approximate method gives

$$[OH^-] \approx \sqrt{K_b C} = \sqrt{1.10 \times 10^{-6} \times 3.00 \times 10^{-8}}$$

$$\approx \sqrt{3.30 \times 10^{-14}} = 1.82 \times 10^{-7}\ M$$

$$pOH = -\log[OH^-] \approx 6.74$$

$$pH = 14 - pOH \approx 7.26$$

As we will see in the following discussion, *this answer is wrong*.

At first this result may seem correct, for a pH of 7.26 is appropriate to a solution that is slightly basic; however, further consideration shows that the calculated $[OH^-]$ is more than six times greater than the original amount of weak base present. The approximation is apparently invalid under these conditions. Indeed, a quick estimate, which should have been made at the outset, shows this to be the case: $K/C = 1.1 \times 10^{-6}/3 \times 10^{-6} = 37$, which is far greater than 10^{-4}. With only slight regret at having wasted the time, let us try another approach.

Example 3.13

Use the quadratic method to calculate the pH of the solution in Example 3.12.

Answer: The quadratic expression is

$$[OH^-] = \frac{-K_b + \sqrt{K_b^2 + 4K_bC}}{2}$$

which yields

$$[OH^-] = 2.92 \times 10^{-8}\ M$$

That is, the aniline is almost 100% ionized. This, at least, is not inconsistent with known behavior of weak acids and bases; their extent of ionization increases as the concentration decreases. When we calculate the pH, however, another problem becomes apparent.

$$pOH = -\log\ [OH^-] = -\log(2.92 \times 10^{-8}) = 7.53$$

$$pH = 14 - pOH = 6.47$$

This result is also wrong. The pH of pure water is 7.00, yet the pH of the methylamine solution shows it to be more acid than water! How can the addition of a base to water result in an acid solution? Of course it cannot; our procedure is thus shown to be invalid at very low concentrations.

Example 3.14

Calculate the pH of a 0.050 M solution of methanol, $K_a = 3.2 \times 10^{-16}$. (Methanol, like most other alcohols, is such a poor proton donor that it is not usually classified as an acid; in fact, it is an extremely weak acid.)

Answer: In this case K/C is about 6×10^{-15}, well below 10^{-4}; the approximate form will therefore give the same result as the quadratic form:

$$[H_3O^+] = \sqrt{K_aC} = \sqrt{3.2 \times 10^{-16}(0.050)} = \sqrt{1.6 \times 10^{-17}}$$
$$= 4.0 \times 10^{-9}\ M$$

$$pH = 8.4$$

Another incorrect result.

Once again the result is physically impossible; the addition of a weak acid to pure water has resulted in a solution that is *less acidic* than water itself. The procedure we have relied on up to now has failed again, this time when confronted with an extremely weak acid.

The inference to be drawn from Examples 3.12, 3.13, and 3.14 is that an explicit recognition of the existence of a second source of H_3O^+ or OH^-, namely, the autoionization of the solvent itself, must be included in the development of a valid procedure for calculating the pH of solutions when the acid is either very weak or very dilute. Perhaps the simplest way of demonstrating this is to ask: What is the concentration of H_3O^+ in 10^{-9} *M* HCl? Up to now we would have said that HCl is a strong acid (100% ionized), and therefore the concentration of $[H_3O^+]$ is 10^{-9} *M*. But in pure water $[H_3O^+] = 10^{-7}$; addition of an acid, however weak or strong, in any quantity, cannot decrease the concentration of $[H_3O^+]$ already present in pure water.

One possible approach to the problem of dilute solutions might be to say that there are two equilibria, operating simultaneously:

$$2H_2O \rightleftharpoons H_3O^+ + OH^- \tag{3-35}$$

$$HA + H_2O \rightleftharpoons H_3O^+ + A^- \tag{3-36}$$

The total $[H_3O^+]$ would be the sum of the individual contributions, which could be calculated separately.

Example 3.15

Assume that solvent autoionization is independent of any additional source of H_3O^+ and estimate the pH of a solution of 1.50×10^{-7} *M* HCl.

Answer: In pure water, $[H_3O^+] = 1.00 \times 10^{-7}$ *M*. The strong acid HCl dissociates completely and so contributes 1.50×10^{-7} mol/liter of H_3O^+ to the solution. The sum of these contributions is 2.50×10^{-7} *M*, corresponding to the pH of 6.60. (The true value, calculated by the methods discussed in Appendix VI, is 2.00×10^{-7} *M*, or pH = 6.70.)

Unfortunately, this approach rests on the false assumption that the two sources of H_3O^+ behave independently of one another; this is simply not the case, as a moment's reflection will show. The principle of Le Châtelier, which governs all equilibrium situations, tells us that the additional H_3O^+ released by the acid will diminish the extent to which the autoionization reaction (3-35) proceeds to the right. The amount of H_3O^+ produced by autoionization is less than it would be if there were no other source of H_3O^+. Similarly, if the acid is weak, H_3O^+ produced by autoionization will shift the equilibrium [reaction (3-36)] to the left and decrease the amount of H_3O^+ supplied by acid ionization. This interaction between the two equilibria introduces complications in the problem.

Fortunately, another approach is available, one that is conceptually simple and soundly based on the physical principles of conservation of mass and conservation of electric charge. We still consider both of the equilibria [reactions (3-35) and (3-36)], but we do not need to assume that they contribute *independently* to the aqueous proton concentration. This approach is developed in Appendix VI, and leads to a cubic equation (3-37) by which $[H_3O^+]$ may be calculated from K_a, C, and K_w:

$$\frac{1}{K_a}[H_3O^+]^3 + [H_3O^+]^2 - \left(C + \frac{K_w}{K_a}\right)[H_3O^+] - K_w = 0 \tag{3-37}$$

The equation is best solved by computer, using the BASIC program given in Appendix II.2.

3.10 CONCLUSION

In this chapter we have discussed pH calculations involving strong acids and bases and the incomplete ionization of weak acids and bases. We have developed several approaches to the calculation of pH in solutions of acids or bases. Each of these approaches has limitations, and there is a range of values of K and C in which each of the treatments should be used. The exact treatment is cumbersome and requires access to a computer: it should be used only when the other methods fail to yield a result of sufficient accuracy. (It should be noted, however, that the BASIC program of Appendix II.2 can readily be modified for use on a programmable calculator.) We suggest using the approximate form (3-33) if $K/C < 10^{-4}$; we also recommend use of the exact equation (3-37) if either $C < 10^{-6}$ or $6 < \text{pH} < 8$ (see Appendix VI).

All of these considerations as to the precision of pH calculations must be interpreted in the light of the ionization constant, which may itself be inaccurate or imprecisely known. Ionization constants, whether taken directly from the literature or from compilations such as Appendix III, are rarely known to more than two significant figures. However, this is sufficiently accurate to ensure an error of less than 1% in calculations of $[H_3O^+]$. In a few cases K is not even known to two significant figures. For example, the value of K_a for HCN is known only to one significant figure, $K_a = 7\text{-}9 \times 10^{-10}$ (this can also be written $8 \times 10^{-10} \pm 13\%$). However, the % error in $[H_3O^+]$ is not as large as the error in K_a. Suppose that we calculate the $[H_3O^+]$ in a 0.01 M solution of HCN. Since $K_a/C < 10^{-4}$, the approximate method is valid; we calculate $[H_3O^+] \approx \sqrt{K_aC} = 2.6 \times 10^{-6}\ M$ for $K_a = 7 \times 10^{-10}$ and $\sqrt{K_aC} = 3.0 \times 10^{-6}$ for $K_a = 9 \times 10^{-10}$. The result can be expressed as $2.8 \pm 0.2 \times 10^{-6}$, that is, an accuracy of 5-10%. If an inaccurate K_a value introduces a 10% uncertainty, we should not worry about computational errors of only 1%.

Another factor bearing on the accuracy of pH calculations is temperature. Ionization constants, like all equilibrium constants, are temperature dependent. There are unfortunately few general rules as to the magnitude of the temperature effect. Carboxylic acids, for example, have K_a values that vary only slightly in the range 15 to 30°C, and phosphoric acid is similar. The ionization constant of phenol, on the other hand, decreases by about 30% as the temperature is raised from 20 to 30°C. Base ionization constants also tend to have larger temperature effects. Clearly, the best procedure is to do the work at the temperature at which the ionization constant is known, or to know the ionization constant at the temperature at which the work is done.

Finally, in precise work, solute activities are used rather than concentrations. This approach to pH calculations is outlined in Appendix IV.

PROBLEMS

3.1 Express the following quantities in moles.

(a) 10 g of HCl
(b) 5 g of H_2SO_4
(c) 1500 g of HNO_3
(d) 12.6 g of $Mg(OH)_2$
(e) 1.8×10^{-5} g of CH_3CO_2H
(f) 25.0 g of SF_6
(g) 7.2 g of H_3PO_4
(h) 5.0 g of Na_2HPO_4

3.2 Express the following quantities in grams.

(a) 0.082 mol of $Al(OH)_3$
(b) 6.02 mol of $C_6H_5CO_2H$
(c) 1.38 mol of $Ag(NH_3)_2Cl$
(d) 4.18 mol of Li_3PO_4
(e) 1.99 mol of $NaCl$
(f) 6.63 mol of H_3PO_4
(g) 3.00 mol of HCl

3.3 Calculate the molarity of each of the following solutions:

(a) 15 g of $NaCl$ in 100 mL of solution
(b) 50 g of H_2SO_4 in 500 mL of solution
(c) 0.1 g of phenol (C_6H_5OH) in 1 liter of solution
(d) 0.038 g of barium hydroxide in 50 mL of solution
(e) 0.05 g of iodine in 100 mL of CCl_4

3.4 Calculate the weight of solute required to prepare each of the following solutions:

(a) 350 mL of 0.01 *M* CsI
(b) 50 mL of 3.0 *M* KOH
(c) 500 mL of 0.1 *M* glucose ($C_6H_{12}O_6$)
(d) 15 liters of 0.05 *M* H_3PO_4

3.5 For each of the following solutions, change the concentration from molarity to normality or from normality to molarity, as appropriate.

(a) 0.5 *M* H_2SO_4
(b) 0.1 *M* HBr
(c) 1.5 *M* $Ba(OH)_2$
(d) 0.08 *M* NH_3
(e) 3.5×10^{-3} *M* HNO_3
(f) 0.75 *N* KOH
(g) 0.005 *N* H_2SO_4
(h) 0.39 *N* $Al(OH)_3$
(i) 3.5 *N* HCl
(j) 0.1 *N* H_3PO_4

3.6 Calculate the molarity and the normality of each of the following aqueous solutions. You may assume that each solution has a density of 1.00 g/mL. The concentrations are all expressed as percent by weight.

(a) 0.5% H_2SO_4
(b) 0.2% $NaCl$
(c) 1.27% $Mg(OH)_2$
(d) 5.0% $C_6H_{12}O_6$
(e) 0.035% H_3PO_4
(f) 1.05×10^{-3}% HNO_2

3.7 The density of a 6.0% (by weight) solution of ammonium nitrate is 1.0255 g/mL at 17°C. Calculate its molarity. Repeat the calculation assuming the density to be 1.00 g/mL and compare the answers.

3.8 The density of a 1.5 *M* K_2CO_3 solution at 20°C is 1.169 g/mL. Express the concentration as weight percent K_2CO_3.

3.9 Derive a general formula for converting weight percent to molarity. (Your formula should include density and molecular weight.) Use this formula to check the answer to the preceding problems.

3.10 How much water must be added to 250 mL of a 0.100 *N* solution to prepare a solution of 0.0300 *N*?

3.11 It is desired to prepare 100 mL of a 0.0375 *M* solution, starting from a stock solution of 0.750 *M*. How much stock solution is required? How much water?

3.12 How much water must be added to x milliliters of a 0.400 *N* solution to effect a four-fold dilution (i.e., to decrease the concentration by a factor of 4, from 0.400 to 0.100)?

3.13 One liter of a 0.1 *M* solution is to be concentrated by warming to evaporate part of the water. What will be the final volume when the concentration has increased to 0.4 *M*? How much water will have been driven off?

3.14 At 20, 25, and 30°C the density of water is 0.99823, 0.99707, and 0.99567 g/mL, respectively. Calculate the molarity of water at each of these temperatures.

3.15 At 20°C, a solution containing 40.00 g of NH_4NO_3 per liter of solution has a density of 1.0148 g/mL. Calculate the molarity of NH_4NO_3 and of H_2O.

3.16 A 0.5 *N* solution of H_2SO_4 has a density of 1.0145 g/mL at 20°C. Calculate the percent (by weight) composition and the molarity of water.

3.17 How much potassium dichromate ($K_2Cr_2O_7$) must be used to prepare 375 mL of a 0.250 *M* solution?

3.18 How many grams of solute is required to prepare each of the following solutions?
(a) 500 mL of 0.500 *N* acetic acid
(b) 75 mL of 0.050 *M* alanine [$CH_3CH(NH_2)CO_2H$]
(c) 350 mL of 0.500% NaCl (by weight; assume that the solution density = 1.00 g/mL)
(d) 200 mL of 2.50 *N* H_3PO_4
(e) 50.0 mL of 0.375 *N* $Ba(OH)_2$

3.19 Describe the procedure as well as the recipe (number of grams of NaCl) for preparing 0.500 mL of a 0.30 *N* solution of NaCl. How should the recipe and the procedure change if you were to prepare 500 mL of 2.00% (by weight) NaCl instead?

3.20 Calculate the pH of the following solutions of strong acids.
(a) 5.00×10^{-3} *M* HCl
(b) 0.03 *M* HNO_3
(c) 7.51×10^{-2} *M* HBr
(d) 1.00% HCl (weight percent; assume that the density = 1.00 g/mL).
(e) 0.05% HBr (weight percent; assume that the density = 1.00 g/mL).

3.21 To what HCl molarity do the following pH values correspond?
(a) 3.05 (b) 5.7 (c) 0.0 (d) −1.00 (e) 2.43

3.22 Calculate the pH of the following solutions of strong bases.
(a) 3 *N* NaOH
(b) 5.8×10^{-2} *N* KOH
(c) 0.00018 *M* $Ba(OH)_2$
(d) 1.325×10^{-4} *N* $Ba(OH)_2$
(e) 2.5 g/liter of lithium hydroxide

3.23 Using the data in Appendix I, calculate the pH of pure water at 10, 20, 30, and 40°C. Plot these results and use your graph to estimate the pH at 25°C. Check this result with the exact value from Appendix I.

3.24 If a 0.0375 *N* solution of a weak base is 3.00% ionized, what is the hydroxide-ion concentration?

3.25 Calculate the hydroxide-ion concentration in the following solutions of weak bases, given the percent ionization.
(a) 3.6×10^{-2} *N* (1.5%)
(b) 8.0×10^{-3} *N* (47%)
(c) 0.15 *N* (0.020%)
(d) 0.78 *N* (3.2×10^{-3}%)

3.26 Calculate the degree or extent of ionization in each of the following solutions of weak, monoprotic acids.
(a) 0.015 *M*, $[H_3O^+] = 10^{-4}$ *M*
(b) 7.5×10^{-3} *M*, pH = 5.2
(c) 9.95×10^{-2} *M*, $[H_3O^+] = 1.5 \times 10^{-3}$ *M*
(d) 1.00 *M*, $[H_3O^+] = 1.5 \times 10^{-4}$ *M*
(e) 4.5×10^{-3} *M*, pH = 5.52
(f) 3.75×10^{-3} *M*, $[H_3O^+] = 3.5 \times 10^{-3}$ *M*

3.27 Some tabulations of acid strength give values for pK_a rather than K_a. Are the following pK_a values arranged in order of increasing or decreasing acid strength? 2.43, 3.85, 6.02, 9.13.

3.28 Convert the pK_a values in Problem 3.27 to K_a values.

3.29 If a 0.050 *M* solution of a monoprotic acid has a pH of 3.2, what is its ionization constant?

3.30 The pH of a solution of a weak base is 9.00. If the solution is 1.50×10^{-3} *M*, what is K_b?

3.31 A monohydric base is 5.0% ionized when its concentration is 0.0015 *M*. Calculate pH, pOH, and K_b.

3.32 The extent of ionization is 0.040 in a 0.75 *M* solution of a monoprotic acid. Calculate K_a.

3.33 Calculate the pH of the following solutions of weak monoprotic acids, using both the approximate form and the quadratic form. In which cases is the approximate form accurate to at least 1%? (The K_a values required in this and the following problems may be found in Appendix III.)

(a) 0.035 *M* chloroacetic acid
(b) 1.5×10^{-3} *M* propanoic acid
(c) 0.0075 *M* benzoic acid
(d) 0.0625 *M* nitrous acid

3.34 Calculate the pH of the following solutions of weak monohydric bases, using both the approximate form and the quadratic form. In which cases is the approximate form sufficiently accurate?

(a) 1.00 *M* *N,N*-dimethylaniline
(b) 0.75 *M* ethylamine
(c) 9.2×10^{-2} *M* hydroxylamine
(d) 0.001 *M* hydrazine
(e) 3.62×10^{-3} *M* pyridine

3.35 Nitrous acid has a pK_a of 3.4. Calculate the pH of a 0.065 *M* solution using the approximate method, and estimate the error by equation (VI-7), in Appendix VI.

3.36 Using equation (VI-7) in Appendix VI, estimate the errors incurred by the use of the approximate form in Problem 3.34. Compare this with the actual difference between the results of the quadratic and approximate forms.

3.37 Assuming that answers must be within 1% of the true value, which of the three calculational methods (approximate, quadratic, or exact) is appropriate for each of the following problems? (More than one of the methods may be appropriate, in particular, the exact method always works.)

(a) pOH of 5.00×10^{-3} *M* ethylamine
(b) pH of 10^{-6} *M* HNO_2
(c) pOH of 1.00×10^{-3} *M* dimethylaniline
(d) pH of 6.25×10^{-7} *M* HCl

3.38 Using the appropriate method, obtain numerical answers for the solutions of Problem 3.37.

3.39 Go to your neighborhood computer, key in Programs II.1 and II.2 in Appendix II, and play with them for a while.

Chapter 4

Lewis Acids and Bases

4.1 INTRODUCTION

Although the Brønsted-Lowry concept of acidity is applicable to a wide variety of substances, there are many species that cannot donate or accept a proton and yet exhibit acidic or basic characteristics. For instance, carbon dioxide and sulfur dioxide are not Brønsted-Lowry acids, since they contain no protons; yet their aqueous solutions exhibit properties characteristic of Brønsted-Lowry acids. Solutions of CO_2 or SO_2 are able, for example, to neutralize proton-accepting bases such as sodium hydroxide.

4.2 WHAT IS A LEWIS ACID?

In order to account for the acidic behavior of compounds that do not contain protons, G. N. Lewis suggested (in 1923) a different definition of acid-base behavior.[1] He defined an acid as any atom, ion, or molecule that is "capable of receiving an electron pair" from another atom, ion, or molecule, provided that a chemical bond results from the interaction. He defined a base as "the atom, ion or molecule which has an electron pair that is capable of entering the valence shell of another atom, resulting in a bond." These definitions became a bit more concise over the next few years due to proposals by Sidgwick[2] and others, and now are stated as follows: A *Lewis acid* is an electron-pair acceptor, that is, any atom, ion, or molecule that is capable of receiving an electron pair; a *Lewis base* is an electron-pair donor, that is, any atom, ion, or molecule that is capable of donating an electron pair.

Several reactions that illustrate these definitions are shown in reactions (4-1) through (4-9).

$$\underset{\text{acid}}{BCl_3} + \underset{\text{base}}{:NH_3} \longrightarrow Cl_3B{-}NH_3 \qquad (4\text{-}1)$$

$$\underset{\text{acid}}{Ni} + \underset{\text{base}}{4:C{\equiv}O:} \longrightarrow Ni(CO)_4 \qquad (4\text{-}2)$$

[1] G. N. Lewis, *Valence and the Structure of Atoms and Molecules*, Chemical Catalog Co., New York, 1923.

[2] N. V. Sidgwick, *The Electronic Theory of Valency*, Oxford University Press, Oxford, 1929.

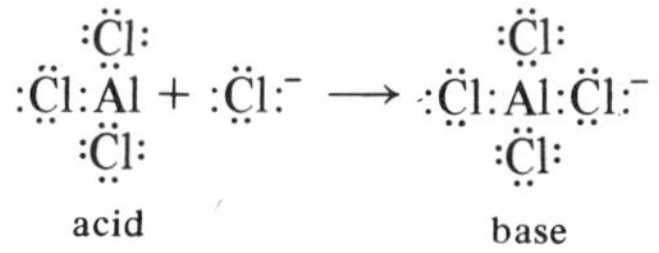

(4-3)

$$\underset{\text{acid}}{Ag^+} + \underset{\text{base}}{2{:}NH_3} \longrightarrow Ag(NH_3)_2^+ \tag{4-4}$$

$$\underset{\text{acid}}{H^+} + \underset{\text{base}}{H{-}\ddot{\underset{\cdot\cdot}{O}}{-}H} \longrightarrow H{-}\overset{+}{\ddot{O}}{-}H \;(\text{with } O{-}H \text{ below}) \tag{4-5}$$

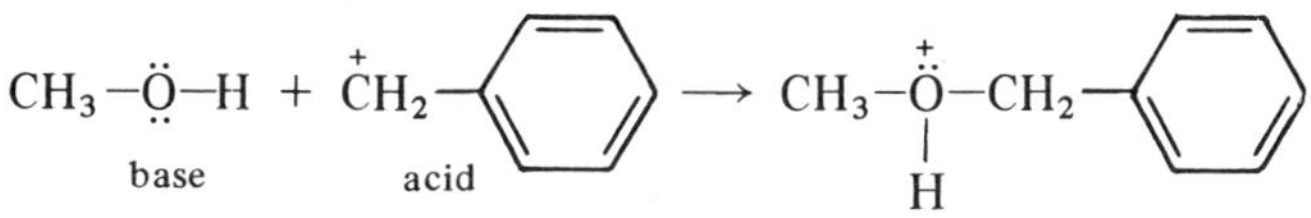

(4-6)

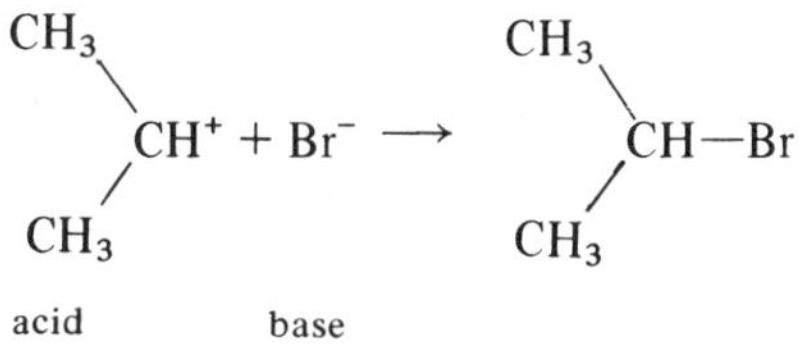

(4-7)

$$\underset{\text{acid}}{Fe^{3+}} + \underset{\text{base}}{3{:}\ddot{\underset{\cdot\cdot}{Cl}}{:}^-} \longrightarrow FeCl_3 \tag{4-8}$$

$$\underset{\text{acid}}{H^+} + \underset{\text{base}}{{:}\ddot{\underset{\cdot\cdot}{O}}{-}H^-} \longrightarrow H{-}\underset{\cdot\cdot}{\ddot{O}}{-}H \tag{4-9}$$

All of these are examples of the reaction of a Lewis acid with a Lewis base, and it is appropriate to refer to them as neutralization reactions. The product of such a reaction is also called an *adduct*, since it results from the addition of two reactant molecules. Adduct formation or neutralization, in the Lewis sense, is the formation of a *coordinate covalent* or "donor-acceptor" bond. A normal covalent bond is formed when each of two atoms provides an electron to form an electron pair, which is then shared by the two atoms. In a coordinate covalent bond, on the other hand, one atom donates *both* members of the electron pair, which is then shared by the two atoms. As illustrated in reaction (4-10),

$$A + {:}B \longrightarrow {}^{\delta-}A{-}B^{\delta+} \tag{4-10}$$

there may be considerable separation of charge (and thus a large dipole moment) in the resulting species. An unsymmetrical distribution of charge is conveniently denoted by the symbols $\delta+$ and $\delta-$, which indicate regions of low and high valence electron density, respectively.[3] Coordinate covalent bonds are not unique in this regard; almost all normal

[3] There are several other systems in use for labeling regions of high and low valence electron density. One of the most common is the system of "formal charges," which we discuss in Section 5.3. The "δ" is omitted in the formal charge system, and the product of reaction (4-10) would be written as $^-A{-}B^+$.

covalent bonds exhibit charge asymmetry to a greater or lesser extent. In this text we do not use the $\delta+$, $\delta-$ notation unless the specific location of positive or negative charge is important to the discussion.

4.3 STRENGTH OF LEWIS ACIDS

The strength of a Lewis acid is defined in terms of its tendency to accept an electron pair. If the tendency is large, the Lewis acid is strong. If we wish to determine which is the stronger of two Lewis acids, we compare the extent to which they react with the same Lewis base. This is illustrated by the reactions of Fe(II) and Fe(III) with chloride ion in reactions (4-11) and (4-12), in both of which an iron ion reacts as a Lewis acid by

$$Fe^{2+} + 2Cl^- \rightleftharpoons FeCl_2 \tag{4-11}$$

$$Fe^{3+} + 3Cl^- \rightleftharpoons FeCl_3 \tag{4-12}$$

accepting electron pairs. If these two Lewis acids differ in acid strength, one or the other will have the greater tendency to accept electron pairs; its reaction with the same base will proceed farther to the right. In fact, the stronger of the two acids is Fe^{3+}, because it has the higher charge and thus more strongly attracts the negative electron pairs than does the Fe^{2+} ion. In general, the more electron deficient a species is, the greater its ability to accept an electron pair and the greater its Lewis acid strength.

It is not necessary for a species to have a positive charge in order to be a Lewis acid. All that is necessary is for the substance to be capable of accepting an electron pair; that is, it must be electron deficient. Many common Lewis acids are neutral molecules. A reaction of one such acid, aluminum chloride, is shown in equation (4-3). In aluminum chloride the aluminum atom is bonded to three chlorine atoms and is uncharged. Two features of aluminum chloride account for its Lewis acidity. First, the aluminum atom is sharing a total of three electron pairs, and as a result has only six electrons in its valence shell (three of its own and one from each of the chlorine atoms to which it is bonded). Since the valence shell can accommodate eight electrons, the aluminum in $AlCl_3$ is electron deficient. Second, since chlorine is highly electronegative, the electrons in the Al–Cl bonds are pulled toward the chlorine atoms; this creates a further deficiency of electrons at the aluminum atom. This electronegativity argument applies to all compounds of formula MX_n (where M is a metal and X is a more electronegative atom); all such compounds may act as Lewis acids.

In general, any factor that increases the positive nature of (withdraws electron density from) a site within a molecule will also serve to increase the acidity of that site in the Lewis sense.

4.4 STRENGTH OF LEWIS BASES

A Lewis base is an electron-pair donor. The strength of a Lewis base depends on its ability to share an electron pair. If an atom, ion, or molecule has a strong attraction for its own electron pairs, these electrons will not be easily available for sharing with another species. Such a species is not a good Lewis base. Any factor that allows an atom, ion, or molecule to have a stronger hold on its electrons will decrease the availability of the electrons and thus decrease the basicity of the species.

Ammonia and amines (substituted ammonia) are very common bases. Reactions (4-13) and (4-14)[4] illustrate the ability of nitrogen to share its nonbonded electron pair (i.e., to

$$:NH_3 + H_2O \rightleftharpoons H:\overset{+}{N}H_3 + OH^- \tag{4-13}$$

$$:NR_3 + H_2O \rightleftharpoons H:\overset{+}{N}R_3 + OH^- \tag{4-14}$$

act as a Lewis base). An amine can be a stronger or weaker base than ammonia, depending on the nature of the R groups. If R represents an atom or group of atoms that is less electronegative than hydrogen, then the amine is a stronger base than ammonia. Having smaller electronegativity means that the R group has a lesser tendency to withdraw valence electron density from the nitrogen atom. The result is that the nitrogen atom has larger electron density; this in turn causes the nonbonded electron pair on the nitrogen atom to be less tightly held and to be more available for sharing with a Lewis acid. On the other hand, if the R groups are better able than hydrogen to withdraw electron density from nitrogen, the amine is a weaker Lewis base than ammonia. Among the factors that affect the ability of a substituent group to donate or withdraw electrons from another atom are inductive effects, electronegativity differences, resonance, and steric hindrance. Several of these effects are treated in more detail in Chapter 5.

4.5 TYPES OF LEWIS ACIDS

Lewis acids, which are defined as electron-pair acceptors, may be thought of in more general terms as electron-deficient species; most can be placed in one of the following categories.

A. Positive Ions

This class of Lewis acids contains the metal cations, all of which have unoccupied valence orbitals available for bonding. Heavy-metal cations such as Ni^{2+}, Cu^{2+}, Fe^{3+}, and Zn^{2+} are relatively strong Lewis acids. The alkaline earth ions Mg^{2+} and Ca^{2+} are weaker Lewis acids, and the alkali metal ions (Li^+, Na^+, etc.) rarely act as Lewis acids. Polyatomic cations, such as the carbocations,[5] also behave as Lewis acids. Some typical reactions of carbocations are shown in (4-6) and (4-7).

B. Compounds Containing an Atom Having Less Than a Full Octet of Electrons in the Valence Shell

Many important Lewis acids are electron deficient because they contain an atom with an incompletely filled valence shell. Common examples of this category are the boron

[4]Note that these reactions can be discussed equally well in Brønsted–Lowry terms. This is true of many acid–base reactions. This should not be thought of as a contradiction, rather as an additional way of understanding acid–base chemistry.

[5]In a carbocation (carbo-cat′-ion) the positive charge is located primarily on the carbon atom.

trihalides, in which boron has only three electron pairs in its valence shell. Since the valence shell can accommodate four pairs, boron is electron deficient and reacts as a Lewis acid. Typical reactions of boron trihalides are shown in equations (4-1) and (4-15).

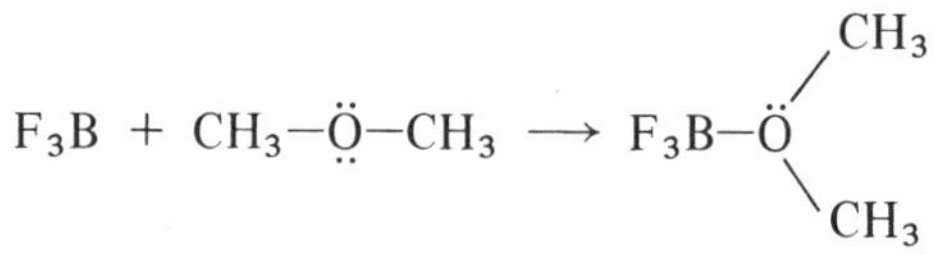
(4-15)

Aluminum chloride is another Lewis acid of the unfilled valence shell type [reaction (4-3)]. It is extensively used in organic syntheses as a catalyst in the alkylation of aromatic compounds (see Section 4.7).

C. Molecules with Polar Double Bonds

Molecules with polar double bonds often behave as Lewis acids. Common examples are molecules containing a carbonyl group, that is, a carbon-oxygen double bond. The rather large difference in the electronegativities of carbon and oxygen induces a shift of electrons away from the carbon. This polarization creates a deficiency of electrons at carbon that is great enough to permit reaction with electron-donating species (i.e., Lewis bases).

As an illustration, consider the formation of a cyanohydrin in the reaction of a ketone with HCN. In the first step [reaction (4-16)] the Lewis base CN^- reacts with the acidic end of the double bond, forming an adduct. The adduct is itself electron rich, and reacts in the second step [reaction (4-17)] with the Lewis acid H^+.

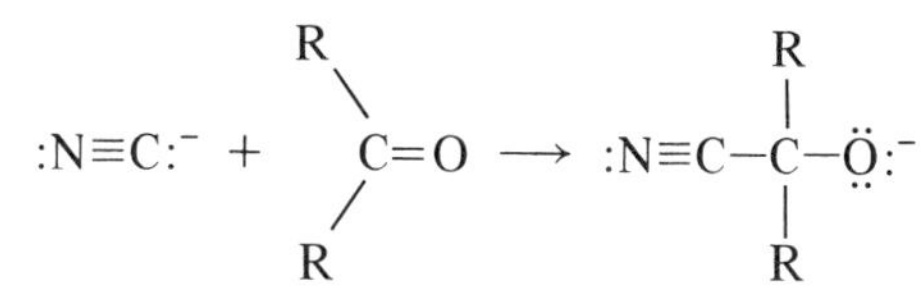
(4-16)

a ketone

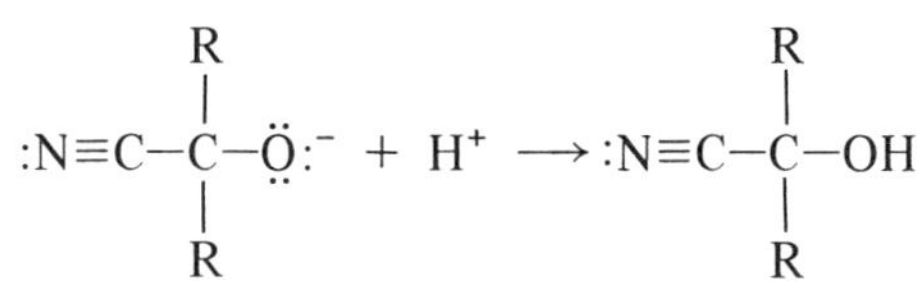
(4-17)

a cyanohydrin

Although a carbon-carbon double bond is inherently symmetrical, it can become polarized if electron-withdrawing or electron-donating groups are incorporated in the molecule (see Section 5.4). For example, 1,1-dicyanoethylene is sufficiently polar to undergo anion addition reactions analogous to (4-16).

(nonpolar)

1,1-dicyanoethylene
(polar)

D. Compounds Containing Atoms with Expandable Valence Shells

Atoms in the third row (and higher) of the periodic table are capable of sharing more than eight electrons, through involvement of *d* orbitals in the bonding. When this occurs, these atoms are said to have expanded their valence shells. If such an atom, acting as a Lewis acid, accepts an electron pair and expands its valence shell, it accepts that pair into a hybrid orbital with *d* character. Atoms must of course have *empty d* orbitals in order to undergo reactions of this type. Reaction (4-18) illustrates the formation of an adduct with an expanded valence shell. The hybrid orbitals used by the central tin atom in the reactant $SnCl_4$ are sp^3, whereas those used by tin in the product, $SnCl_6^{2-}$, are d^2sp^3.

$$SnCl_4 + 2Cl^- \longrightarrow SnCl_6^{2-} \tag{4-18}$$

4.6 TYPES OF LEWIS BASES

Lewis bases are conveniently divided into three types, listed below. Of course all are, by definition, electron-pair donors or electron-rich species. Any atom, ion, or molecule with a nonbonded electron pair may act as a Lewis base. However, an atom can have such a tight grip on an electron pair that the pair is not readily shared; such species are very weak bases.

A. Negative Ions

Examples of anionic Lewis bases (Cl^-, Br^-, and OH^-) are given in reactions (4-3), (4-7), (4-8), and (4-9). Any anion is able to function as a Lewis base provided that the reaction partner is a sufficiently strong Lewis acid.

B. Neutral Species with at Least One Unshared Pair of Electrons

The most common examples of this type of substance are ammonia, amines, and water. Typical reactions are shown in reactions (4-1), (4-4), (4-5), (4-13), and (4-14). Reactions of other neutral bases are shown in (4-2), (4-6), and (4-15).

C. Molecules That Contain Carbon-Carbon Double Bonds

Carbon-carbon double bonds (π bonds) are regions of high electron density and thus are capable of acting as Lewis bases. (This category is not to be confused with the polar double bond of carbonyl groups discussed above as one type of Lewis acid.) The carbon-carbon double bond Lewis base activity depends not on bond polarity but on the high electron density associated with π bonds. In reaction (4-19) the π electron pair in

$$CH_2{=}CH_2 + H^+ \longrightarrow \underset{\text{(an unstable carbocation)}}{CH_3{-}CH_2^+} \tag{4-19}$$

the double bond is used to form the C–H bond in the carbocation product. Reactions of this type occur as the first step in many organic reaction sequences. Note that in the product of reaction (4-20), the silver atom is not bonded to a specific atom. Instead, it is bonded to a region of high electron density, the π cloud of the alkene.

$$\underset{\text{ethene}}{\begin{array}{c}CH_2\\ \| \\ CH_2\end{array}} + Ag^+ \longrightarrow \left[Ag-\begin{array}{c}CH_2\\ \| \\ CH_2\end{array}\right]^+ \tag{4-20}$$

4.7 DISPLACEMENT REACTIONS

Most of the reactions of Lewis bases considered thus far may be classified as *adduct-forming* neutralizations. In such reactions, generalized by equation (4-21), a species

$$B: + A \longrightarrow B\text{—}A \tag{4-21}$$

having a nonbonded electron pair forms a coordinate covalent bond with an electron-deficient substance to give the adduct B–A. In addition to neutralization there is another important class of Lewis acid-base interactions, called *displacement reactions.* In displacement reactions, illustrated by the general reaction (4-22), a base B:′ reacts with an acid-base adduct A–B to form a different adduct, A–B′. In the process the base B: is displaced from the adduct by the base B:′.

$$B:' + A\text{—}B \longrightarrow B'\text{—}A + B: \tag{4-22}$$

As an example of the displacement reaction, let us consider the reaction between $NH_3(aq)$ and the weak acid HF(aq):

$$H_3N: + HF \longrightarrow NH_4^+ + F^- \tag{4-23}$$

In the Arrhenius view the reaction occurs in several steps: HF dissociates to H^+ and F^-, NH_3 reacts with water to form $NH_4^+ + OH^-$, and finally H^+ and OH^- neutralize one another [reactions (4-24) through (4-26)]. There are several defects in this point of view.

$$HF \longrightarrow H^+ + F^- \tag{4-24}$$

$$NH_3 + H_2O \longrightarrow NH_4^+ + OH^- \tag{4-25}$$

$$H^+ + OH^- \longrightarrow H_2O \tag{4-26}$$

$$\text{overall reaction} \quad HF + NH_3 \longrightarrow NH_4^+ + F^- \quad (4\text{-}23)$$

First, it presents as complex an interaction that is actually quite simple. Second, it involves a free or bare proton which somehow finds its way from HF to OH^-, and as we have seen, bare protons do not exist in aqueous solution.

In the Brønsted-Lowry point of view, reaction (4-23) is a transfer of a proton from HF to NH_3; the role of valence electrons in the interaction is not addressed. In the Lewis view (4-27), the base ammonia *displaces* the base F^- from the adduct HF, in the

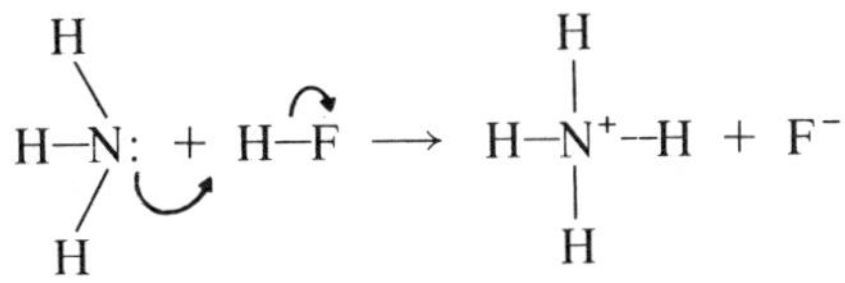

(4-27)

process forming a new adduct with a new coordinate covalent bond. The Lewis view stresses the movement of electrons from the old to the new bond configuration. In reaction (4-27) we have used arrows to indicate the movement of valence *electron pairs*; two electrons from the nitrogen atom form the new N—H bond, and both electrons from the H—F bond go with the F^- ion. Note that the arrows point toward where the electrons will be in the product. Reaction (4-27) also introduces a modification of the $\delta+$, $\delta-$ notation (Section 4.2) used to indicate regions of positive or negative charge. If the species is an ion rather than a neutral species, it is customary to omit the "δ" and simply write a "+" or a "–" near the symbol of the atom with the most pronounced charge. Thus in reaction (4-27) the notation indicates that the positive charge of the NH_4^+ ion resides primarily (but not exclusively) on the nitrogen atom. The notation used in reaction (4-29) means that the nitrogen atom is the most positively charged atom in the species $(CH_3NH_3)^+$, and the oxygen atom is the most negatively charged in the species $(HSO_4)^-$. The combination of this with the arrow notation is very helpful in visualizing the movement of charge during chemical reactions.

Reactions (4-28) through (4-31) are additional examples of acid-base reactions that occur via a displacement mechanism. In each case a stronger base displaces a weaker base from an adduct.

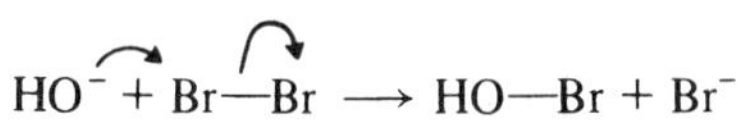

(4-28)

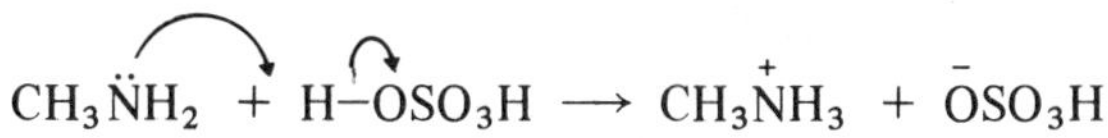

(4-29)

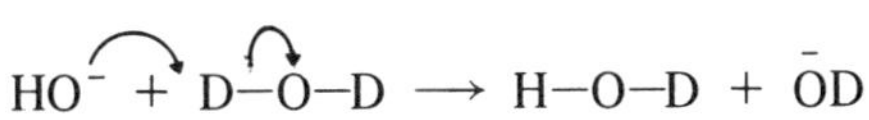

(4-30)

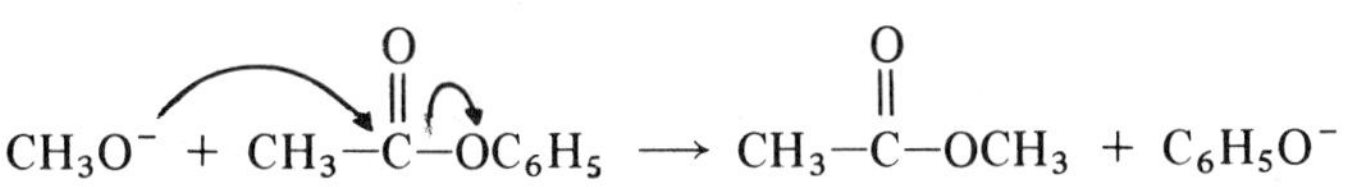

(4-31)

Reaction (4-31) is one of the many organic reactions that proceed via a Lewis base displacement mechanism. This type of reaction is also referred to as a *nucleophilic substitution*, because one nucleophile (CH_3O^-) is substituted for another ($C_6H_5O^-$) in the molecule. A nucleophile (nucleus-lover) is a reagent which is attracted to a nucleus, or in general to regions of positive charge. In this case the region of positive charge is the carbon end of the carbonyl group (see Section 4.5). Equation (4-31) represents an overall reaction process which is the sum of the two sequential steps (4-32) and (4-33).

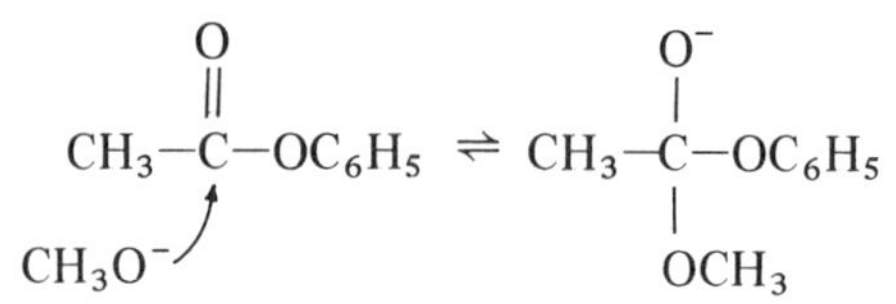

(4-32)

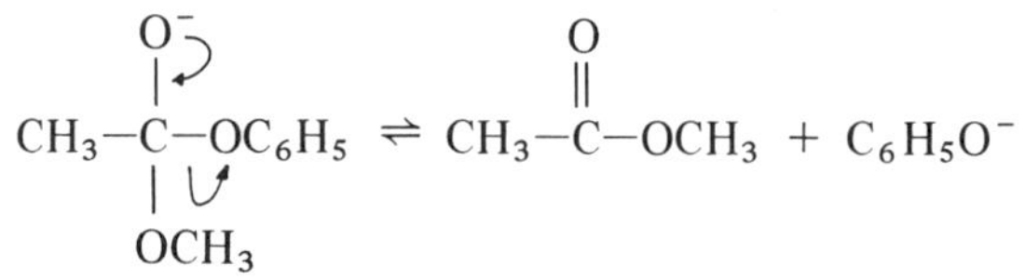

(4-33)

In the first step, methoxide ion attacks the electron-deficient carbon atom of the carbonyl group. The result of this Lewis acid-base addition reaction is the formation of an unstable tetrahedral intermediate. In the second step the intermediate decomposes, which it may do in either of two ways. If it decomposes by loss of CH_3O^- ion, the reaction is the reverse of (4-32) and the original starting materials are re-formed. In fact, this reaction is reversible, and this does occur part of the time. If, however, the intermediate decomposes by loss of $C_6H_5O^-$ ion (4-33), the product methyl acetate is formed. Since both steps of the mechanism are reversible, the overall reaction (4-31) does not go to completion; a position of equilibrium is reached. Just as in Brønsted-Lowry acid-base reactions, the equilibrium favors the side with the weaker base; that is, the stronger base displaces the weaker base. In this case the attacking base (CH_3O^-) is quite a bit stronger than the leaving base ($C_6H_5O^-$), so the reaction produces the desired ester in good yield.

Several other examples of nucleophilic substitution are shown in reactions (4-34) through (4-36). All of them involve the initial attack of a Lewis base (a nucleophile) on

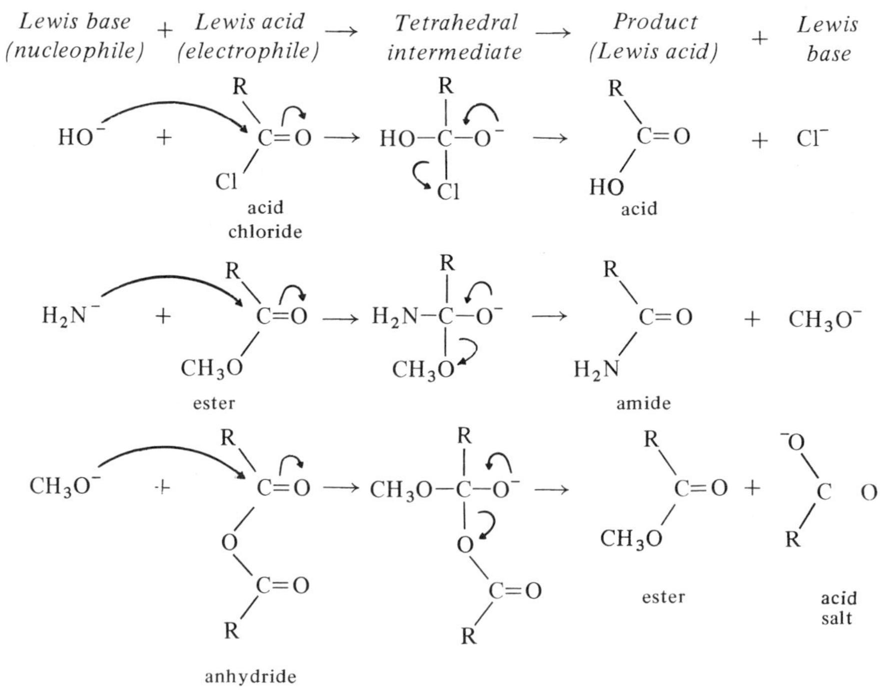

(4-34)

(4-35)

(4-36)

the carbon atom (a Lewis acid) of the carbonyl group and all of them proceed via the same displacement mechanism as that discussed for reaction (4-31). These reactions are frequently encountered in synthetic organic chemistry.

Nucleophilic substitution reactions can also take place at a carbon atom that does not have a double bond. Several examples are shown in reactions (4-37), (4-38), and (4-39). Note that in each of these reactions a strong Lewis base is displacing a weaker Lewis base.

$$HO^- + CH_3{-}I \longrightarrow CH_3OH + I^- \qquad (4\text{-}37)$$

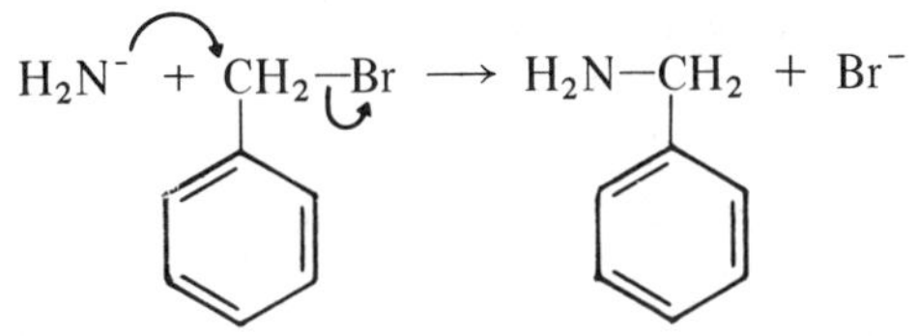

(4-38)

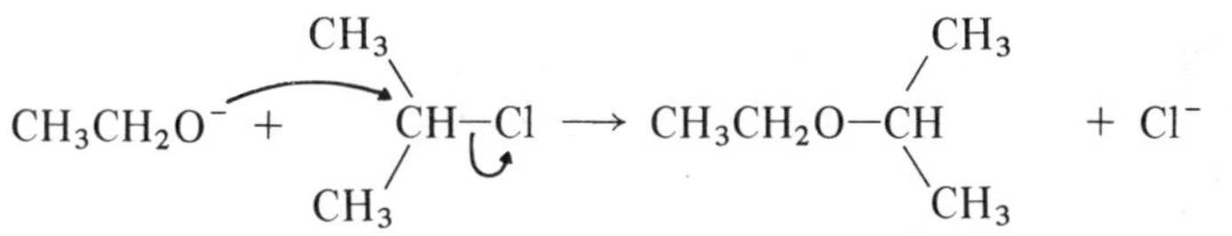

(4-39)

There are also reactions that take place via an *electrophilic substitution* mechanism. An electrophile ("electron-lover") is a species that is attracted to a negatively charged center. These reactions differ from nucleophilic substitutions in that the attacking and leaving groups are Lewis acids rather than Lewis bases. Electrophilic substitution is often encountered in organic syntheses; one example is the Friedel–Crafts reaction, in which an alkyl group is incorporated in an aromatic compound. In the first step (4-40) the Lewis acid $AlCl_3$ reacts by electrophilic displacement with an organic halide to produce the intermediate carbocation R^+. This intermediate is also a Lewis acid, and it is strongly electrophilic. In the second step the electrophile R^+ reacts with (for example) benzene, producing a different intermediate carbocation. In the final step the intermediate transfers a proton to the base $AlCl_4^-$, producing the substituted benzene and HCl, and regenerating the $AlCl_3$ used in the first step. In the overall reaction (4-43) the electrophile R^+ has displaced the electrophile H^+ from benzene; the Lewis acid $AlCl_3$ is a catalyst that participates, but is not consumed in the process.

$$R{-}Cl + AlCl_3 \longrightarrow R^+ + AlCl_4^- \qquad (4\text{-}40)$$

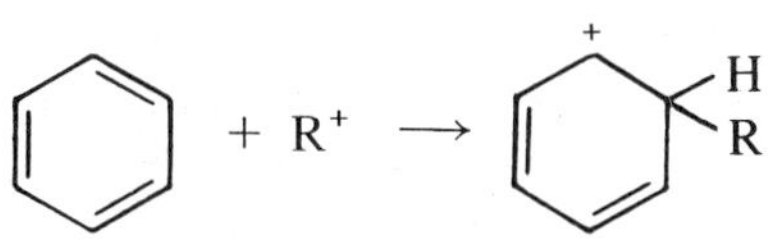

(4-41)

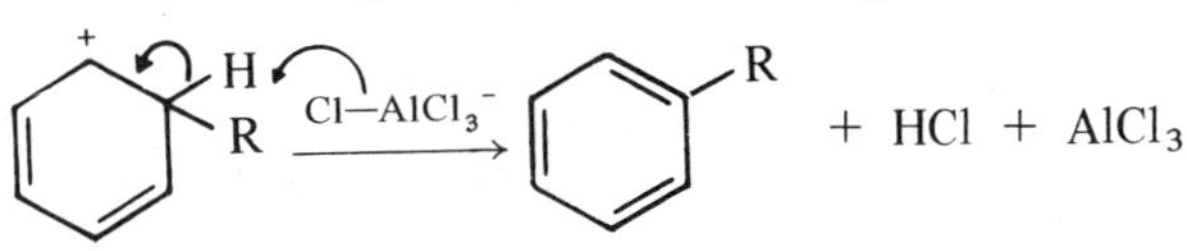

(4-42)

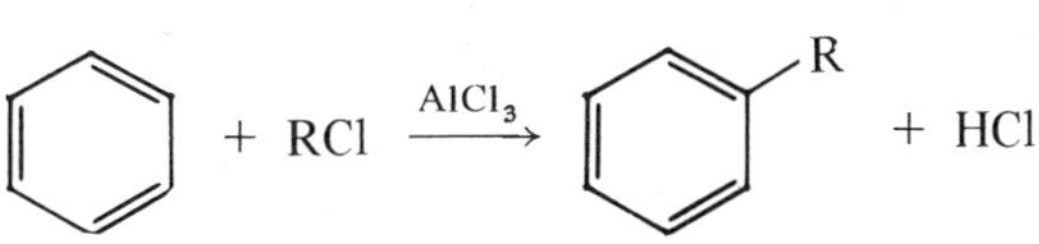

(4-43)

4.8 COMPARISON OF LEWIS AND BRØNSTED-LOWRY VIEWS

Two difficulties arise when the Lewis theory is compared to the Brønsted-Lowry theory. The first of these concerns Brønsted-Lowry acids such as HNO_3, HCl, or H_2CO_3. These proton-donating species are not electron deficient in any sense and are therefore not Lewis acids. Instead, an acid like HCl is a Lewis acid-Lewis base adduct in the same sense that $Cl_3B{-}NH_3$ is an adduct [reaction (4-1)]. Lewis theory would probably be easier to learn if the term "Lewis acid" included all Arrhenius and Brønsted-Lowry acids as well as some additional species such as $AlCl_3$, but unfortunately it does not. Brønsted-Lowry acids should be regarded as *sources* of a Lewis acid because H^+ *is* a Lewis acid.

$$\underset{\substack{\text{Brønsted-Lowry}\\ \text{acid, and/or}\\ \text{Lewis adduct}}}{HCl} \longrightarrow \underset{\text{Lewis acid}}{H^+} + \underset{\text{Lewis base}}{Cl^-} \qquad (4\text{-}44)$$

The second and more serious difficulty involves the correlation of relative acid or base strengths in the Brønsted-Lowry sense with relative strengths in the Lewis sense. In some cases such comparisons lead to contradictory results. For example, Zn^{2+} is a Lewis acid and as such reacts readily with either OH^- or CN^-. Both of these Lewis bases are good electron-pair donors. In reactions where both OH^- and CN^- are available, Zn^{2+} bonds preferentially with CN^-. From this we conclude that CN^- forms a stronger bond with Zn^{2+} than does OH^-; therefore, in this case CN^- is the stronger Lewis base. However, with respect to accepting a proton from an acid such as HCl, that is, functioning as a Brønsted-Lowry base, the relative strength of these two bases is reversed. Hydroxide ion is a stronger base than cyanide ion (i.e., a better proton acceptor). Relative base strengths cannot be carried over from the Brønsted-Lowry to the Lewis concept. Relative strengths in the series of bases may even vary depending on the electron pair acceptor (Lewis acid) used to establish the ranking. Difficulties of this type have led to the development of the *hard and soft acid-base concept*. There are several excellent review articles available in this area, and students who are interested in exploring these concepts are encouraged to read these reviews.[6]

4.9 USANOVICH ACID-BASE THEORY

A very general theory of acids and bases has been set out by Usanovich.[7,8] In this scheme an acid is a substance that produces or donates a cation, or one that combines with (accepts) an anion or an electron. A base is a substance that produces (donates) an anion, or one that combines with (accepts) a cation or other electron-deficient species. These definitions encompass the three other definitions, Arrhenius, Brønsted-Lowry and Lewis. However, this consolidation of concepts is a rather small advantage, since few chemists have difficulty in using Arrhenius, Brønsted-Lowry, and Lewis theories separately. (Indeed, there is a definite conceptual advantage to keeping them separate.) The advantage of the Usanovich definitions is that they include reactions in which *any*

[6] R. G. Pearson, *J. Chem. Ed.*, *45*, 581 and 643 (1968).

[7] M. Usanovich, *Zh. Obschei Khim.*, *9*, 182 (1939).

[8] J. E. Huheey, *Inorganic Chemistry: Principles of Structure and Reactivity*, Harper & Row, New York, 1972.

number of electrons is transferred. We are no longer limited to electron-pair interactions, and may even include single electron transfers in our thinking. Thus, using the Usanovich concept, we can apply the powerful methods of acid–base theory to two important categories of reactions that would otherwise seem to be unrelated to acid–base behavior. These are free-radical reactions, involving a single electron, and redox reactions, which involve transfer of one or more electrons.

Some examples of free-radical reactions are (4-45), a displacement reaction that occurs in the photochlorination of methane, and (4-46), an addition reaction that takes place in the production of polyethylene. Free-radical addition polymerizations, of which reaction (4-46) is a typical example, are of great industrial importance.

$$:\ddot{\underset{\cdot\cdot}{Cl}}\cdot + CH_4 \longrightarrow ClCH_3 + H\cdot \qquad (4\text{-}45)$$

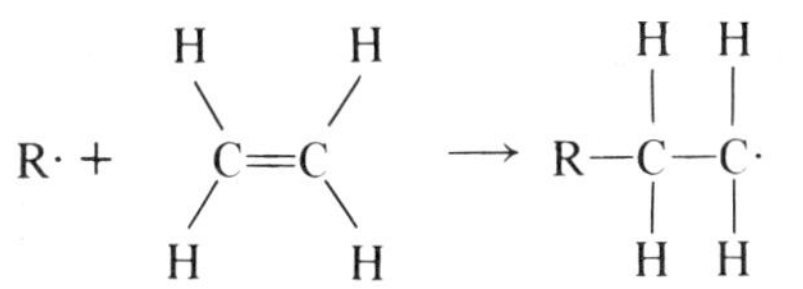

(4-46)

Reactions such as (4-47) and (4-48) are examples of redox reactions. Reaction (4-48) is a convenient laboratory test for the presence of an aldehyde group (—CHO) in a molecule. Aldehydes (RCHO) react with Fehling's solution (alkaline $CuSO_4$) to give a reddish precipitate of Cu_2O; the copper ion acts as an acid by accepting electrons from the aldehyde. The closely related ketones ($R_2C{=}O$) do not react in this way.

$$Zn(s) + Sn^{4+}(aq) \longrightarrow Zn^{2+}(aq) + Sn^{2+}(aq) \qquad (4\text{-}47)$$

$$CH_3{-}\overset{\overset{\displaystyle O}{\|}}{C}{-}H + 2Cu^{2+} + 5OH^- \longrightarrow CH_3{-}\overset{\overset{\displaystyle O}{\|}}{C}{-}O^- + Cu_2O + 3H_2O \qquad (4\text{-}48)$$

4.10 SOME GENERALIZATIONS

It is very useful to think about chemical species and their interactions in terms of the concepts developed in the various acid–base theories. Although there is some overlap, each theory emphasizes and is appropriate to different aspects of chemistry. Arrhenius theory is simple, and suitable for proton acids and hydroxide bases and for their reactions in aqueous solutions. It is also suitable for calculations involving pH, percent ionization, buffer solutions, solubilities, and so on. Brønsted–Lowry theory is appropriate to all of the above, and in addition is more suitable for nonhydroxide bases, for acid–base behavior in nonaqueous solvents, and for chemical reactions involving the transfer of a proton from one species to another. Usanovich theory can be used for oxidation–reduction reactions and for free-radical interactions. Lewis theory has wide applications in both inorganic chemistry and organic chemistry. It is suitable for discussion of any species that contains or is capable of forming a coordinate covalent bond, for displacement reactions, and generally for electron-rich and electron-deficient species. In the remainder of this text we discuss many topics in chemistry, both qualitative and quantitative in nature. One of our important goals is to develop an appreciation for the central importance of electron-deficient-electron-rich character in chemistry.

PROBLEMS

4.1 Complete and balance each of the following equations.

(a) $AlCl_3 + Br^- \longrightarrow$

(b) $BF_3 + CH_3-O-CH_3 \longrightarrow$

(c) $CO_2 + H_2O \longrightarrow$

(d) Sulfur dioxide + $Na_2O \longrightarrow$

(e) $(CH_3)_3N + HI \longrightarrow$

(f) 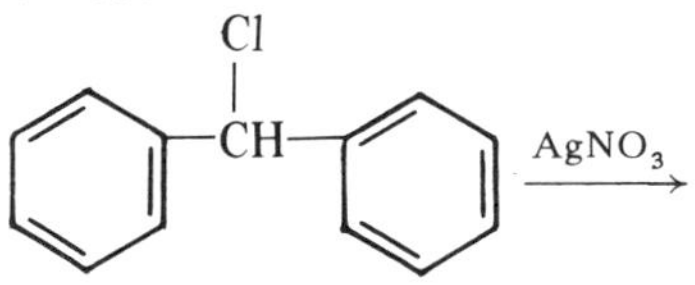

(g) $CH_3ONa + CH_3NH_3Cl \longrightarrow$

4.2 Define or explain each of the following terms. Write out a complete answer, then check your text or other sources.

(a) Lewis acid

(b) Lewis base

(c) Usanovich acid

(d) Usanovich base

(e) Coordinate covalent bond

4.3 List the four types of Lewis acids and show one example of each type.

4.4 List the three types of Lewis bases and show one example of each type.

4.5 Explain why Brønsted-Lowry (proton-donating) acids do not fit the Lewis definition of acids.

4.6 Using the Usanovich acid-base definition, label as acids or bases the reagents on the left side of each of the following equations. Think about your answers, then go argue with a friend.

(a) $NaCl \xrightarrow{H_2O} Na^+(aq) + Cl^-(aq)$

(b) $3H_2 + N_2 \longrightarrow 2NH_3$

(c) $NaH + H_2C{=}O \longrightarrow H_3CONa$

(d) $2Cl^- \longrightarrow Cl_2 + 2e^-$

(e) $2H^+ + 2e^- \longrightarrow H_2$

(f) 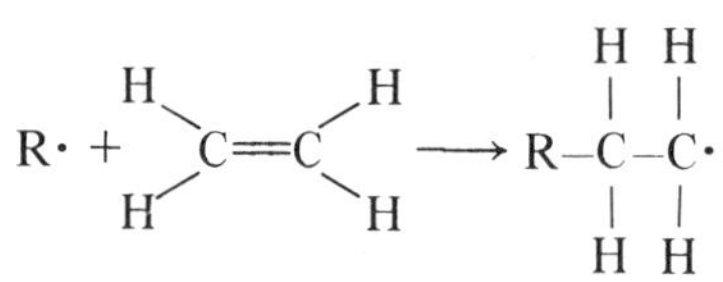

(g) $Zn(s) + Sn^{4+}(aq) \longrightarrow Zn^{2+}(aq) + Sn^{2+}(aq)$

Chapter 5

The Effects of Structure on Acid-Base Strength

5.1 INTRODUCTION

The ability of a species to act as an acid or a base depends on the existence, within the molecule, of regions of electron deficiency or abundance. The number and kind of atoms in a molecule are of course influential, but the way in which they are bonded together, that is, the structure of the molecule, is an even more important factor in determining acidity or basicity. Any structural features that increase the electron deficiency in any part of the molecule will increase the ability of the species to accept an electron pair, that is, increase the Lewis acid strength. Some structural features facilite the loss of protons from a molecule, thereby increasing the Brϕnsted-Lowry acid strength. In the case of electron-rich species, any structural feature that increases the availability of electrons and/or the species' willingness to donate the electrons increases the Lewis base strength of the species. In this chapter we examine several features of molecular structure in order to achieve a general understanding of the effects of structure on acidity and basicity.

5.2 ELECTRONEGATIVITY

Electronegativity is a measure of the tendency of an atom in a compound to attract valence electrons to itself. Electronegativity can have a profound effect on the acid-base strength of compounds. The effect is illustrated by the second-period binary hydrides methane through hydrogen fluoride. In the second period, electronegativity increases smoothly from C through F; corresponding to this increase, the acid strength of the hydrides also increases: $CH_4 < NH_3 < OH_2 < FH$. As the electronegativity of the central atom (the second-period element) increases, the electron density in the sigma bond(s) to the hydrogen atom(s) is shifted toward the central atom. This weakens the bond(s) and increases the tendency to cleave heterolytically, producing a proton and the conjugate base.

Another way of understanding the observed acidity order is to consider the stability of the conjugate base formed once the proton is lost. The bond strength and the conjugate base stability are both consequences of the electronegativity and have an inverse relationship to one another: the weaker the bond strength, the greater the stability of the conjugate base. If it is more convenient or easier to think in terms of base stability (rather than in terms of bond strength) when making predictions of relative acid strength, this should be done; both lines of reasoning lead to the same result. The stability of the conjugate bases of the second-period binary hydrides increases in the following order: $CH_3^- < NH_2^- < OH^- < F^-$. As electronegativity increases in this series, so does the atom's ability to accomodate and hold on to an extra electron.

These generalizations work quite well for comparisons within a given horizontal row of the periodic table. However, atomic size also exerts an influence on acid-base strength, and this factor is more influential than electronegativity when comparisons are made within a given vertical column of the periodic table (see Sections 5.5 and 5.7).

5.3 RESONANCE

A. Bonding and Charge Delocalization

If a particular site within a molecule has high electron density (e.g., the oxygen atom of a carbonyl group), that site is capable of acting as a Lewis base. Any structural feature that spreads the excess electron density out over a larger portion of the molecule (delocalizes it) will decrease the Lewis base strength of that site.

One mechanism for the delocalization of electrons within a molecular or ionic species is known as *resonance*. Resonance may be defined as a distribution of electrons throughout a system of overlapping *p* orbitals.[1] The atoms involved in a resonance system are usually sp^2 hybridized. As shown in Figure 5.1, the three equivalent sp^2 hybrid orbitals are directed toward the corners of an equilateral triangle. An sp^2 hybridized atom also has a *p* orbital that is perpendicular to the plane of the triangle; it is this *p* orbital that is involved in resonance stabilization.

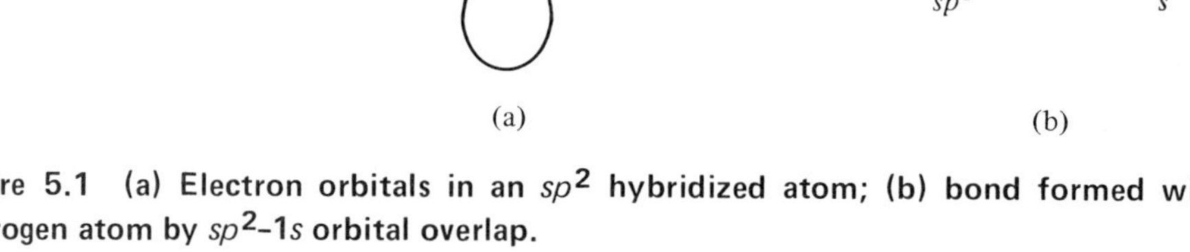

Figure 5.1 (a) Electron orbitals in an sp^2 hybridized atom; (b) bond formed with a hydrogen atom by sp^2–1s orbital overlap.

If two adjacent sp^2 hybridized carbon atoms are oriented so that their *p* orbitals are parallel, it is possible for overlap (i.e., bond formation) to occur in the *p* orbitals as well as the sp^2 hybrid orbitals. This is illustrated by the bonding in the ethene molecule

[1] For elements heavier than second-period elements, *d* orbitals may also be involved.

(Figure 5.2). The C—H bonds in ethene are formed by the overlap (not shown) of the 1*s* orbital of hydrogen with an sp^2 hybrid orbital of carbon. These C—H bonds are *σ bonds*. An important characteristic of a σ bond is that the region of overlap is centered on the internuclear axis (the line joining the nuclei of the two atoms participating in the bond). A C—C σ bond also results from the end-on overlap of two sp^2 hybrid orbitals of the adjacent carbon atoms. The *p* orbitals of the adjacent carbon atoms are in position to interact by a *side-to-side* overlap. The result of this interaction is an additional bond, called a *π bond*. Molecules containing a π bond have several important characteristics. First, the geometry in the neighborhood of the π bond is planar: the five σ bonds define a plane, and the nuclei of all six atoms (four hydrogen atoms and two carbon atoms) joined by these bonds lie in the plane. Since this is a characteristic of π bonds, it is true not only in the $CH_2{=}CH_2$ molecule but also in larger molecules containing a carbon-carbon double bond. A second important characteristic is that the π bond, formed by side-by-side overlap of the atomic *p* orbitals, lies both above and below (but not in) the plane.

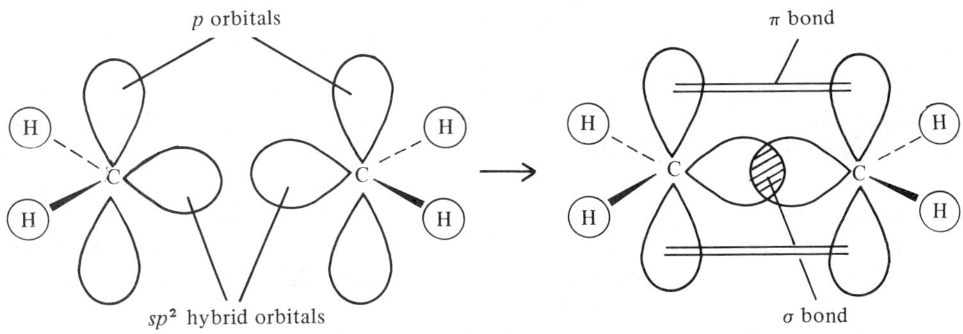

Figure 5.2 Double-bond formation in ethene, C_2H_4.

In general, the greater the degree of overlap between two orbitals, the stronger the resulting bond. The end-on orientation of the sp^2 hybrid orbitals permits greater overlap than does the side-to-side overlap of the *p* orbitals, and the C—C σ bond is stronger than the C—C π bond in ethene. To understand resonance, it is necessary to be familiar with hybridization and double-bond formation: the student is referred to any general chemistry or organic text for a more extensive discussion of these concepts.

We have defined resonance as the delocalization or spreading out of electrons into a *system* of overlapping *p* orbitals. For resonance to occur, the system must involve more than two adjacent atoms with *p* orbitals. If the *p* orbitals of only two atoms overlap, the result is not resonance; it is simply a double bond. Resonance occurs whenever the structure is such that two or more satisfactory Lewis "electron dot" structures may be drawn for a molecule or ion. For sulfur trioxide, for example, three equally probable dot structures can be drawn:

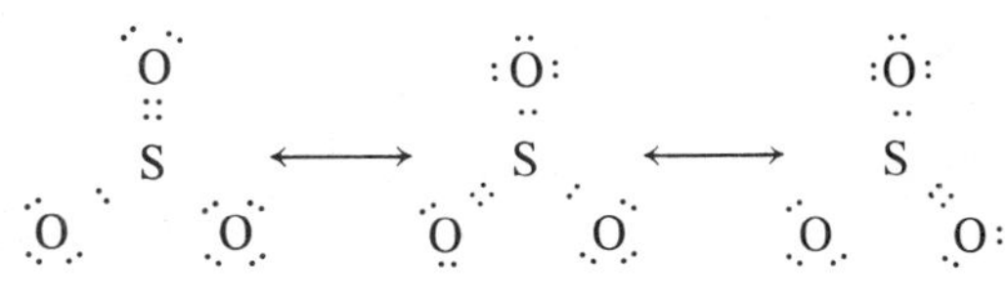

(5-1a)

This situation can also be depicted by less cluttered structures, (5-1b), in which nonbonding electrons are not shown. In both of these representations the double-headed

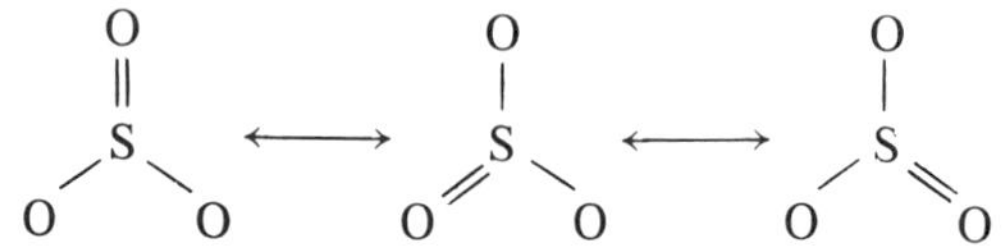

(5-1b)

arrows mean that resonance occurs. In each of these structures there are three σ bonds and one π bond. The *actual* structure of SO_3 is a blend, intermediate among these three; it is called a *resonance hybrid*, and it cannot be depicted by any one of these simple diagrams. The actual structure has three *identical* bonds which are more than single and less than double in character. There are six σ electrons, localized pairwise in the three σ bonds; the two π electrons are not localized, but are distributed throughout the entire molecule. We say that the π electrons are "delocalized throughout the π *system*," that is, into the system of overlapping *p* orbitals.

The following structure is a way of emphasizing these points:

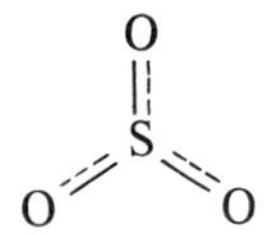

(5-1c)

The molecule has *one* structure, not three different ones. Resonance in a molecule cannot be depicted in a conventional dot structure; the π electrons are delocalized in a π-orbital system. As indicated by the dashed lines, the bonds have partial double-bond character; the dashed lines also delineate the region of electron delocalization.

Structures such as those appearing in (5-1a) and (5-1b) are called *resonance contributors*. A "major contributor" has a structure closely similar to the true structure and a "minor contributor's" structure is less similar. The position of the nuclei is the same in all resonance contributors, which differ only in the placement of valence electrons when the dot structures are drawn.

B. Formal Charges

It is often helpful to keep track of the "location"[2] of electrons in a molecule, ion, or resonance contributor, and a scheme known as *formal charges* has been developed to do so. This scheme is widely used in organic chemistry, although less so in other branches of chemistry. It is especially useful in acid-base chemistry, which is greatly influenced by charge distribution.

The formal charge scheme is based on the notion that molecules consist of neutral atoms, and that the bonds which hold the atoms together consist of pairs of electrons from the valence shells of the constituent atoms. Each atom in a molecule can be assigned a formal charge on the basis of how its valence electrons are used. A neutral atom has neither gained nor lost electrons, and has a formal charge of zero. When an atom shares one of its electrons with another to form a normal covalent bond it loses one-half of a negative charge; by the same token it gains one-half of a negative charge by sharing the electron that the other atom contributes to the bond. There is no net effect, so that:

[2]Remember that electrons do not truly have locations within a molecule. Negative charge is not distributed pointwise but is continuous and extends throughout the entire molecule.

1. The formal charge of an atom is not changed by normal covalent bonding. However, in a coordinate bond both electrons come from one atom. Since the electrons are shared, the acceptor atom gains ½ + ½ = 1 unit of negative charge and the donor atom loses 1 unit. Thus
2. The formal charge of an atom changes by +1 for each coordinate bond for which it is the donor, and changes by −1 for each coordinate bond for which it is the acceptor.

These principles lead to the following rules for calculating the formal charge of any atom in either a neutral molecule or a polyatomic ion such as SO_4^{2-}.

First, draw the Lewis dot structure. Then assign the formal charges.

$$\text{formal charge} = \text{number of valence electrons (group number of atom)} - \text{number of nonbonding electrons} - \tfrac{1}{2}\,\text{number of bonding electrons}$$

When these rules are applied correctly, the sum of the formal charges equals the charge on a polyatomic ion or is zero for a neutral molecule. For example, the resonance contributors of SO_3 [(5-1a) and (5-1b)] have formal charges as shown in (5-1d).

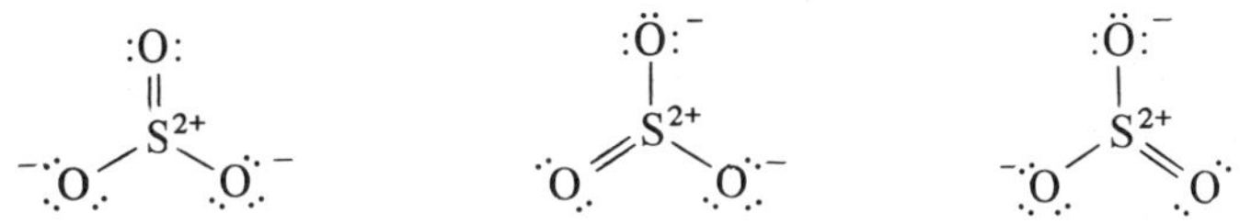

(5-1d)

The charges are calculated as follows:

sulfur: 6 valence electrons
0 nonbonded electrons
½ of 8 bonding electrons

6 − 0 − 4 = formal charge of +2

doubly bonded oxygen: 6 valence electrons
4 nonbonded electrons
½ of 4 bonding electrons

6 − 4 − 2 = formal charge of zero

singly bonded oxygens: 6 valence electrons
6 nonbonded electrons
½ of 2 bonding electrons

6 − 6 − 1 = formal charge of −1

The sum of the formal charges on all four atoms (+2, 0, −1, −1) is zero, as it should be for a neutral molecule.

The dashed-line representation of the hybrid structure (5-1c) is not compatible with our definition of formal charge, and the δ notation should be used to suggest charge distribution.

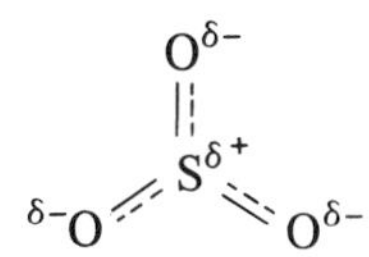

(5-1e)

Formal charge assignments in a molecule or ion are not unique. For example, two of several possible formal charge assignments in the sulfate ion are:

 (5-2)

It must be emphasized that formal charges do not indicate actual, measurable charges resident on an atom; we are simply doing bookkeeping on the electrons in Lewis dot structures. However, dot structures and formal charges do indicate charge distribution in a qualitative way. An atom having a negative formal charge is more negative than an atom with a formal charge of zero. Carefully used, formal charges can help in understanding acid-base interactions. Since it may be easier to memorize a few instances of formal charges than to apply the rules each time, a partial listing of formal charges in common bonding situations is given in Table 5.1.

Table 5.1 Formal Charges in Some Common Structures[a]

Element	*Structure*	*Formal Charge*
Hydrogen	—H	0
Boron	>B— (three bonds)	0
	—B— (four bonds)	−1
Carbon	—C— (four bonds), >C=, or —C≡	0
	—C: (three bonds)	−1
	—C (three bonds)	+1
Nitrogen	—N̈— (three bonds), =N̈—, or ≡N:	0
	—N— (four bonds)	+1
Oxygen	—Ö— or =Ö:	0
	=Ö— or —Ö— (three bonds)	+1
	—Ö: (three lone pairs)	−1
Fluorine	—F: (three lone pairs)	0

[a]A line, such as –H, indicates a covalent bond to *any* atom. All nonbonding electrons are shown explicitly in these dot structures.

In this text we indicate formal charges only when they are important to the discussion: they are usually important in resonance contributors because they are clues to the charge distribution in the true structure, the resonance hybrid.

C. Lewis Acid Strength

Resonance, which spreads out charge density, strongly affects Lewis acid-base strength. If a negative charge is delocalized, the Lewis base strength of the species is weakened. If a positive charge is delocalized, the Lewis acid strength of the species is decreased. In addition, the existence of resonance confers extra stability on a molecule or ion, compared to similar species lacking resonance.

The relative acidity of carboxylic acids and alcohols provides an excellent illustration of the effect of resonance. The ionization reactions are (5-3) for ethanol and (5-4) for

$$H_2O + \underset{\text{ethanol}}{CH_3CH_2OH} \rightleftharpoons \underset{\text{ethoxide ion}}{CH_3CH_2O^-} + H_3O^+ \qquad (5\text{-}3)$$

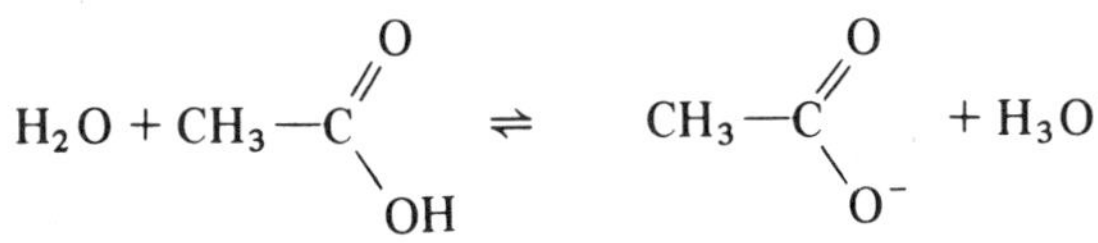

acetic acid — acetate ion (5-4)

acetic acid. Acetic acid is a much stronger acid than ethanol, and it is generally true that carboxylic acids are stronger acids than alcohols. The ionization constant for acetic acid at 25°C is 1.8×10^{-5}, and the ionization constant for ethanol is approximately 1×10^{-16}. Since the ionization reaction (the heterolytic fission of an O—H bond) is the same in both compounds, the difference in acidity must be due to more subtle, structural differences between the two molecules. In acetic acid the OH is bonded to a carbonyl group, which contains a π bond, whereas in ethanol only σ bonds are present. Acetate ion is substantially more stable than ethoxide ion because of resonance delocalization of charge (5-5).

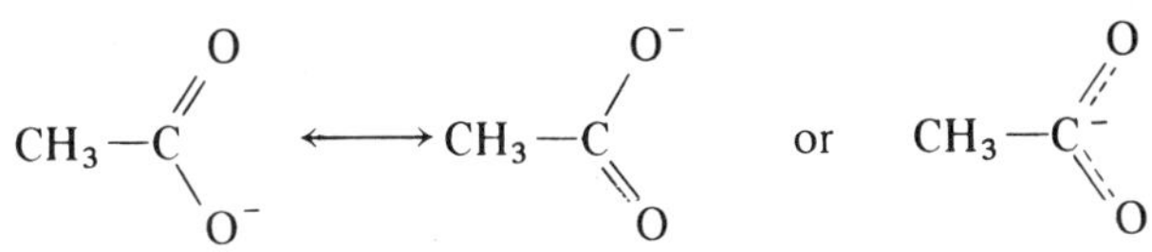

(5-5)

Each of the three atoms in the $—CO_2^-$ part of the acetate anion has a p orbital for π bonding. There are two identical Lewis structures, and the negative charge is distributed throughout the O—C—O π system. X-ray diffraction studies corroborate this resonance hybrid structure, by showing that the two carbon-oxygen bonds in the acetate ion are exactly the same length. In comparison, in nonionized acetic acid two different carbon-oxygen bond lengths are observed; one corresponds to a single bond and one to a double bond. There is essentially no resonance, no stabilization, and no delocalization in the nonionized acid. The fact that the acetate anion is stabilized relative to the nonionized molecule means that it is formed more readily than the ethoxide ion, which is not resonance stabilized; that is, the strength of the acid is increased. All carboxylate anions are resonance stabilized.

Two further examples of resonance stabilization are the allyl carbocation, with two resonance contributors, and the cyclopentadienyl anion, with five contributors. These are shown in Figure 5.3. The basic character of the cyclopentadienyl anion is decreased (relative to an anion such as $CH_3CH_2^-$) because of resonance delocalization of the negative charge. In the allyl carbocation positive charge is delocalized, and it is therefore the electron-acceptor (Lewis acid) strength that is decreased.

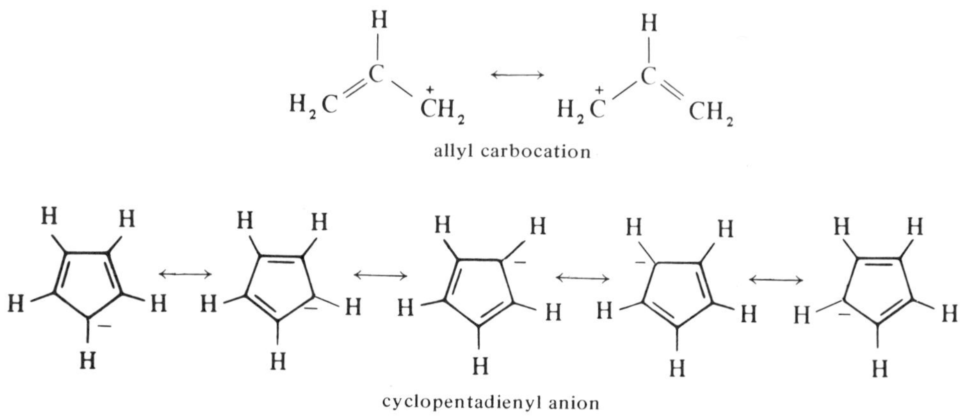

Figure 5.3 Resonance contributors of the allyl and cyclopentadienyl ions.

The resonance species discussed in this section are redrawn in Figure 5.4, using the dashed-line representation. As discussed above, this method emphasizes the intermediacy of the structure, the partial double-bond character which is a consequence of resonance, and the region in which the charge is delocalized.

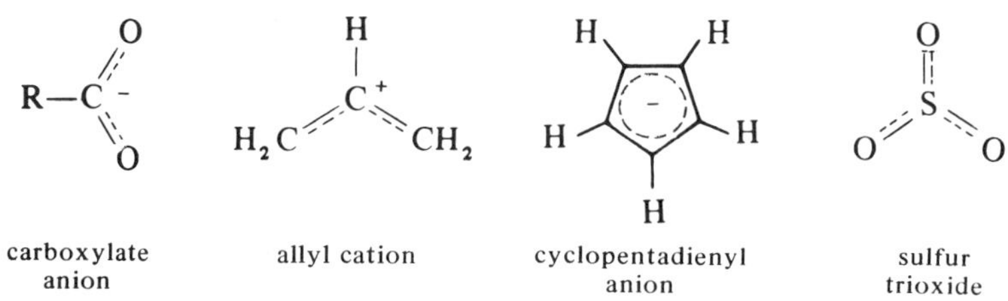

Figure 5.4 Alternative representation of resonance.

There are many other systems in which resonance stabilization helps to account for acid-base strength; some of these are discussed in Chapter 6.

5.4 INDUCTIVE EFFECT

The greater acidity of acetic acid relative to ethanol is due almost entirely to resonance stabilization of the acetate anion; the effect is so strong that it overshadows a quite different effect. In ethanol the substituent is the ethyl group, $—C_2H_5$. All alkyl groups, including ethyl, are mild electron donors; an alkyl group is less electronegative than the oxygen to which it is bonded. The effect of this mild electron donation is to intensify the negative charge on the oxygen, making heterolytic fission more difficult; this decreases the acidity of ethanol. In acetic acid the substituent is the acetyl group

(CH_3CO); this type of substituent has an electron-withdrawing effect. In acetic acid, the electrons in the O—H bond are shifted strongly toward the acetyl group. The O—H bond becomes more polar, making heterolytic cleavage to $-O^-$ and H^+ easier. The acidity of acetic acid is thus enhanced by the electron-withdrawing group.

If an atom or group of atoms of high electronegativity is part of a larger molecule, its presence will affect the electron distribution not only in the immediate neighborhood, but throughout the entire molecule. As an example, consider chloromethane, CH_3Cl. The high electronegativity of chlorine causes the electron density in the C—Cl σ bond to be closer to the chlorine atom, leaving a residual partial positive charge on the carbon atom. This in turn causes a (somewhat weaker) shift of electron density toward carbon in each of the three C—H σ bonds, *even though carbon and hydrogen have about the same electronegativity.*

$$\begin{array}{c} H \\ \downarrow \\ H \rightarrow C \rightarrow Cl \\ \uparrow \\ H \end{array} \qquad (5\text{-}6)$$

This persistence of the effect of an electronegative atom (or group) beyond its immediate vicinity is called the *inductive effect*, and it has an effect on the acidity or basicity of a molecule. The inductive withdrawal of electron density from a site in a molecule will increase the acidity (or weaken the basicity) of that site.

As an example of the inductive effect on acidity, consider a substituted acetic acid in which one of the hydrogen atoms on the $-CH_3$ group has been replaced by the group G. The acid ionization reaction of this generalized carboxylic acid is given by (5-7). In Section 5.3 we described the increased acidity of carboxylic acids due to resonance stabilization of the conjugate base; this effect increases the strength of all carboxylic acids by about the same amount, independent of any influence of the substituent G. The inductive effect of G works in addition to the effect of resonance.

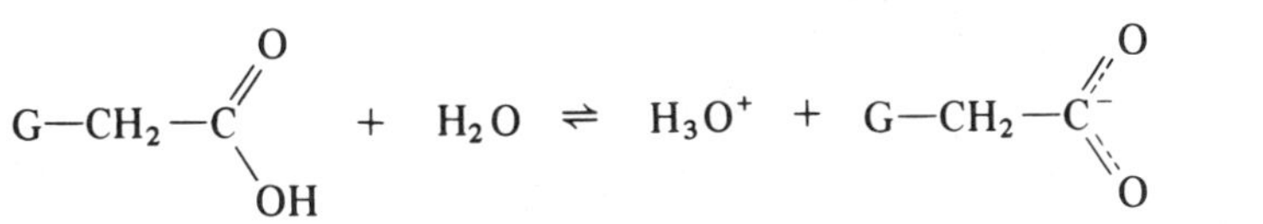

(5-7)

If G is a good electron-withdrawing group (a highly electronegative atom or group of atoms), it has the effect of increasing the acidity. Electron density is shifted toward G, and consequently withdrawn from the O—H bond. This weakens the bond and promotes ionization. Alternatively, we can consider the effect of an electron-withdrawing group on the stability of the anion. If G withdraws electrons, the negative charge on the oxygen will be distributed throughout the molecule. Any mechanism by which the charge on a highly charged atom can be reduced will increase the stability of a species. If the anion is made more stable in this way, its tendency to form from the acid is increased.

If, on the other hand, G is an electron-releasing group (an atom or group of lower electronegativity than the carbon to which it is bonded), electron density is shifted toward the carboxyl group; this *strengthens* the O—H bond and *decreases* ionization. The shift of electron density toward the carboxyl group destabilizes the carboxylate anion and thus makes ionization less favorable. The inductive effect can have a positive or a negative effect on acidity. Several specific examples of inductive effects are given in Table 5.2.

Table 5.2 Inductive Effect of Selected Substituents on Acidity

Structure	*Name*	pK_a
$CH_3CH_2CH_2CO_2H$	Butanoic acid	4.8
$CH_3CH_2CH(Cl)CO_2H$	2-Chlorobutanoic acid	2.8
$CH_3CH(Cl)CH_2CO_2H$	3-Chlorobutanoic acid	4.0
$Cl{-}CH_2CH_2CH_2CO_2H$	4-Chlorobutanoic acid	4.5
CH_3CO_2H	Acetic acid	4.8
$Cl{-}CH_2CO_2H$	Chloroacetic acid	2.9
$Cl{-}CH(Cl)CO_2H$	Dichloroacetic acid	1.3
$Cl{-}C(Cl)_2{-}CO_2H$	Trichloroacetic acid	0.7
$CH_3CH_2CO_2H$	Propanoic acid	4.9
$CH_3CH(Cl)CO_2H$	2-Chloropropanoic acid	2.8
$Cl{-}CH_2CH_2CO_2H$	3-Chloropropanoic acid	4.1
CH_3CO_2H	Acetic acid	4.8
$CH_3CH_2CO_2H$	Propanoic acid	4.9
$(CH_3)_2CH{-}CO_2H$	2-Methylpropanoic acid	4.8
$CH_3{-}C(CH_3)_2{-}CO_2H$	2,2-Dimethylpropanoic acid	5.0
CH_3CO_2H	Acetic acid	4.8
$F{-}CH_2CO_2H$	Fluoroacetic acid	2.6
$Cl{-}CH_2CO_2H$	Chloroacetic acid	2.9

Table 5.2 (cont'd)

Structure	*Name*	pK_a
$Br—CH_2CO_2H$	Bromoacetic acid	2.9
$I—CH_2CO_2H$	Iodoacetic acid	3.1
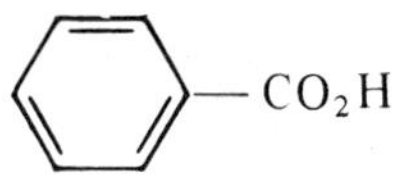	Benzoic acid	4.2
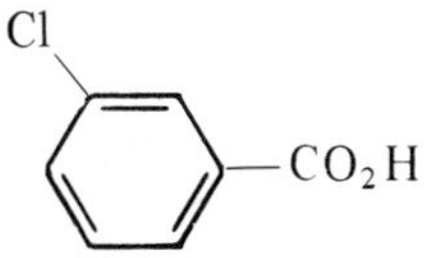	*m*-Chlorobenzoic acid[a]	3.8
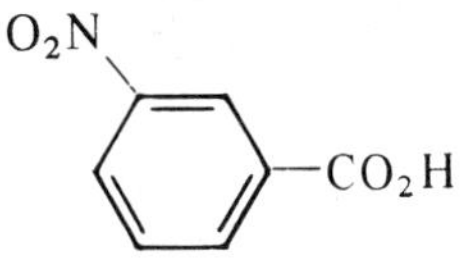	*m*-Nitrobenzoic acid	3.4
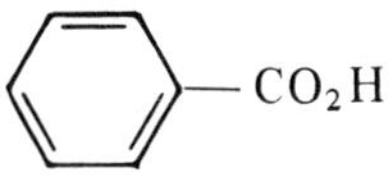	Benzoic acid	4.2
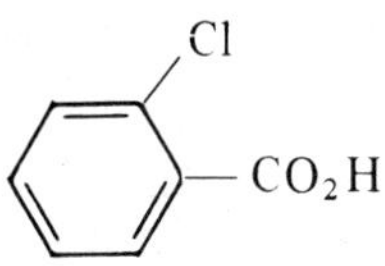	*o*-Chlorobenzoic acid[a]	2.9
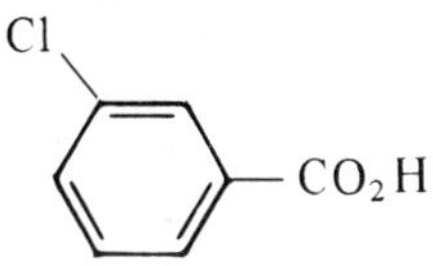	*m*-Chlorobenzoic acid	3.8
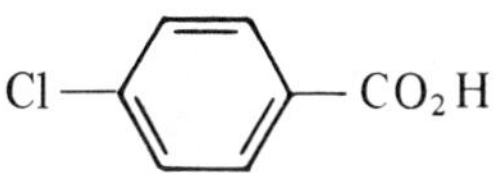	*p*-Chlorobenzoic acid[a]	4.0
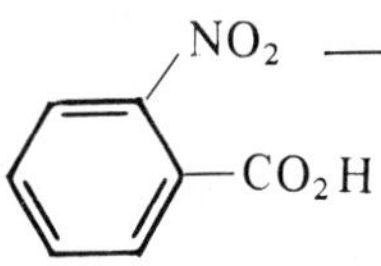	*o*-Nitrobenzoic acid	2.2
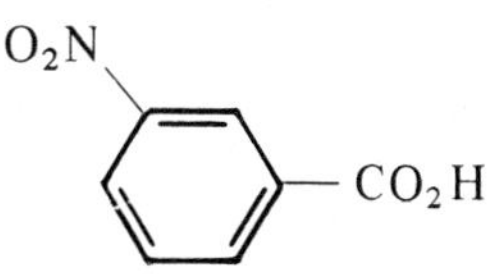	*m*-Nitrobenzoic acid	3.5
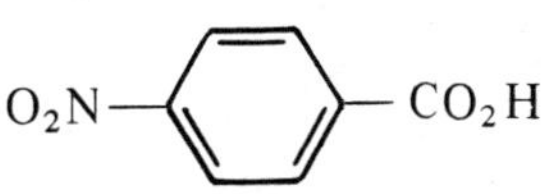	*p*-Nitrobenzoic acid	3.4

[a]The prefixes *o* (*ortho*), *m* (*meta*), and *p* (*para*) denote the three possible positional isomers of a distubstituted benzene.

Source: Some of the data are taken from A. L. Ternay, Jr., *Contemporary Organic Chemistry*, W. B. Saunders, Philadelphia, 1979.

Table 5.3 Inductive Effect of Selected Substituents on the Basic Strength of Substituted Amines

Structure	*Name*	pK_b
$:NH_2—OH$	Hydroxylamine	8.03
$:NH_3$	Ammonia	4.73
$:NH_2—CH_3$	Methylamine	3.38
$:NH_2—CH_2CH_3$	Ethylamine	3.37
$:NH(CH_3)_2$	Dimethylamine	3.23
$:NH(C_2H_5)_2$	Diethylamine	2.90
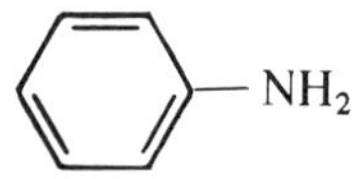	Aniline	9.42
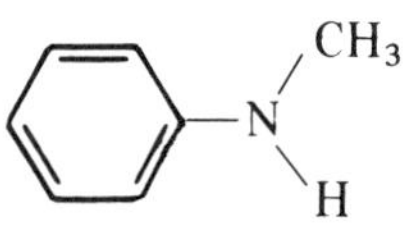	*N*-Methylaniline	9.15
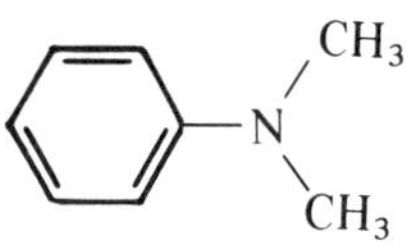	*N*,*N*-Dimethylaniline	8.93
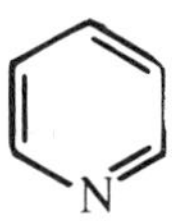	Pyridine	8.89
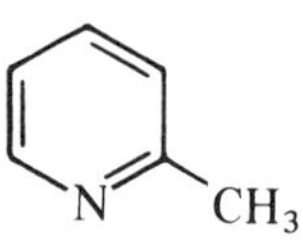	2-Methylpyridine	8.03
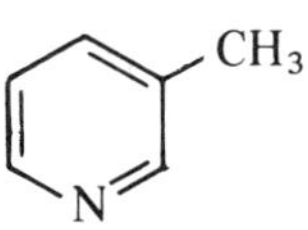	3-Methylpyridine	8.32
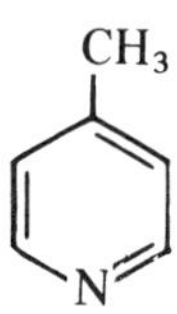	4-Methylpyridine	7.98

Consider the various monochlorinated butanoic acids, for which the order of acid strengths is as follows:

Cl on C-2	Cl on C-3	Cl on C-4	
$CH_3CH_2\overset{\overset{\textstyle Cl}{|}}{C}HCO_2H$	$CH_3\overset{\overset{\textstyle Cl}{|}}{C}HCH_2CO_2H$	$\overset{\overset{\textstyle Cl}{|}}{C}H_2CH_2CH_2CO_2H$	$CH_3CH_2CH_2CO_2H$
$pK_a = 2.8$	4.0	4.5	4.8
$K_a = 1.6 \times 10^{-3}$	1.0×10^{-4}	3.0×10^{-5}	1.6×10^{-5}

These compounds are structural isomers and vary only in the positioning of the cholorine atom. As the highly electronegative halogen is moved away from the carboxyl group, the acidity falls off noticeably. There is only a slight difference in acidity between 4-chlorobutanoic and the unsubstituted butanoic acid. Inductive effects fall off rapidly with distance and can be effectively transmitted through only four or five bonds. In 4-chlorobutanoic acid it is transmitted through six bonds, and the effect on acidity is quite small.

The magnitude of an inductive effect is also dependent on the electronegativity of the group. For example the inductive effect of a fluorine atom is greater than that of a CH_3 (methyl) group. Comparison of the acidities shows that 2-fluoroacetic acid (pK_a = 2.6) is more than 100 times stronger than 2-methylacetic acid (pK_a = 4.9).

Inductive effects also influence the acidity of inorganic acids and bases. The order of acidity in the hypohalous acids is determined by the inductive effect of the halogens bonded to the oxygen. As the electronegativity of the halogen increases, so does the acidity of the hypohalous acid.

↑ acid strength increases			conjugate base strength increases ↓	
	$HOCl + H_2O \rightleftharpoons H_3O^+ + OCl^-$	$pK_a = 7.5$		(5-8)
	$HOBr + H_2O \rightleftharpoons H_3O^+ + OBr^-$	$pK_a = 8.7$		(5-9)
	$HOI + H_2O \rightleftharpoons H_3O^+ + OI^-$	$pK_a = 11$		(5-10)

The base strength of a compound is also influenced by the inductive effect; basicity is weakened by withdrawal of electron density from the basic site. The student is invited to consider the examples of bases given in Table 5.3, and to explain the observed pK_b values in terms of the inductive effect.

5.5 SIZE OF THE ION

Any mechanism that operates to distribute charge over a larger region in general increases the stability of an ion. Energy is required to pack electrical charge into a smaller volume because of the mutual repulsion of like charges. A smaller ion has a higher energy than a larger ion, and is more likely to react in such a way as to alleviate the high charge density, that is, via acid-base interaction with a species of opposite charge. Spreading out of charge in a large cation decreases its tendency to attract negative charge, and hence decreases its acidity. In an anionic species spreading out of charge decreases the basicity. Examples of this trend are found in most groups of the periodic table.

In group I the acid strength of the cations decreases in the order $Li^+ > Na^+ > K^+$, while the ionic radius increases in the order $Li^+ < Na^+ < K^+$. In group II Mg^{2+} is a stronger acid than the larger Ca^{2+}.

The Lewis base strength of anions is also affected by ionic size and charge density. In group VII, both the Lewis and the Brønsted-Lowry base strength of the halide ions decreases in the order $F^- > Cl^- > Br^- > I^-$, while the ionic radius increases in the same order, $F^- < Cl^- < Br^- < I^-$.

5.6 MAGNITUDE OF THE CHARGE

The effect of ionic charge on Lewis acid-base strength is quite straightforward. As the positive charge increases so does the Lewis acid strength; in the second-period elements, Al^{3+} is a stronger acid than Mg^{2+}, which in turn is stronger than Na^+. As the negative charge increases, so does the Lewis base strength; nitride ion (N^{3-}) is a stronger base than oxide (O^{2-}), which in turn is stronger than fluoride (F^-).

The effect of the magnitude of the charge is also illustrated by the series OH^-, H_2O, H_3O^+:

$$\xrightarrow{\text{positive charge increases}}$$
$$OH^- < H_2O < H_3O^+$$
$$\xrightarrow{\text{acid strength increases}}$$

in which the acid strength increases markedly with the incorporation of each additional H^+.

5.7 APPLICATIONS: HYDROGEN HALIDES AND BORON HALIDES

A. Hydrogen Halides

In this chapter we discussed some features of molecular structure and their effect on the relative strength of acids and bases. Application of these principles is not always straightforward; the information available from experiment must be supplemented by careful reasoning in order to provide satisfactory explanations for observed acidity trends. A case in point is the relative acidity of the binary hydrides which we will discuss using the hydrogen halides as examples. The acidity of the hydrogen halides increases regularly in the order HF < HCl < HBr < HI. The pK_a values, as well as other relevant data, are collected in Table 5.4.

Of the influential factors discussed in this chapter, only electronegativity and anionic size are applicable. Resonance delocalization of charge does not occur in the binary hydrides. Since there are no "distant parts of the molecule," the inductive effect is the same as electronegativity. The binary hydrides are neutral compounds, so we need not consider varying magnitude of charge.

Table 5.4 Hydrogen Halides

Acid	pK_a	*Electronegativity of Halogen Atom*	*Anionic Radius of Halide Ion (nm)*
HF	3	4.0	0.136
HCl	−7	3.0	0.181
HBr	−9	2.8	0.195
HI	−10	2.5	0.216

Electronegativity and anionic size influence the acidity of the hydrogen halides in opposite ways. Since the electronegativity of the halogen atom decreases in the order F > Cl > Br > I, we would expect the acidity of the hydrogen halides to *decrease* in the order HF > HCl > HBr > HI (see Section 5.2). On the other hand, anionic size increases in the order $F^- < Cl^- < Br^- < I^-$, so we would expect the acidity to *increase* in the order HF < HCl < HBr < HI (see Section 5.5). Since the latter is the observed order of acidities, we conclude that, within group VII, anionic size exerts a more powerful influence on acidity than does electronegativity.

This behavior is true in other groups as well. In group VI the electronegativity decreases in the order O > S > Se > Te, while anionic size increases in the order $HO^- < HS^- < HSe^- < HTe^-$. Acidity increases in the order $H_2O < H_2S < H_2Se < H_2Te$; that is, it follows the trend in size rather than electronegativity. In group V the trends are: electronegativity, N > P > As > Sb; size, $NH_2^- < PH_2^- < AsH_2^- < SbH_2^-$; acidity, $NH_3 < PH_3 < AsH_3 < SbH_3$. Here too size is more important than electronegativity.

When acidity comparisons are made within a *period* rather than a group, a different pattern emerges. In the second period electronegativity increases in the order C < N < O < F so we would expect the acidity to increase in the order $CH_4 < NH_3 < H_2O < HF$. The size of the conjugate base decreases slightly in the order $CH_3^- > NH_2^- > OH^- > F^-$, and on this basis we would expect a different effect on acidity, namely $CH_4 > NH_3 > H_2O > HF$. Since the observed acidity order is $CH_4 < NH_3 < H_2O < HF$, we conclude that within the second period, electronegativity rather than size is the dominant factor. Analogous trends hold in the third, fourth, and fifth periods.

For the binary hydrides, then, (1) *within a period relative acidity is determined by electronegativity*, and (2) *within a group acidity is determined by anionic size.* This behavior is not surprising, because elements in the same period have similar sizes while differing greatly in chemical properties. Elements in the same group, however, share similar chemical properties while differing significantly in size.

B. Boron Trihalides

The boron trihalides BX_3 (X = F, Cl, or Br) form another group of compounds in which two of the structural features discussed in this chapter work in opposite ways to affect acidity. These compounds are very strong Lewis acids, and readily form adducts with ammonia and other electron donors.

$$H_3N{:} + BF_3 \longrightarrow H_3N{-}BF_3 \qquad (5\text{-}11)$$

Boron (a group III element) has three valence electrons which occupy three sp^2 hybrid orbitals, as shown in Figures 5.1 and 5.5. Figure 5.5 also shows the three σ bonds in a

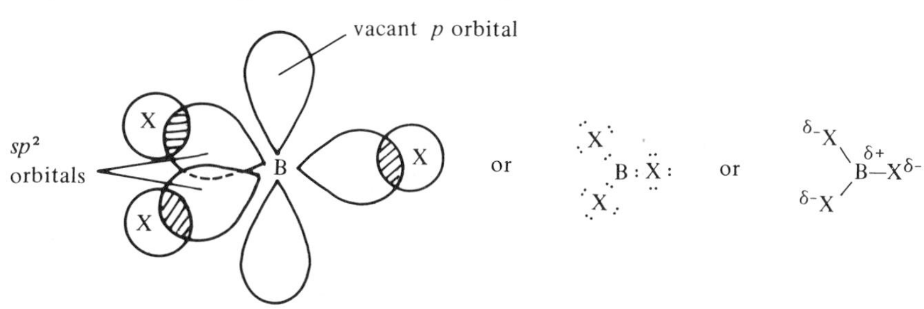

Figure 5.5 σ bonding in boron trihalides.

boron trihalide. These are normal covalent bonds; that is, each of the bonded atoms contributes one electron to the shared pair. The dot representation of the σ bonding is also shown in Figure 5.5. Note that the boron atom has only six electrons in its valence shell. It is therefore electron deficient, which accounts for its Lewis acidity. Since the halogens are all more electronegative than boron, the B—X bonds are polar and as a result the B atom has an enhanced positive charge.

Since enhancement of positive charge increases the Lewis acidity of a site, and since halogen electronegativity increases in the order $Br < Cl < F$, we would expect the Lewis acidity of the boron trihalides to increase in the order $BBr_3 < BCl_3 < BF_3$. In fact, the opposite order is observed, and we conclude that another factor is at work.

The additional factor is resonance. The p orbital on the boron atom is vacant and unused in the σ bonds, and the adjacent halogen atoms have p orbitals in the proper position for orbital overlap to occur to form a π bond. This is shown in Figure 5.6; note the similarity to Figure 5.2. The π bond is a coordinate bond, since both of its electrons come from the halogen atom. The boron atom acquires a share of two additional electrons and has a formal charge of (−1). This completes its valence-shell octet, and it cannot form any additional bonds.

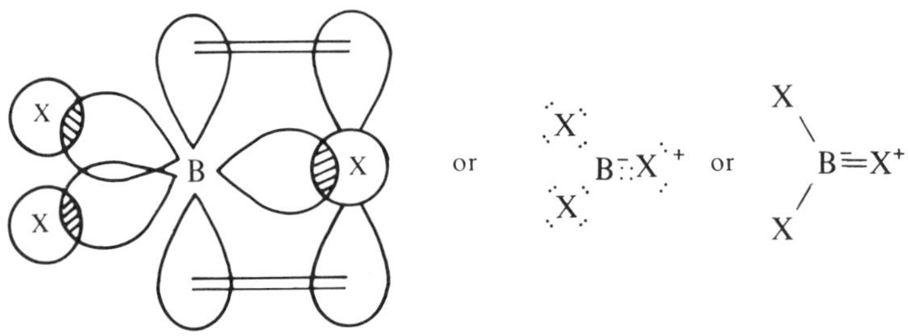

Figure 5.6 π **bonding in boron trihalides.**

Three equivalent dot structures can be drawn, and, as pointed out in Section 5.3, this means that resonance occurs. The resonance contributors are:

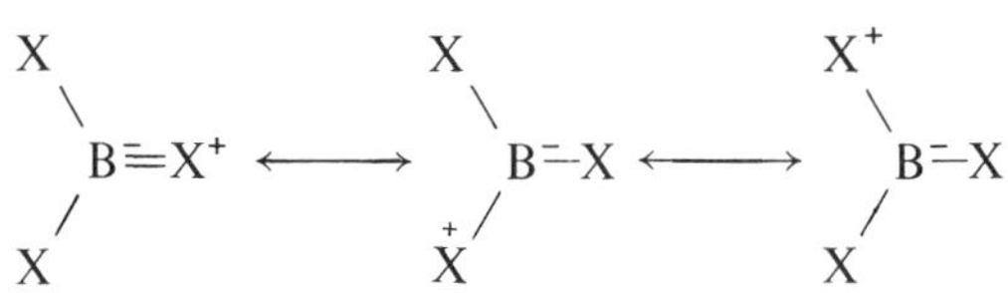

The structure can also be represented by the hybrid,

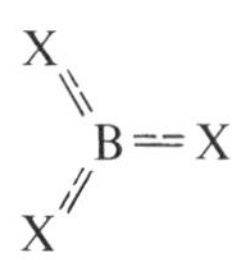

The boron atom is not electron deficient in the π-bonded contributors; the effect of resonance is to decrease the electron deficiency, and hence decrease the Lewis acidity of the trihalides. The extent to which the π-bonded forms contribute to the hybrid structure is greatest for BF_3 and least for BBr_3; the effect is great enough to reverse the order of acidity expected from electronegativity alone. The observed order of acidity is $BBr_3 > BCl_3 > BF_3$.

We can summarize the factors bearing on boron trihalide acidity as follows.

1. With only six electrons in the valence shell, the boron atom in the trihalide is an electron acceptor. All boron trihalides are strong Lewis acids.
2. The greater electronegativity of halogens, compared to boron, enhances the acidity.
3. The effect of π-bonded resonance contributors decreases the acidity; in boron trihalides this effect is more important than electronegativity in determining the relative acidities.

PROBLEMS

5.1 Draw a Lewis dot structure for the nitro group and explain why it is an excellent electron-withdrawing group, through both resonance and the inductive effect.

5.2 Without looking up the relevant pK_a values, arrange each of the following sets of compounds in order of decreasing acidity, and explain your reasoning.

(a) $Br{-}CH_2{-}CH_2{-}CO_2H$, $Br{-}CH_2{-}CH(Cl){-}CO_2H$, $Br{-}CH(Cl){-}CH_2{-}CO_2H$

(b) FCH_2CO_2H, F_2CHCO_2H, F_3CCO_2H

(c) $CH_3{-}C({=}O){-}CH_2{-}CO_2H$, $O_2N{-}CH_2CO_2H$, CH_3CO_2H

(d) 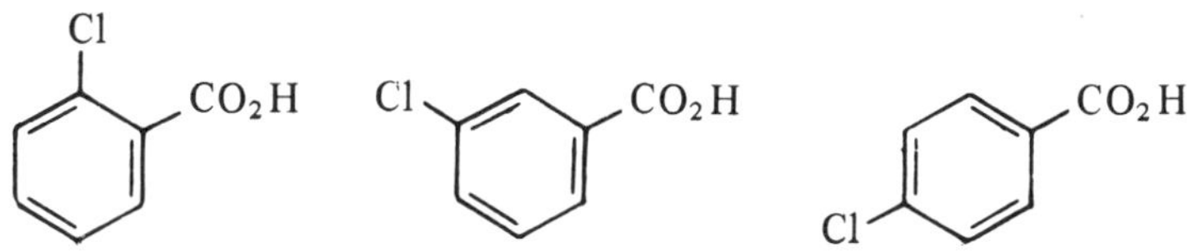

(e) CH_3CH_2OH 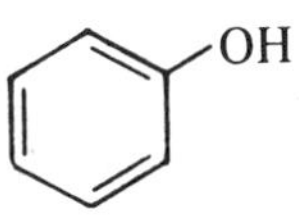

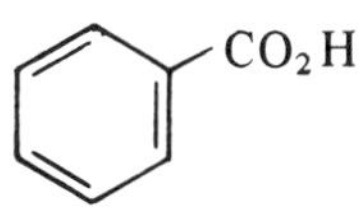

5.3 Without looking up the relevant pK_b values, arrange each of the following sets in order of decreasing basicity, and explain your reasoning.

(a) CH_3O^-, $CH_3CH_2O^-$, $(CH_3)_2CHO^-$

(b) NF_3, NCl_3, NH_3, NBr_3

(c) $CH_3CO_2^-$, $FCH_2CO_2^-$, $BrCH_2CO_2^-$

(d) 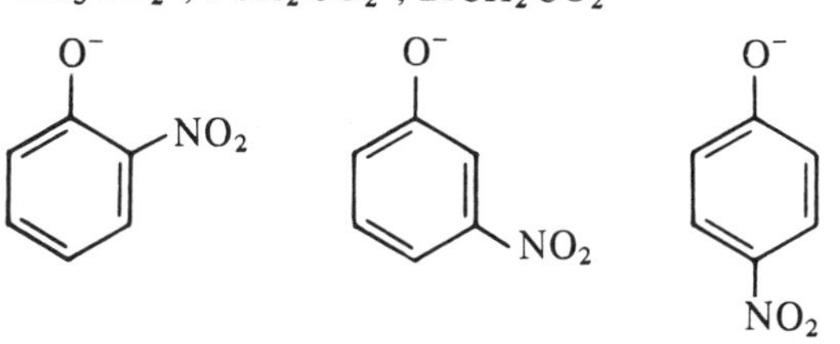

5.4 Draw the major contributing resonance forms for each of the following species.
(a) The anion formed when benzoic acid loses its most acidic proton.
(b) The anion formed when the species $CH_2{=}CH{-}\mathbf{CH_3}$ loses one of the bold-faced protons.
(c) The anion formed when the proton is lost from H–N=C=O.
(d) The anion formed when the proton is lost from N≡C–O–H.

5.5 In problem 5.4, parts (c) and (d), you were asked to draw contributing resonance structures for two related acids. Examine these resonance forms carefully. Are they the same or different? What effect might they have on the relative acidity of these two compounds?

5.6 Calculate the hydronium ion concentration for each of the following solutions. Refer to Table 5.2 for any necessary information.
(a) 0.010 *M* 2-chlorobutanoic acid
(b) 0.001 *M* benzoic acid
(c) A 0.17% (by weight) solution of chloroacetic acid
(d) 0.003 *M* hydrochloric acid

5.7 Calculate the hydroxide ion concentration for the solutions listed in Problem 5.6.

5.8 Which of the following is the stronger Lewis acid, $(CH_3)_3B$ or Cl_3B? Explain your choice.

5.9 Deduce the formal charge on each of the atoms (other than hydrogen) in the following structures.

(a) NO_2^-
(b) N_3^-
(c) SO_2
(d) 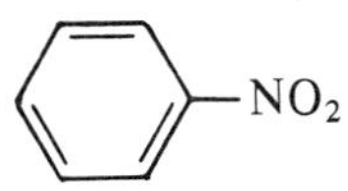
(e) 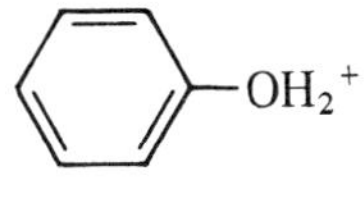
(f) CH_3CO_2H
(g) $CH_3CO_2^-$
(h) O_3
(i) CN^-
(j) 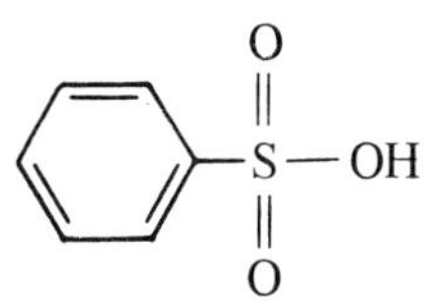
(k) ClO_3^-
(l) $CH_3NH_3^+$
(m) 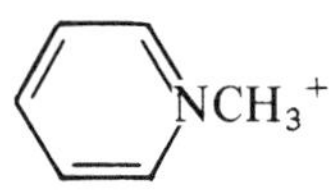

Chapter 6

Applications and Examples in Organic Chemistry

6.1 INTRODUCTION

A complex chemical transformation can be analyzed in terms of a sequence of simple steps, referred to as the *mechanism* of the reaction. Understanding the mechanism permits better control of the process and allows a reasoned approach to the synthesis of other molecules. The vast majority of organic reactions involve the interaction of an electron-rich species with an electron-deficient material, which, as we have seen, is another name for acid-base chemistry. In this chapter we discuss the application of acid-base principles to some reaction mechanisms which are important to the study and practice of organic chemistry.

6.2 HYDROCARBONS: THE EFFECT OF HYBRIDIZATION

Hydrocarbons, compounds made up of only hydrogen and carbon, may be separated into three categories: (1) *saturated hydrocarbons* (alkanes and cycloalkanes); (2) *unsaturated hydrocarbons* (alkenes and alkynes); and (3) *aromatic hydrocarbons* such as benzene. Compounds in each category exhibit similar chemical reactivity; the reactivity differences between compounds in different categories are due, to a significant extent, to different hybridizations of the carbon atoms.

A. Alkanes

Alkanes are compounds having the molecular formula C_nH_{2n+2}, where n is any integer. Examples are methane, CH_4; ethane, CH_3CH_3; propane, $CH_3CH_2CH_3$; and butane, $CH_3CH_2CH_2CH_3$. The carbon atoms in alkanes are sp^3 hybridized; the C—H bonds in alkanes result from the overlap of a singly occupied sp^3 hybrid orbital on carbon with a singly occupied 1s orbital on a hydrogen atom. The simplest alkane, methane (CH_4),

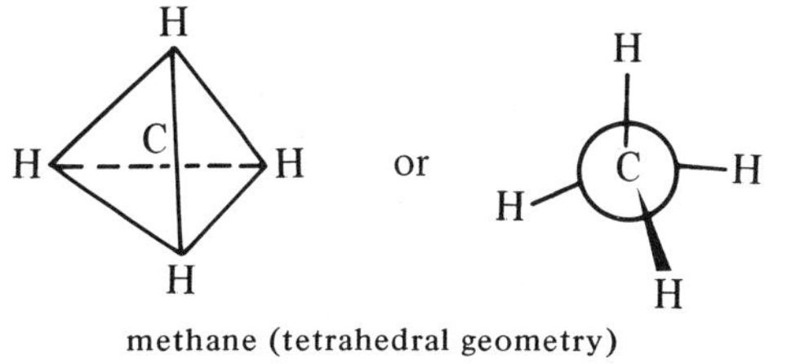

methane (tetrahedral geometry)

is a symmetrical molecule in which each C—H bond is directed toward one of the corners of a regular tetrahedron. The H—C—H bond angles are 109.5°. When acid ionization occurs, as shown for methane in reaction (6-1), one of the C—H bonds is heterolytically

$$CH_4 \longrightarrow H^+ + H_3C\,(:)^- \qquad (6\text{-}1)$$

cleaved. Both electrons of the former C—H bond become nonbonding and reside on the conjugate base. This nonbonded pair still occupies an sp^3 orbital.[1]

B. Alkenes

Alkenes, like alkanes, are compounds containing only the elements carbon and hydrogen. Unlike alkanes, they contain at least one carbon-carbon double bond. Examples are ethene (also called ethylene), $CH_2{=}CH_2$; propane, $CH_3{-}CH{=}CH_2$; 1-butene, $CH_2{=}CH{-}CH_2{-}CH_3$; and 2-butene, $CH_3{-}CH{=}CH{-}CH_3$. In alkenes the doubly bonded carbon atoms are sp^2 hybridized. The C—H bonds at an sp^2 carbon are formed by the overlapping of a singly occupied sp^2 orbital on carbon with a singly occupied 1*s* orbital on hydrogen (see Figure 5.2). The geometry about doubly bonded carbon atoms is illustrated below for the simplest alkene, ethene:

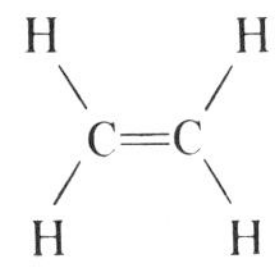

ethene

The bond angles of the H—C—H and the C—C—H bonds are close to 120°, and all six atoms lie in the same plane. Acid ionization by heterolytic cleavage of a C—H bond leaves a nonbonded pair of electrons on the conjugate base [reaction (6-2)]. In alkenyl

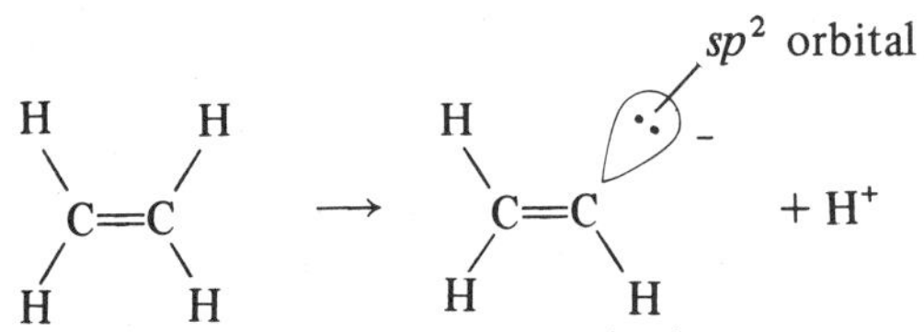

(6-2)

ethene — ethenyl carbanion (conjugate base of ethene)

carbanions this lone pair resides in a carbon sp^2 orbital, and the 120° bond angles are almost unchanged from those of the alkene. Loss of a proton from an sp^2 hybridized carbon atom in benzene, which is characterized by resonance delocalization of electron density, also gives a conjugate base in which the lone electron pair resides in an sp^2 orbital [reaction (6-3)].

[1] The shape of the methyl carbanion is a triangular pyramid in which the H—C—H bond angles have decreased slightly to approximately 108°. This structure is, as one would expect, almost the same as that of ammonia, NH_3. Ammonia also has a nonbonded ("lone") pair of electrons occupying an sp^3 orbital.

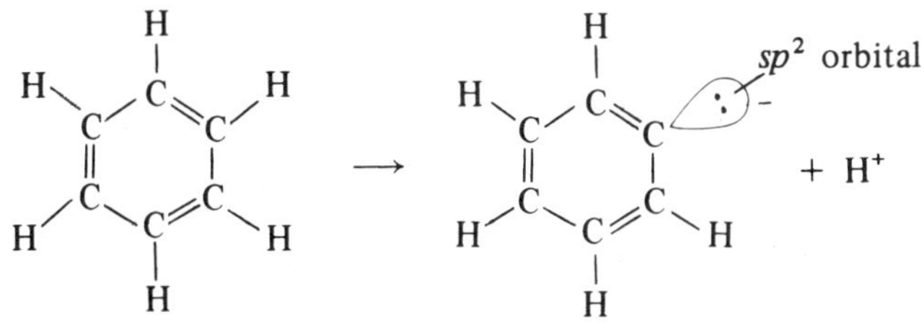

benzene phenyl carbanion (6-3)

C. Alkynes

Hydrocarbons containing one or more carbon-carbon triple bonds are called alkynes. Examples are acetylene, H—C≡C—H, and propyne, H—C≡C—CH_3. The carbon atoms in alkynes have a third type of hybridization. Triply bonded carbon atoms form bonds to each other and to hydrogen using *sp* hybrid orbitals. The acid ionization of the simplest alkyne, acetylene (ethyne), is shown in reaction (6-4). Again, in the heterolytic

$$\underset{\text{acetylene}}{\text{H—C≡C—H}} \longrightarrow \underset{\text{acetylide ion}}{\text{H—C≡C}}\,(:)^{-}\ \ \textit{sp}\text{ orbital} + \text{H}^{+} \qquad (6\text{-}4)$$

cleavage of a C—H bond a nonbonded pair of electrons remains on the conjugate base, the acetylide ion. In an alkyne carbanion this lone pair resides in an *sp* hybrid orbital.

The type of orbital (*sp*, sp^2, or sp^3) in which the nonbonded electron pair of a carbanion resides affects the acid strength of the hydrocarbon (the conjugate acid). The effect is due to the increasing "*s* character" of the hybrid orbitals in the series $sp^3 < sp^2 < sp$.

Hybrid Type	sp^3	sp^2	sp
%s Character	25%	33%	50%

The greater the *s* character of an orbital, the closer the electrons are to the nucleus. The electrons in an *sp* orbital are closer to the positively charged nucleus, and therefore more tightly bound than they are in an sp^2 or an sp^3 orbital. Since, in general, electronegativity is inversely related to orbital size, it is also true that as *s* character increases, so does electronegativity. As *s* character increases, the electrons are less available for sharing and the carbanion is a weaker Lewis base. The acidity-basicity order of the hydrocarbons and their conjugate bases conforms to this principle, as shown below.

		Hybrid Type	pK_a	*Acid Ionization*	pK_b	
s character increases ↓	acid strength increases ↓	sp^3	49	H_3C—H → $H^+ + H_3C:^-$	−35	base strength decreases ↓
		sp^2	44	H_2C=CH—H → $H^+ + H_2C{=}\ddot{C}H^-$	−30	
		sp	25	H—C≡C—H → H^+ + H—C≡C:$^-$	−11	

The effect of *s* character is quite large and accounts for an apparent anomaly in the relative acidities of the second-period binary hydrides. On the basis of electronegativity considerations (see Chapter 5) we expect the following order:

$$\xrightarrow{\text{electronegativity (of the non-H atom) increases}}$$
$$CH_4 < NH_3 < H_2O < HF$$
$$\xrightarrow{\text{acid strength increases}}$$

and this is indeed observed. If electronegativity were the only factor determining acidity, all hydrocarbons would be weaker acids than ammonia. In fact, they are not: the increased *s* character of the *sp* orbitals in acetylene so enhances the acidity that acetylene is a stronger acid than ammonia.

$$CH_4 < CH_2{=}CH_2 < NH_3 < HC{\equiv}CH < H_2O < HF$$

Evidence for the relative acidities comes from data on equilibria such as reaction (6-5),

$$NH_2^- + CH_3{-}C{\equiv}C{-}H \rightleftharpoons CH_3{-}C{\equiv}C{:}^- + NH_3 \qquad (6\text{-}5)$$

which proceeds almost entirely to the right. The formation of substituted acetylide ions by reaction with the NH_2^- ion is of great practical value in organic syntheses.

6.3 ORGANIC OXYGEN ACIDS

There are many organic compounds which are more acidic than hydrocarbons. In one class of such compounds the acidic hydrogen is bonded to oxygen, and the general category is called *oxygen acids*. The three major types of oxygen acid are (1) sulfonic acids; (2) carboxylic acids; and (3) phenols. Typical ionization reactions of these are shown in reactions (6-6), (6-7), and (6-8). In these reactions water acts as the solvent as well as the base. In general, sulfonic acids are much more acidic than carboxylic acids, which are in turn much more acidic than phenols.

$$\underset{\text{methanesulfonic acid}}{CH_3SO_3H} + H_2O \rightleftharpoons H_3O^+ + CH_3SO_3^- \qquad pK_a \cong -1.2 \qquad (6\text{-}6)$$

$$\underset{\substack{\text{acetic acid}\\ \text{(a carboxylic acid)}}}{CH_3{-}C(=O){-}OH} + H_2O \rightleftharpoons H_3O^+ + CH_3{-}C(=O){-}O^- \qquad pK_a = 4.8 \qquad (6\text{-}7)$$

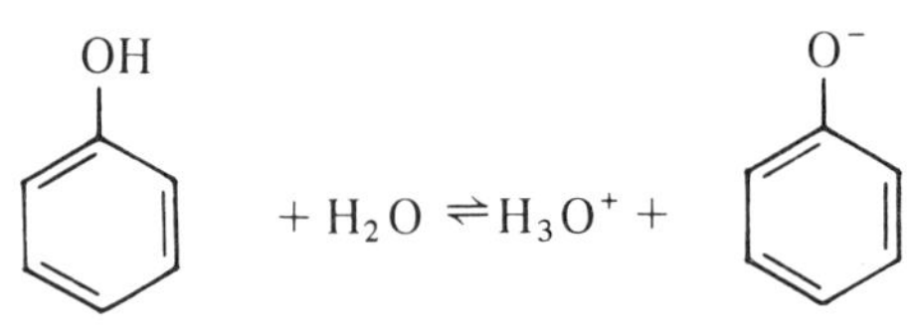

$pK_a = 10.0$ (6-8)

phenol

A. Sulfonic Acids

The acidity of sulfonic acids is in many cases nearly as great as that of common mineral acids. They are essentially 100% ionized in dilute aqueous solution. Sulfonic acids should be viewed as derivatives of sulfuric acid, in which a hydroxy group has been replaced by a carbon-containing substituent.

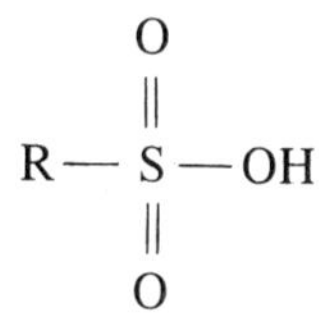

The factors responsible for the strongly acidic nature of H_2SO_4 are also responsible for the high acidity of sulfonic acids. The most important factor is the resonance delocalization of the negative charge in the anion. There are three resonance contributors to the

Table 6.1 Acidities of Selected Sulfonic Acids

Name	*Structure*	K_a	pK_a
Methanesulfonic acid	CH_3SO_3H	15.8	−1.2
Naphthalenesulfonic acid	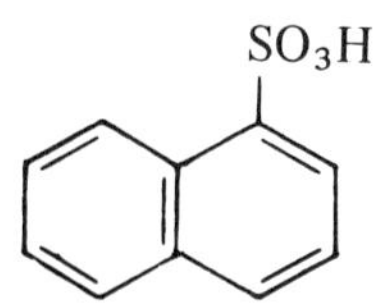	2.7×10^{-1}	0.57
Benzenesulfonic acid	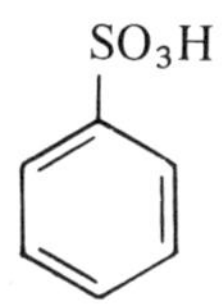	2.0×10^{-1}	0.70
o-Aminobenzenesulfonic acid	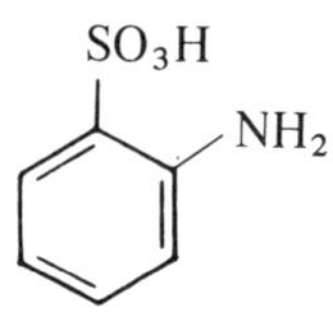	3.3×10^{-3}	2.48
m-Aminobenzenesulfonic acid	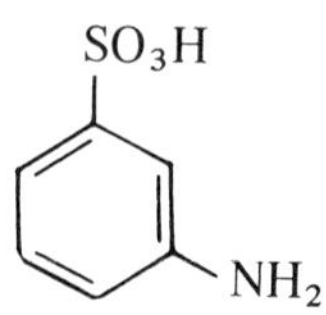	1.85×10^{-4}	3.73
p-Aminobenzenesulfonic acid	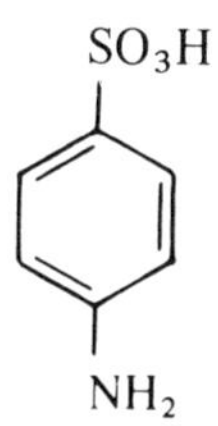	5.81×10^{-4}	3.24

hybrid structure of the anion (6-9). Extensive charge delocalization enhances the sta-

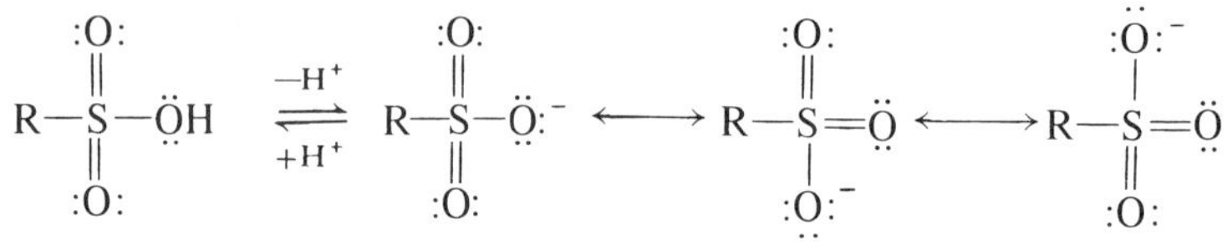

acid — resonance forms of conjugate base (6-9)

bility of this anion, and the ionization equilibrium (6-9) is pulled to the right. There are many sulfonic acids, and the pK_a values for several of them are given in Table 6.1.

B. Carboxylic Acids

Carboxylic acids are derivatives of carbonic acid, in which one of the hydroxy groups has been replaced by an organic group.

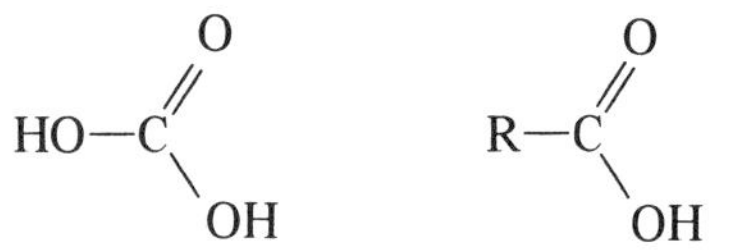

carbonic acid — a carboxylic acid

Table 6.2 Acidities of Substituted Acetic and Benzoic Acids

Substituent Y	pK_a at 25°C			
		Y—C$_6$H$_4$COOH		
	Y—CH$_2$COOH	*ortho*	*meta*	*para*
Hydrogen (H)	4.74	4.20	4.20	4.20
Methyl ($-CH_3$)	4.87	3.91	4.27	4.38
Ethyl ($-CH_2CH_3$)	4.82	3.79	4.27	4.35
Fluoro (F)	2.59	3.27	3.86	4.14
Chloro (Cl)	2.85	2.92	3.83	3.97
Bromo (Br)	2.90	2.85	3.81	3.97
Iodo (I)	3.18	2.86	3.85	4.02
Cyano (—CN)	2.47	3.14	3.64	3.55
Trifluoromethyl ($-CF_3$)	3.06		3.77	3.66
Hydroxy (—OH)	3.83	2.98	4.08	4.57
Methoxy ($-OCH_3$)	3.57	4.09	4.09	4.47
Phenyl ($-C_6H_5$)	4.31	3.46	4.14	4.21
Nitro ($-NO_2$)		2.21	3.49	3.42

Source: Reproduced with the permission of Macmillan Publishing Company, from *Introduction to Organic Chemistry*, 2nd ed., 1981, by A. Streitwieser, Jr., and C. H. Heathcock.

They are considerably weaker than sulfonic acids, and can be described as moderately weak acids. Since carbon is less electronegative than sulfur, and since there are fewer resonance forms to contribute to the stability of the conjugate base, carbonic acid is weaker than sulfuric acid. As with sulfuric acid and the sulfonic acids, the acidity of carbonic acid and the carboxylic acids is enhanced by resonance stabilization of the conjugate base (6-10); the negative charge is delocalized throughout the π system of the

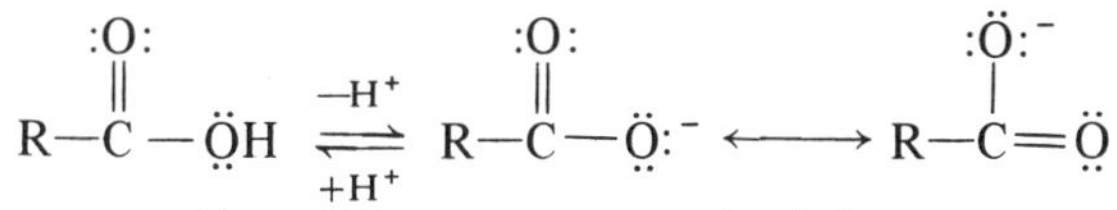

acid conjugate base (6-10)

carboxy group. In addition to resonance, inductive effects also contribute to the acidity of carboxylic acids.

A list of representative carboxylic acids, including both substituted benzoic and acetic acids, is presented in Table 6.2. Study this table and note the nature of the various subtituents and their effect on acidity. Many of these substituents are highly electronegative, and thus enhance the acidity. Other groups are electron donors (less electronegative than the groups to which they are attached) and thus decrease the acidity. The proximity of the group to the acid-base site also affects the acid-base strength. As the group is moved farther away from the acidic site, its inductive effect becomes weaker.

C. Phenols

Phenols are rather weak acids. They are *less* acidic than the carboxylic acids, but *more* acidic than the closely related alcohols.

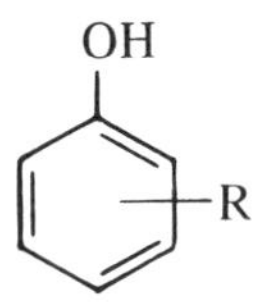

a substituted phenol[2]

$R{-}CH_2{-}OH$

an alcohol

The only important factor influencing the acidity of alcohols is the inductive effect, whereas the acidity of phenols [reaction (6-11)] is determined by both inductive and

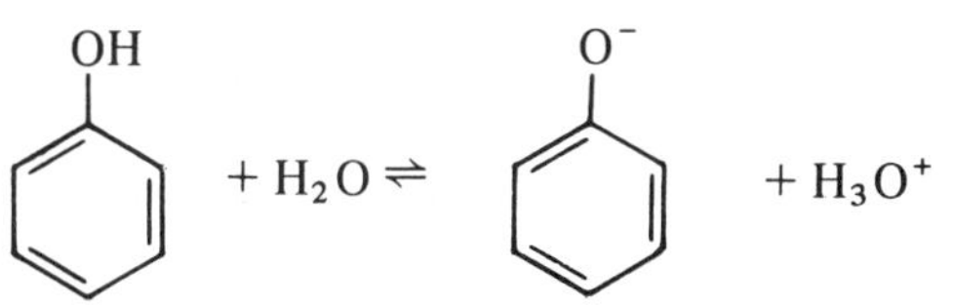

phenol phenoxide ion (6-11)

resonance effects. The benzene ring, with sp^2 hybridized carbon atoms, is more electronegative than the RCH_2 group, in which the carbon atom is sp^3 hybridized. The conjugate

[2] This structural formula, in which a bond drawn to the center of the ring, means that the position of the substituent, R, with respect to the hydrogen group, is not designated: the compound may be an *ortho-*, *para-*, or *meta-*substituted phenol.

base, phenoxide ion, is thus more stabilized by inductive withdrawal of negative charge by the ring from the oxygen atom than is the anion of an alcohol.

More importantly, the phenoxide ion is stabilized by resonance. The four possible resonance structures **A**, **B**, **C**, and **D** are shown in (6-12). Structure **A** is important, since

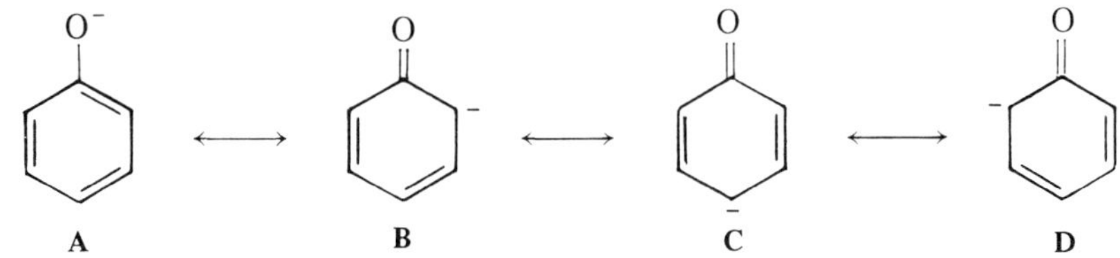

(6-12)

the negative charge resides on the more electronegative atom, oxygen. In forms **B**, **C**, and **D** the negative charge is delocalized into the ring: as always, delocalization of charge enhances stability.

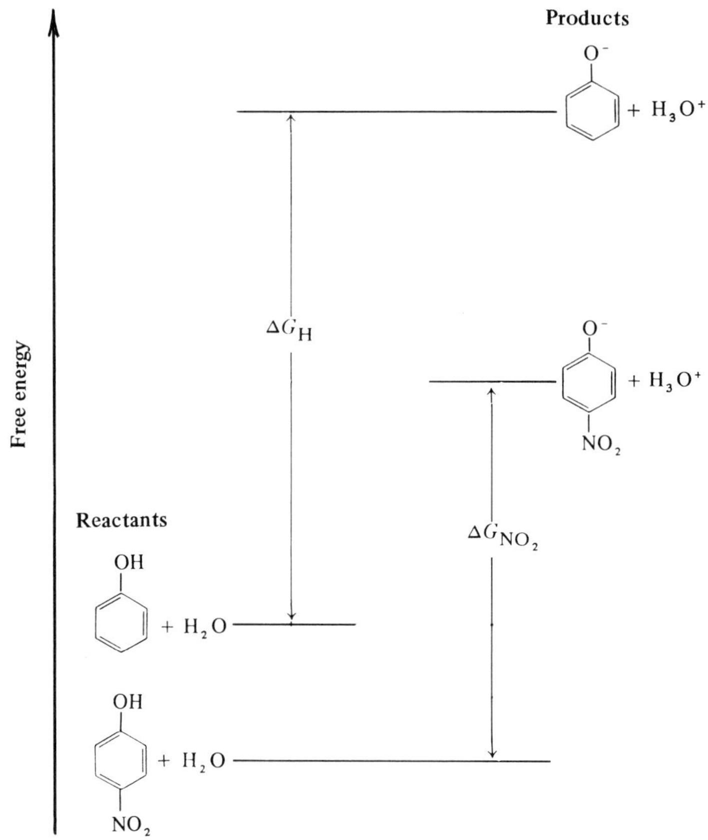

Figure 6.1 When a *para*-nitro substituent is introduced into the phenol molecule, the energies of both the neutral acid and the conjugate base are decreased. What determines the effect of substitution on acidity is the change in the free energy difference between neutral and anionic forms. Since ΔG_{NO_2} is smaller than ΔG_H, *p*-nitrophenol is ionized more easily than phenol and is a stronger acid than phenol.

Similar considerations apply to the stability of the conjugate bases of substituted phenols. In the case of p-nitrophenol (6-13) the resonance forms involving the ring (**F**, **G**,

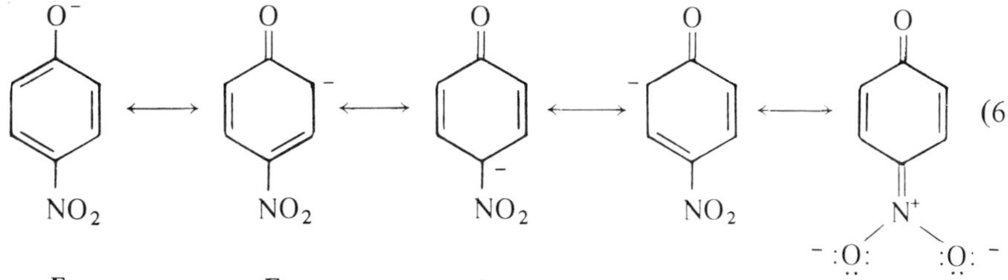

(6-13)

E F G H I

H, and **I**) are even more important contributors to the hybrid structure than is the case with phenoxide ion. This is a consequence of the great electron-withdrawing power of the nitro group. The additional structure, **I**, in which the charge resides on the oxygen atoms of the nitro group, is an important contributor to the resonance hybrid. The p-nitrophenoxide ion, having considerable charge delocalization, is more stable than the phenoxide ion. The relative energies of the nonionized molecules are also important in determining acidities. The p-nitrophenol molecule is more stable than phenol in polar solvents, because of its greater polarity. These relationships are diagrammed in Figure 6.1. The stronger of the two acids, p-nitrophenol, is the one whose ionization requires the smaller free energy input, ΔG_{NO_2}. Normally the solvation energies of ionized species are subject to greater variation than those of their uncharged conjugate acids.

Tables 6.3 and 6.4 list the pK_a values of several phenols. As you study these tables, keep in mind that electron-donating substituents decrease acidity, and electron-withdrawing substituents enhance acidity. Both resonance and inductive effects are important in determining the relative acidities of these materials. Problem 6.25, at the end of the chapter, provides a good opportunity for application of these concepts.

One practical consequence of the weak acidity of phenols is the ease with which they may be separated from carboxylic and sulfonic acids. All three types of acid react with strong bases, such as aqueous sodium hydroxide, to form the conjugate base and water. Hydroxide ion is such a strong base that these reactions, (6-14), (6-15), and (6-16), go essentially to completion. However, if these same materials are treated with a weaker

$$R{-}SO_3H + OH^- \longrightarrow R{-}SO_3^- + H_2O \qquad (6\text{-}14)$$

$$RCO_2H + OH^- \longrightarrow RCO_2^- + H_2O \qquad (6\text{-}15)$$

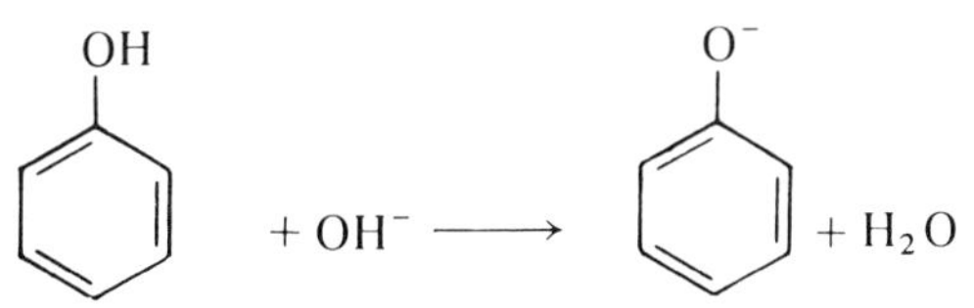

(6-16)

base, such as bicarbonate ion, only the reactions of the stronger acids (RSO_3H and RCO_2H) go to completion. This is illustrated in reactions (6-17) and (6-18). Phenol does

$$R{-}SO_3H + HCO_3^- \xrightarrow{H_2O} R{-}SO_3^- + H_2CO_3 \qquad (6\text{-}17)$$

$$RCO_2H + HCO_3^- \xrightarrow{H_2O} RCO_2^- + H_2CO_3 \qquad (6\text{-}18)$$

Table 6.3 Acidity of Selected Nitrophenols

Compound	*Structure*	pK_a
Phenol	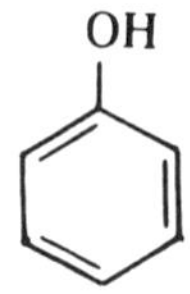	10.00
o-Nitrophenol	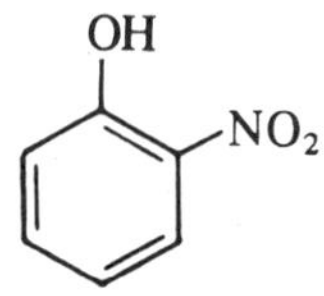	7.22
m-Nitrophenol	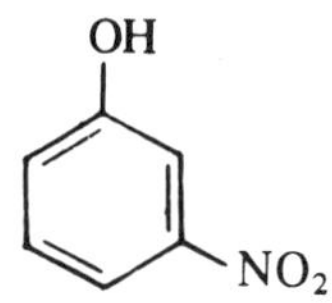	8.39
p-Nitrophenol	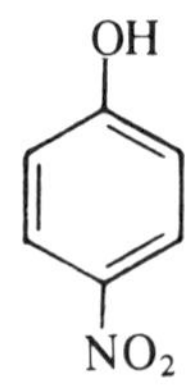	7.15
2, 4-Dinitrophenol	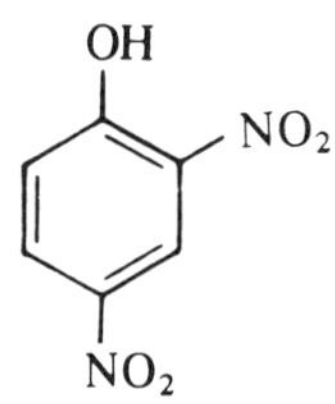	4.09
2, 4, 6-Trinitrophenol (picric acid)	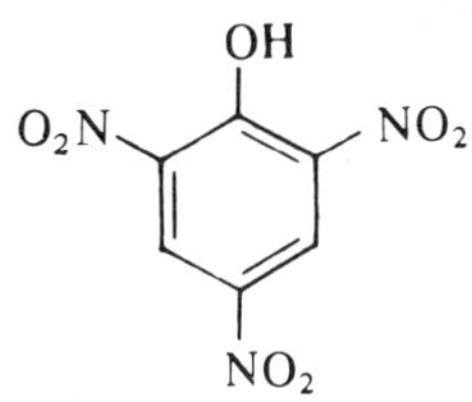	0.25

react [reaction (6-19)], but the equilibrium lies far to the left. Phenol can therefore be

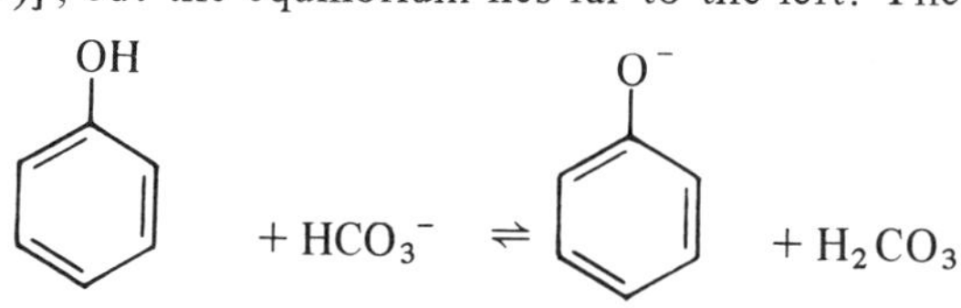

(6-19)

Table 6.4 Effect of *ortho*, *meta*, and *para* Substituents on the Acidity of Phenol

	pK_a (25°C)		
Substituents	*ortho*	*meta*	*para*
Hydrogen (H)	10.00	10.00	10.00
Methyl ($-CH_3$)	10.29	10.09	10.26
Fluoro (F)	8.81	9.28	9.81
Chloro (Cl)	8.48	9.02	9.38
Bromo (Br)	8.42	8.87	9.26
Iodo (I)	8.46	8.88	9.20
Methoxy ($-OCH_3$)	9.98	9.65	10.21
Methylthio ($-SCH_3$)		9.53	9.53
Cyano ($-CN$)			7.95
Nitro ($-NO_2$)	7.22	8.39	7.15

Source: Reproduced with the permission of Macmillan Publishing Company from *Introduction to Organic Chemistry*, 2nd ed., 1981, by A. Streitwieser, Jr., and C. H. Heathcock.

separated from a mixture containing all three types of acid. The mixture is dissolved in some organic solvent that is not soluble in water (diethyl ether, for example) and the solution is then extracted with an aqueous solution of $NaHCO_3$. The two stronger acids are converted to the anionic forms RSO_3^- and RCO_2^- and pass into the aqueous layer, in which they are readily soluble. The relatively unreactive phenol remains behind in the ether layer, since in its nonionized form it is only slightly soluble in water.

6.4 BASIC FUNCTIONAL GROUPS

One of the most common basic substituents in organic molecules is the amino group, $-NH_2$. Molecules containing an amino group are related to ammonia; if one or more of the hydrogen atoms of ammonia is replaced by an organic group, the compound is an *amine*.

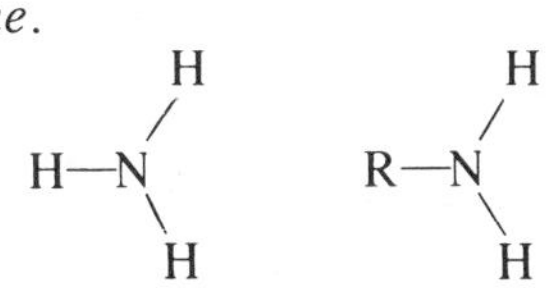

ammonia an amine

$R{-}NH_2$ primary amine R_2NH secondary amine R_3N tertiary amine

These compounds, mono-, di-, or trisubstituted, are known as primary, secondary, and tertiary amines, respectively.

The base ionization reaction of amines is given by reaction (6-20).

$$R{-}NH_2 + H_2O \rightleftharpoons R{-}NH_3^+ + OH^- \qquad (6\text{-}20)$$

As we saw in Chapter 4, the basicity of these materials is due to the ease of donation of the nonbonded electron pair on nitrogen to an electron-pair acceptor. The substituents on nitrogen affect the availability of the electron pair. If these substituents are electron donating (less electronegative than nitrogen), they will enhance the availability of the pair of electrons and hence the basicity of the species. If the R group is electron-withdrawing (more electronegative than nitrogen), the electron pair is more tightly bound, less available for sharing, and the basicity of the amine is decreased. Table 5.2 gives pK_b values of some substituted amines and should be referred to at this point.

Resonance greatly affects the basicity of amines. Compare the relative basicities of aliphatic amines, in which the nitrogen atom is bonded to an sp^3 carbon atom, with aniline, in which the substituent is a phenyl group. Aniline is a much weaker base (pK_b = 9.42) than methylamine (pK_b = 3.81). The nonbonded electron pair on the nitrogen atom of aniline is more tightly held due to both inductive and resonance effects. Inductive withdrawal is greater by the phenyl group (sp^2 hybridized carbon) than by methyl (sp^3 hybridized carbon). In addition, the electrons on the nitrogen atom of aniline are further delocalized by resonance interaction with the ring. The resonance contributors have the structures **A**, **B**, **C**, and **D**:

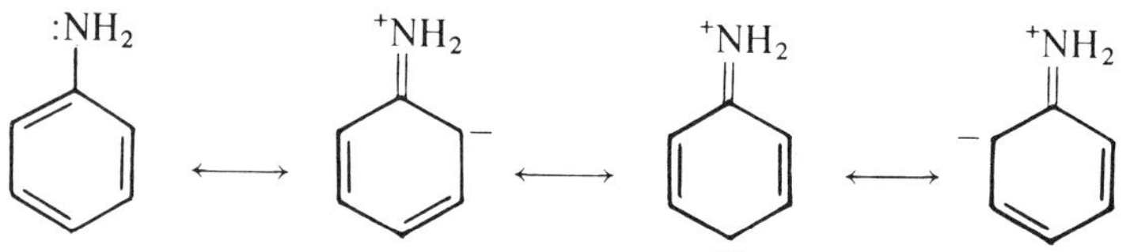

(6-21)

Note that in forms **B**, **C**, and **D**, the nitrogen atom has no nonbonding electrons. To the extent that these forms contribute to the overall resonance hybrid, the availability of the electron pair on nitrogen for Lewis base activity is substantially reduced. In addition, the electron density of the carbon atoms *ortho* and *para* to the amino group is increased, so that these atoms can also act as Lewis bases. The increased basicity of the *ortho* and *para* carbons leads to certain reactions of appropriately substituted benzenes, called *electrophilic substitutions*, which are discussed later in this chapter.

Electron-withdrawing substituents on the benzene ring, especially if *ortho* or *para* to the NH_2 group, can have a pronounced effect on the basicity of aniline. This phenomenon is illustrated by the nitro-substituted anilines shown below. In this sequence the order of

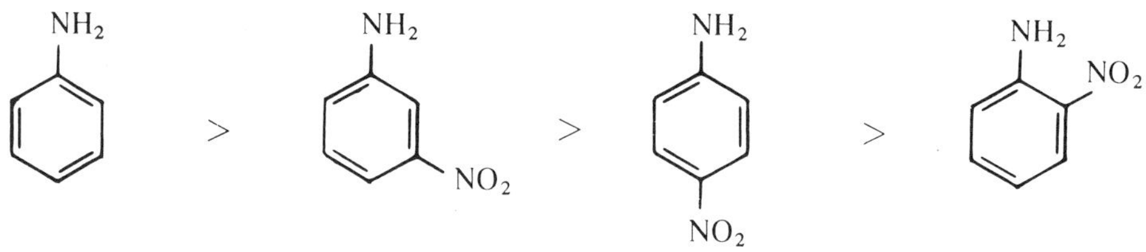

$K_b = 3.8 \times 10^{-10}$	$K_b = 2.8 \times 10^{-12}$	$K_b = 1.0 \times 10^{-13}$	$K_b = 5 \times 10^{-15}$
pK_b = 9.42	pK_b = 11.6	pK_b = 13.0	pK_b = 14.3
aniline	*m*-nitroaniline	*p*-nitroaniline	*o*-nitroaniline

decreasing base strength is *meta* > *para* > *ortho*. Once again both resonance and inductive effects are involved. As we have seen in the discussion of phenols, the nitro group has a powerful electron-withdrawing effect. In the *ortho* and *para* isomers, electron withdrawal by the nitro group from the amine leads to the resonance contributors shown in (6-22).

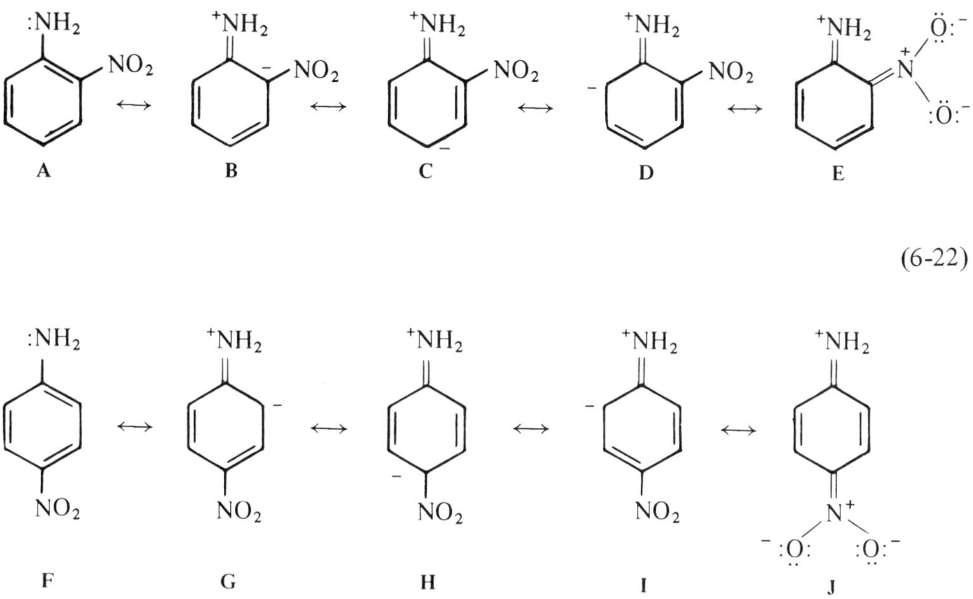

(6-22)

The resulting delocalization of negative charge weakens the basicity of the amine group to about the same extent in both the *ortho* and *para* isomers. The *ortho* isomer is the weaker base of the two, however, because of the additional, *inductive* withdrawal of electrons of the NO_2 group. In the *ortho* isomer the nitro group is closer to the basic site and thus its effect is more strongly felt. The *meta* isomer is the strongest base in this series because the electron pair of its amine nitrogen atom is the least delocalized by resonance. The only resonance forms that contribute to the *meta* isomer are shown in

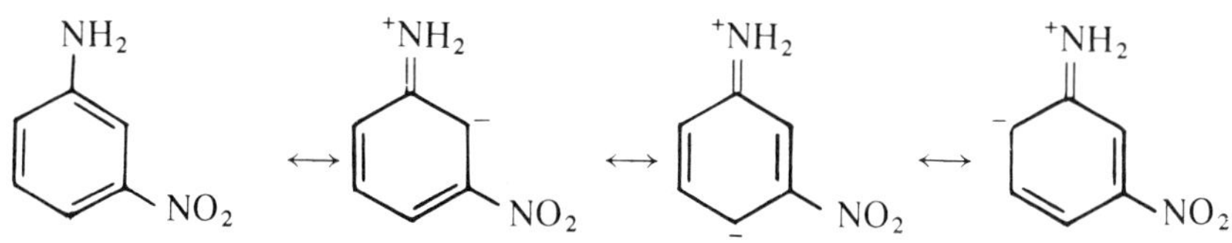

(6-23)

(6-23); none of these allows delocalization of charge into the nitro group. It is instructive to compare the resonance forms shown in (6-21), (6-22), and (6-23) for aniline and substituted anilines with those shown for phenol and its derivatives in (6-12) and (6-13).

The K_b and pK_b values of some substituted anilines are given in Table 6.5. The effect of the various substituents on basicity should be carefully noted. A comparison of the data found in Tables 6.1 through 6.5 should provide further insight into the electron-donating or electron-withdrawing power of a considerable number of substituents.

Table 6.5 Strength of Substituted Anilines[a] (25°C)

Name	*Structure*	K_b	pK_b
Aniline	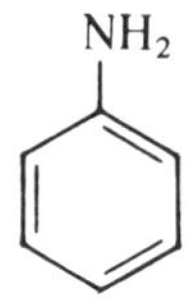	3.82×10^{-10}	9.42
o-Chloroaniline	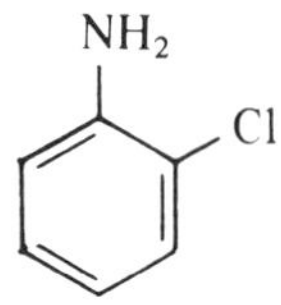	4.4×10^{-12}	11.36
m-Chloroaniline	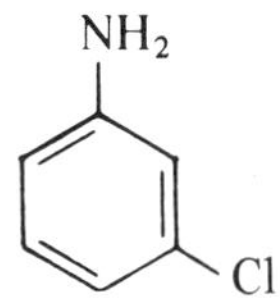	2.88×10^{-11}	10.54
p-Chloroaniline	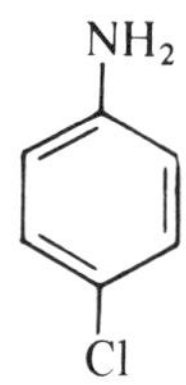	1×10^{-10}	10.00
p-Fluoroaniline	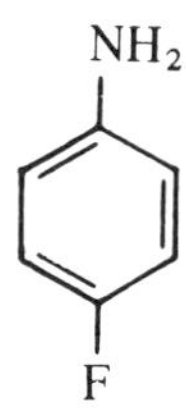	4.3×10^{-10}	9.37
p-Bromoaniline	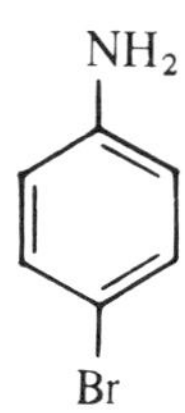	7×10^{-11}	10.15
p-Iodoaniline	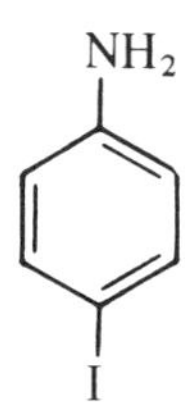	6×10^{-11}	10.22

Table 6.5 (cont'd)

Name	*Structure*	K_b	pK_b
p-Methoxyaniline	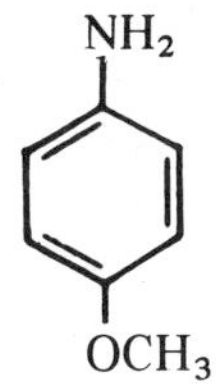	1.9×10^{-9}	8.71
o-Nitroaniline	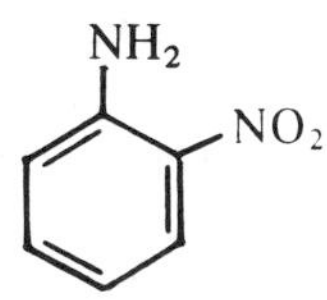	5.1×10^{-15}	14.29
m-Nitroaniline	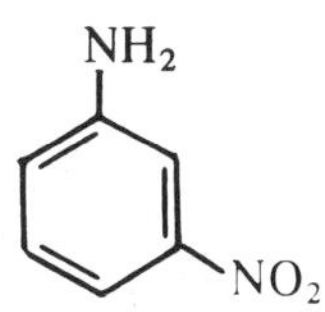	2.8×10^{-12}	11.55
p-Nitroaniline	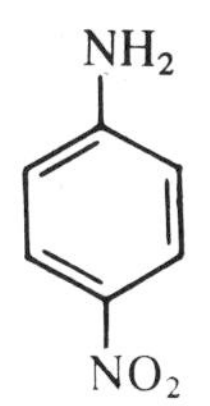	1×10^{-13}	13.00
p-Toluidine	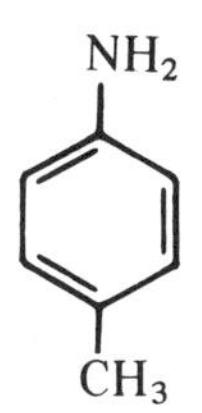	1.18×10^{-9}	8.92
Diphenylamine (*N*-phenylaniline)	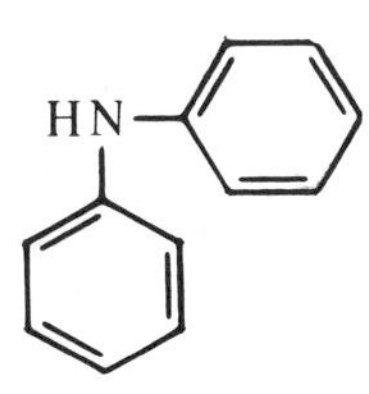	7.6×10^{-14}	13.12
N-Methylaniline	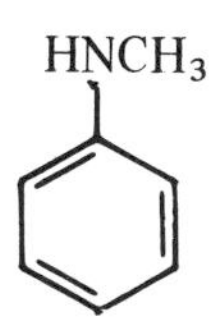	7.1×10^{-10}	9.15

[a]See also Table 5.2.

6.5 AMIDES AND IMIDES

If one of the hydrogen atoms of ammonia is replaced with a carbonyl (>C=O) group, the resulting compound is called an *amide*. If two of ammonia's hydrogen atoms are replaced by carbonyl groups, the compound is called an *imide*. Both of these classes

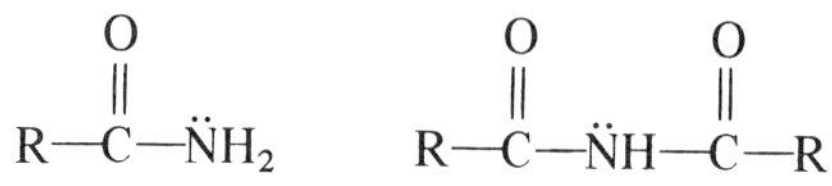

an amide — an imide

of compounds are substantially less basic than the amines. The electron pair on the nitrogen atom in both amides and imides is essentially unavailable for sharing with acids.

The high electronegativity of oxygen withdraws considerable electron density from the nitrogen atom; the molecules are also stabilized by resonance. In several of the contributing resonance structures, shown in (6-24) and (6-25), the nitrogen atom bears a

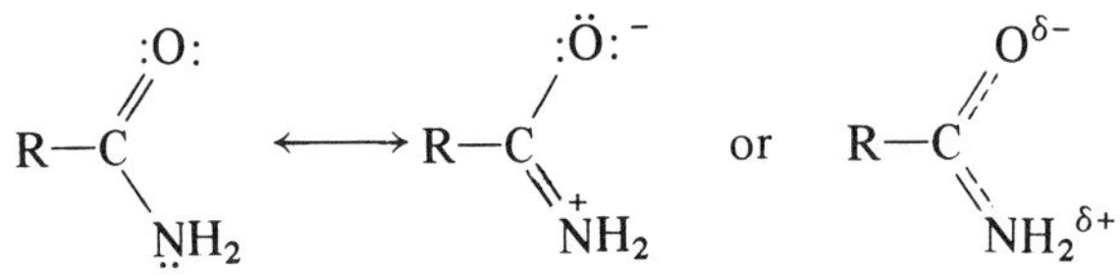

(6-24)

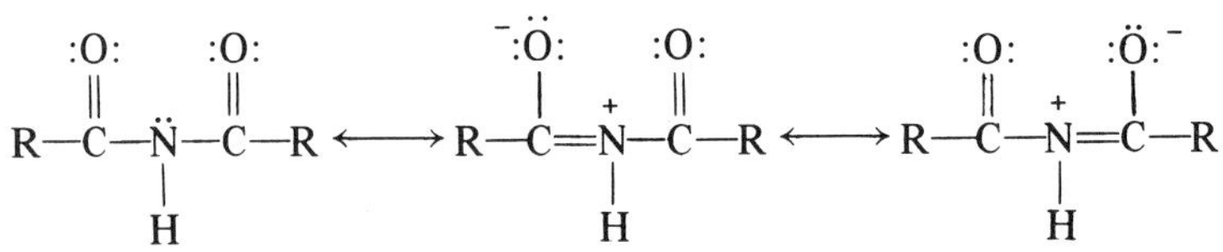

(6-25)

positive charge. In imides there are two carbonyl groups which can accommodate a negative charge, so that the nitrogen atom's electron pair is more delocalized than in amides. The very weakly basic properties of amides and imides[3] are due to protonation of oxygen rather than nitrogen, as illustrated in reaction (6-26). (See problem 6.26.)

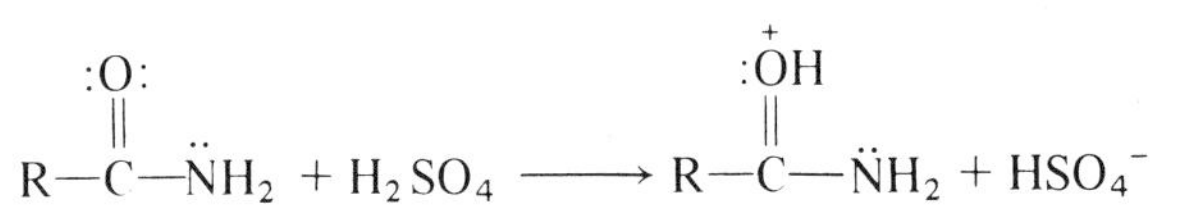

(6-26)

An interesting and important consequence of the presence of two carbonyl groups on a single nitrogen atom is that imides can act as Brønsted *acids*. The proton on the nitrogen atom is so acidic that it can easily be removed by hydroxide ion (6-27). The conjugate

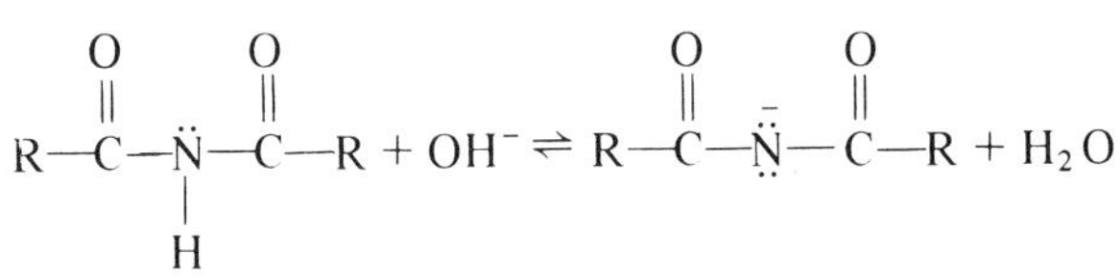

(6-27)

base is greatly stabilized by resonance (6-28). The enhanced stability of the anion pulls the equilibrium (6-27) to the right (i.e., makes the imide a stronger acid). The acidic

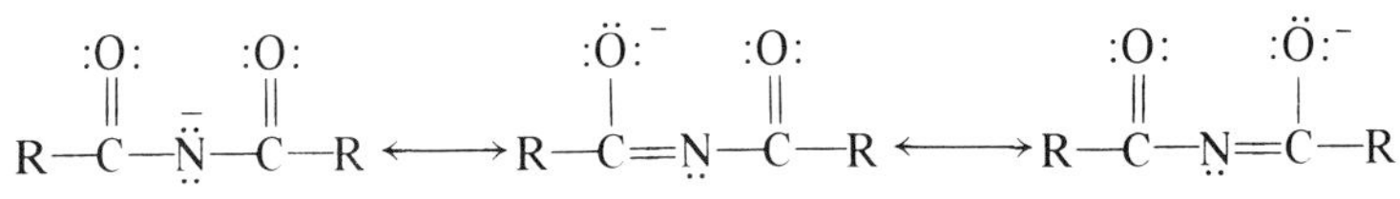

(6-28)

[3]Values of pK_b for amides range from 14 to 18, depending on the nature of the substituent.

nature of imides is exploited in the synthesis of primary amines. For example, in the Gabriel synthesis, primary alkyl halides are treated with phthalimide in the presence of base. In the first step (6-29) the acidic proton of phthalimide ($K_a = 5 \times 10^{-9}$) is abstracted by hydroxide ion to give the conjugate base. The basic anion produced in the first step displaces[4] a halide ion from the alkyl halide, forming an *N*-alkylphthalimide (6-30). In the final step (6-31) the *N*-alkylphthalimide is hydrolyzed in dilute acid (or, less frequently, in dilute base) to produce a primary amine.

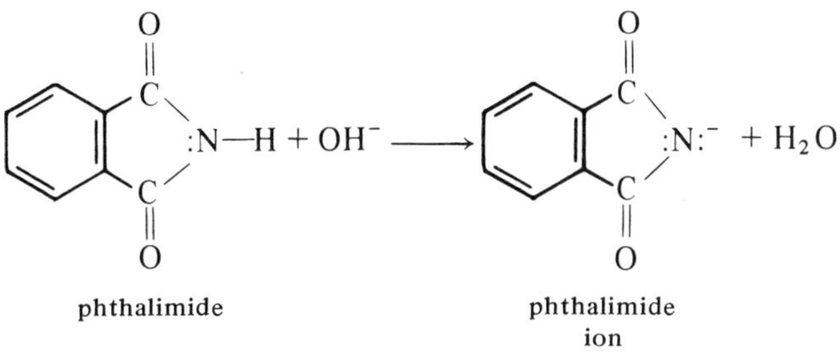

(6-29)

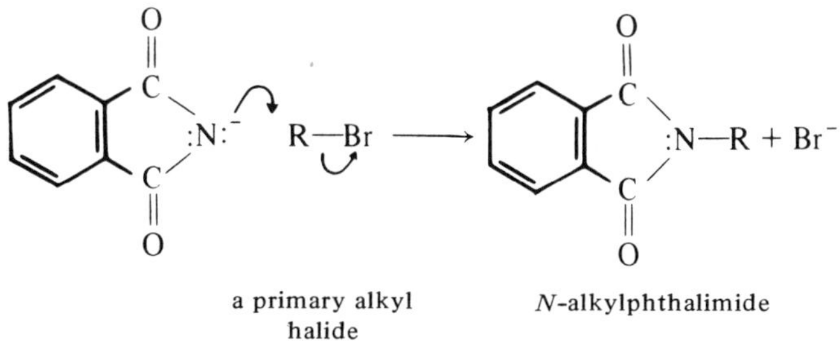

(6-30)

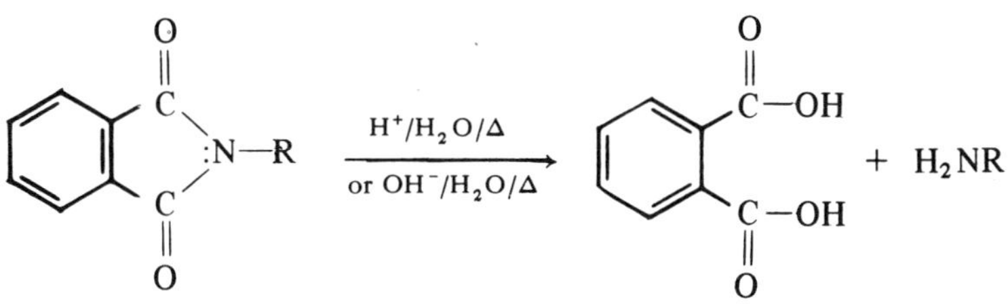

(6-31)

6.6 SULFONAMIDES AND THE HINSBERG TEST

If an organic compound is insoluble in water, but soluble in dilute HCl (5%), the compound is a base and is quite probably an amine. The *Hinsberg test*, which involves the formation of a sulfonamide derivative of the amine, can be used to establish whether the amine is primary, secondary, or tertiary. Compounds containing the functional group $-SO_2N<$ are called *sulfonamides*. Before describing the Hinsberg test in detail we will discuss the nature of the sulfonamide group. The sulfonyl group

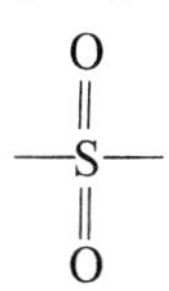

[4] This is a nucleophilic displacement (see Section 4.7).

is a powerful electron-withdrawing substituent because the contributing resonance structures **B**, **C**, and **D** (6-32) all contain a positively charged sulfur atom. A sulfonyl

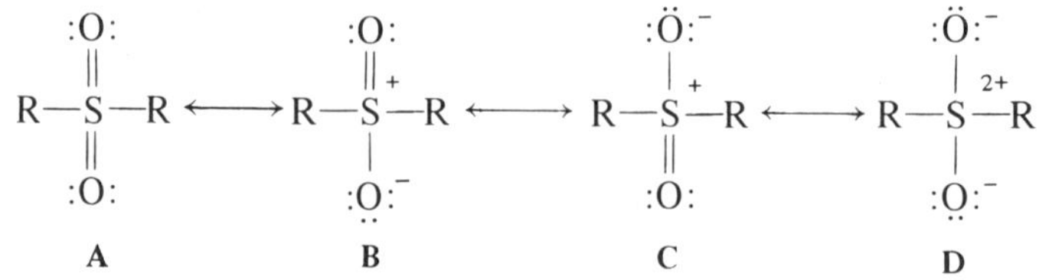

(6-32)

group, when bonded to an >N—H group, substantially increases the acidity of the site. [Compare this with the effect of the sulfonyl group on an adjacent —OH group in sulfuric and sulfonic acids (Section 6.3) and the acidity of imides discussed in Section 6.5.] As a result, the proton of RSO_2NHR can be removed by hydroxide ion (6-33).

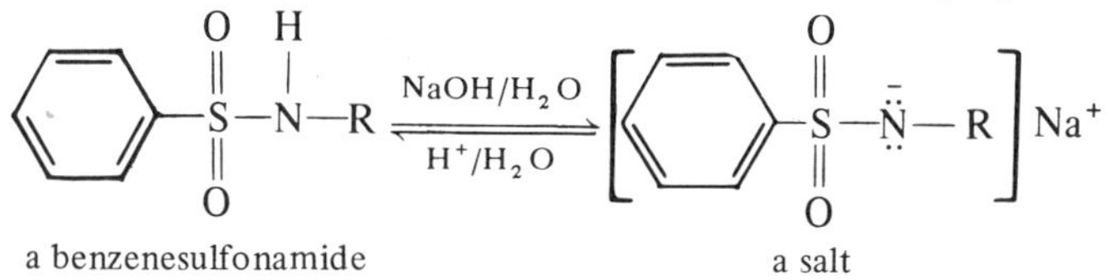

(6-33)

The Hinsberg test consists of treating an amine first with benzenesulfonyl chloride and base, and then with acid. The three types of amine respond differently to this treatment. A primary amine, when treated with benzenesulfonyl chloride and aqueous potassium hydroxide, forms the (soluble) potassium salt of the sulfonamide (6-34). Acidification of the solution with dilute HCl yields the benzenesulfonamide, which is insoluble. A precipitate thus confirms the initial presence of a primary amine.

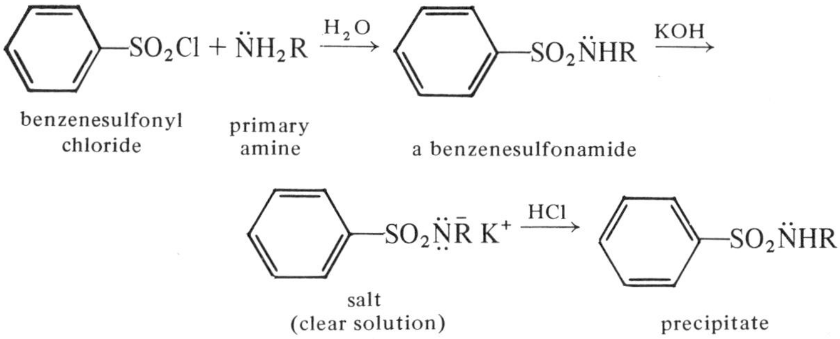

(6-34)

The benezenesulfonamide derived from a secondary amine (6-35) has no acidic proton and therefore cannot form a soluble salt. The precipitation of a solid from the alkaline solution is a positive test for a secondary amine; addition of HCl has no effect on the mixture.

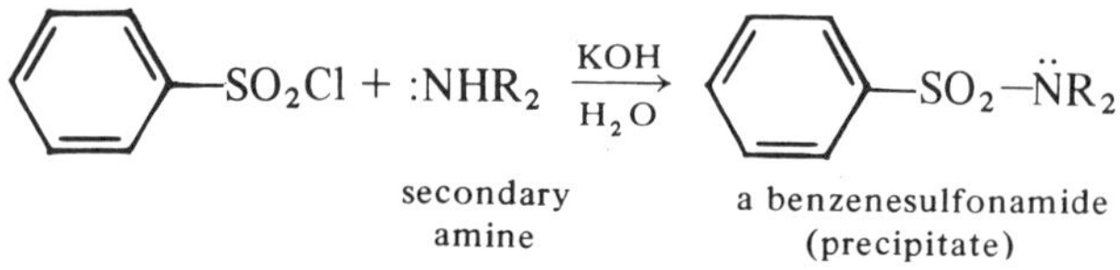

(6-35)

With a tertiary amine the process (6-36) is more complicated. Benzenesulfonyl chloride reacts with the amine to form an unstable salt. The salt is hydrolyzed by hy-

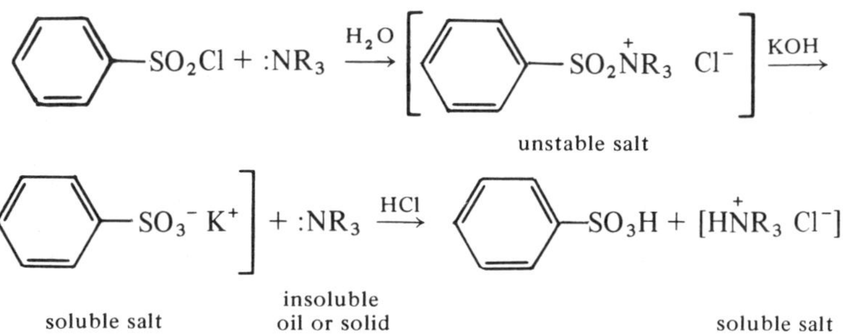

(6-36)

droxide ion, yielding benzenesulfonic acid and regenerating the original tertiary amine. The benzenesulfonic acid dissolves in the excess base, forming the salt. To the experimenter it often appears as if nothing has happened; the tertiary amine ends up as it began, as a solid or an insoluble oil. Addition of HCl causes the amine to dissolve, yielding a clear solution.

6.7 CARBANIONS: THEIR USE IN ORGANIC SYNTHESIS

The strongest bases commonly encountered in organic chemistry are the carbanions. The basicity of these negatively charged carbon species generally exceeds that of oxygen bases (R–O:$^-$) which are in turn stronger than nitrogen bases (R_3N:). As with other bases, electron-withdrawing substituents decrease the basicity of carbanions and electron-donating substituents increase basicity. Saturated alkyl groups are mild electron donors and, when bonded to a negatively charged carbon, increase its basicity. This accounts for the observed order of base strengths of primary, secondary, and tertiary carbanions.[5]

$$H_3C{:}^- < RCH_2{:}^- < R_2CH{:}^- < R_3C{:}^-$$

methyl carbanion	primary carbanion	secondary carbanion	tertiary carbanion

——— basic strength increases ———→

The use of carbanions in organic syntheses is widespread, and in this section we discuss two examples.

A. Grignard Reagents

A Grignard reagent is prepared by reaction of an organic halide with magnesium metal. Example reactions are shown for alkyl halides in reaction (6-37) and for aryl halides in

$$CH_3CH_2CH_2Br + Mg \xrightarrow{Et_2O} CH_3CH_2CH_2MgBr \qquad (6\text{-}37)$$

n-propyl bromide (1-bromopropane) → *n*-propylmagnesium bromide (an alkyl Grignard reagent)

[5]Carbon atoms with sp^3 hybridization are classified as primary, secondary, or tertiary, as follows: A primary carbon atom is bonded to only one other carbon atom; a secondary carbon atom is bonded to two other carbon atoms; a tertiary carbon atom is attached to three other carbon atoms. This terminology is not applicable to carbon atoms with sp^2 or sp hybridization.

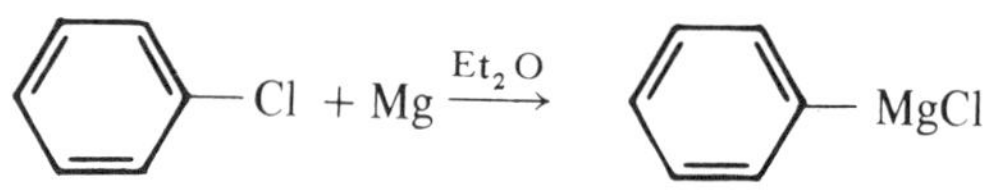

(6-38)

chlorobenzene — phenylmagnesium chloride (an aryl Grignard reagent)

reaction (6-38).[6] The carbon and magnesium atoms in a Grignard reagent are linked by a highly polar covalent bond: the carbon bears a partial negative charge and, in the presence of acids, can behave as if it were a free carbanion.

$$\overset{\delta-}{R} - - - \overset{\delta+}{MgX}$$

One synthetic use of Grignard reagents is the conversion of an alkyl halide to a hydrocarbon (i.e., the replacement of a halogen with a hydrogen atom). This is illustrated in reaction (6-39). The formation of propane from the Grignard reagent [the second step

$$\underset{\text{2-bromopropane}}{CH_3\underset{|\atop CH_3}{CH}-Br} + Mg \xrightarrow{Et_2O} CH_3\underset{|\atop CH_3}{CH}MgBr \xrightarrow[H_2O]{HCl} \underset{\text{propane}}{CH_3\underset{|\atop CH_3}{CH}-H} + MgBrCl \qquad (6\text{-}39)$$

of (6-39)] is an acid-base reaction in which the base $(CH_3)_2CH^-$ accepts a proton from HCl. Grignard reagents, like other strongly basic compounds, react with even very weak acids. Any acid stronger than a carbon acid, that is, a species in which the acidic hydrogen is bonded to carbon, can protonate a Grignard reagent to produce a hydrocarbon. This is readily understandable in terms of the Brϕnsted-Lowry theory of proton transfer reactions. An equilibrium such as (6-40) lies to the right for any acid HA stronger than the

$$R_3C^- + HA \rightleftharpoons R_3C-H + A^- \qquad (6\text{-}40)$$

acid R_3C-H. Since carbon acids are exceedingly weak, there are many compounds that are more acidic. Water, alcohols, ammonia, and amines are all more strongly acidic than carbon acids and therefore readily react with Grignard reagents (6-41) and (6-42).

$$R-MgX + R'OH \longrightarrow R-H + MgX(OR') \qquad (6\text{-}41)$$

$$R-MgX + R'_2NH \longrightarrow R-H + MgX(NR'_2) \qquad (6\text{-}42)$$

Another important use of Grignard reagents is in the preparation of alcohols. A Grignard reagent is first treated with an aldehyde or ketone; the resulting salt is treated with dilute acid to give the product, an alcohol. This sequence is shown in reaction (6-43) for the synthesis of 1-propanol from formaldehyde and ethylmagnesium bromide. The mechanism of this reaction involves several acid-base interactions.

$$\underset{\text{ethylmagnesium bromide}}{CH_3CH_2MgBr} \xrightarrow[\text{(2) dilute acid}]{\text{(1) formaldehyde}} \underset{\text{1-propanol}}{CH_3CH_2CH_2OH} \qquad (6\text{-}43)$$

The carbon atom in a carbonyl group bears a partial positive charge (due to the greater

[6] Et_2O is a common abbreviation for diethyl ether, which is the usual solvent in Grignard reactions.

electronegativity of oxygen), and in the first step (6-44) of the reaction the carbanion-like Grignard reagent attacks this carbon atom. The result of this Lewis acid-base reaction

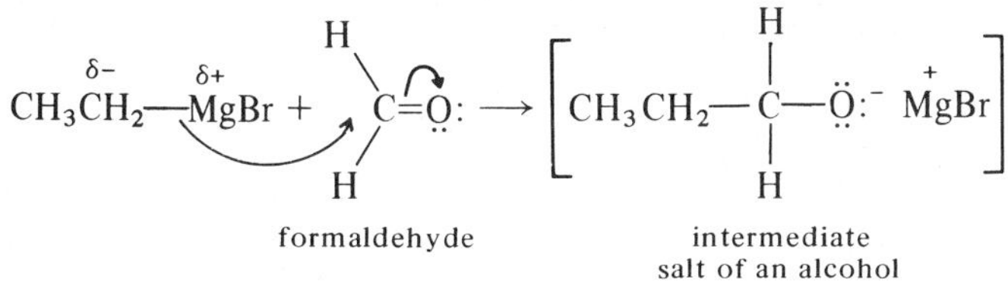

(6-44)

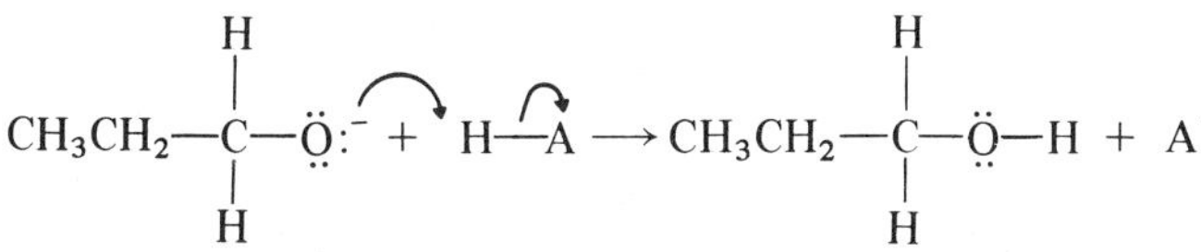

(6-45)

is a magnesium salt of an alcohol. In the second step (6-45) an acid reacts with this strongly basic salt, forming the alcohol (1-propanol).

The reaction of a Grignard reagent and formaldehyde results in a primary alcohol;[7] use of a different aldehyde, or a ketone, yields a secondary alcohol (6-46) or a tertiary alcohol (6-47). In reactions (6-46) and (6-47), X represents Cl, Br, or I, and HA is any Brønsted acid.

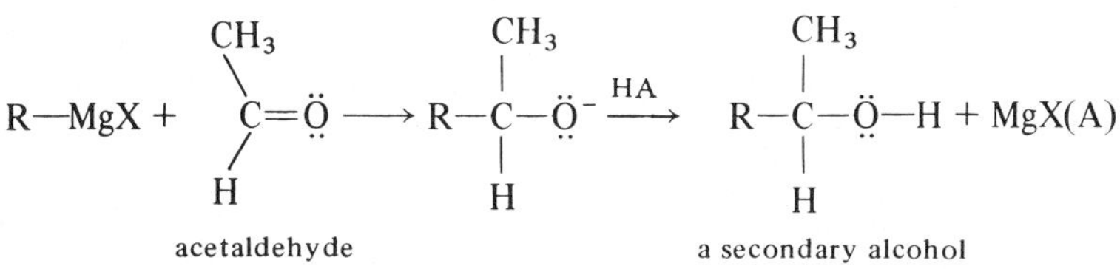

(6-46)

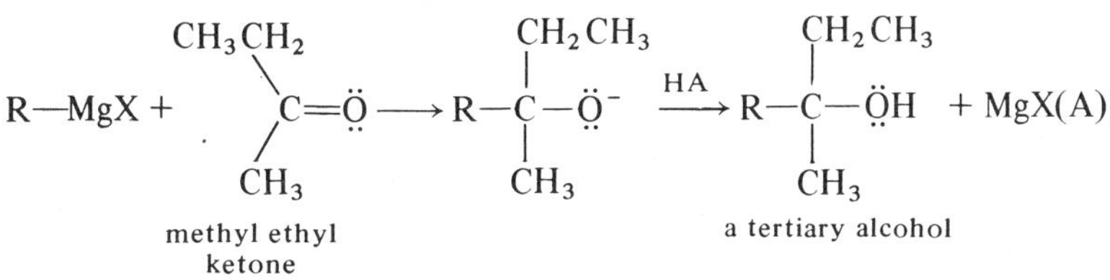

(6-47)

B. The Acidity of α-Hydrogens: The Aldol Condensation

The electron-withdrawing power of the carbonyl group can have a large effect on acid strength. For example, it is responsible for the fact that the acidity of carboxylic acids is much greater than that of alcohols (see Section 5.3).

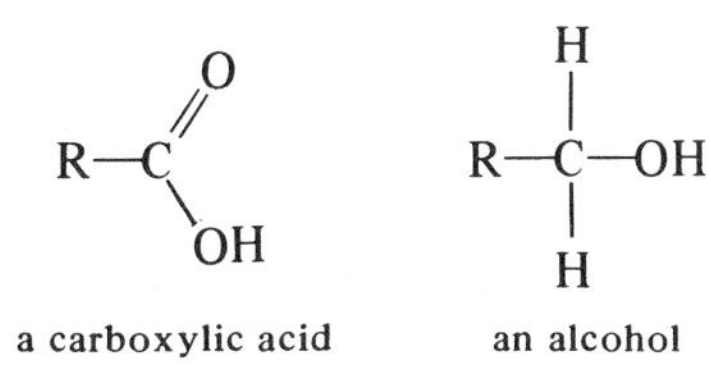

[7]Alcohols are classified as primary, secondary, or tertiary according to whether the carbon atom to which the —OH group is bonded is primary, secondary, or tertiary (see footnote 5). Methanol (CH_3OH) is unique in that it is neither primary, secondary, nor tertiary; its chemical reactivity is very similar to that of primary alcohols.

The carbon atom adjacent to a carbonyl group is referred to as the *α-carbon*, and any hydrogens bonded to it are called *α-hydrogens*. The enhanced acidity of the α-hydrogen is due to resonance stabilization of the anion, the conjugate base. Reactions (6-48) and (6-49) show the similarity between the contributing resonance forms of the carboxylate ion and the anion formed on the carbon α to the carbonyl group. The acidity of these

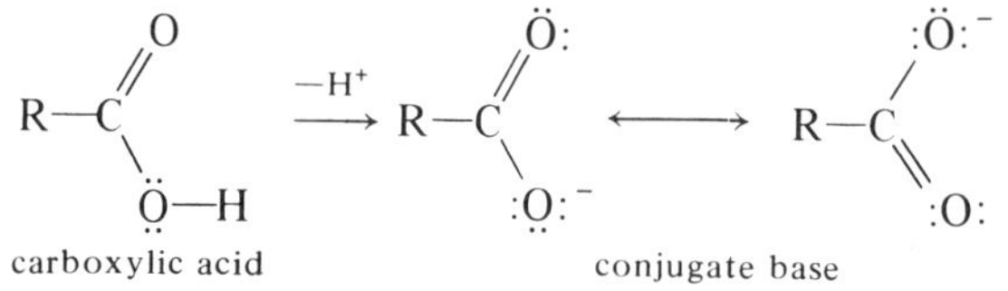

(6-48)

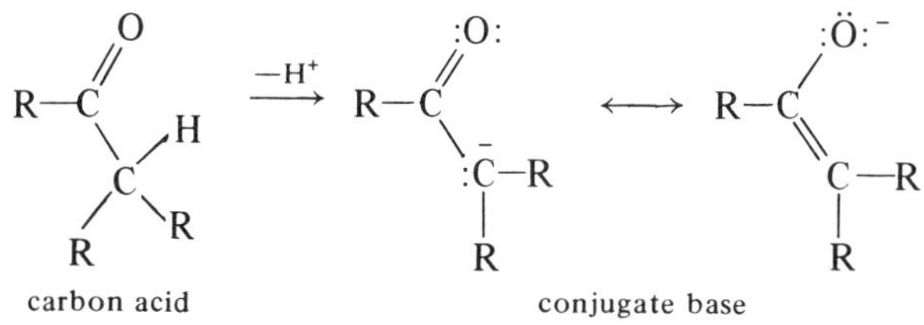

(6-49)

α-hydrogens is small compared to that of mineral acids, carboxylic acids, or sulfonic acids. The pK_a of acetone, for example, is approximately 20; in the presence of hydroxide ion only a small concentration of the α-carbanion is produced (6-50). However, reaction with a stronger base such as RHN^- results in a nearly complete conversion to the anion (6-51).

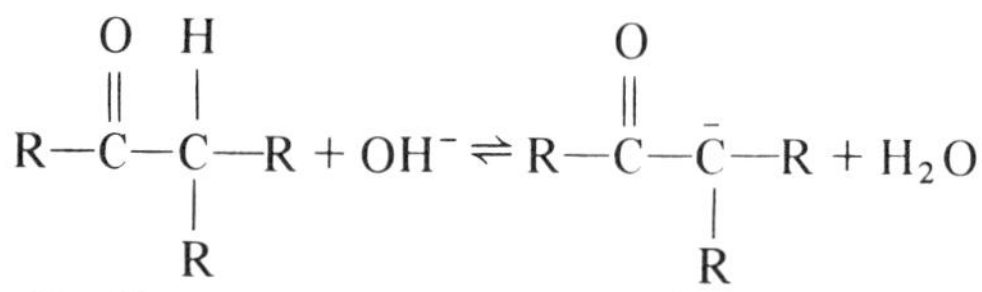
(6-50)

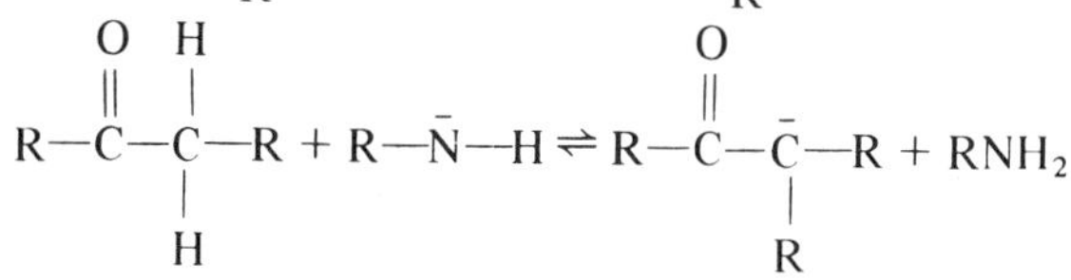
(6-51)

There are many reactions that depend on the acidity of an α-hydrogen; one of these is the aldol condensation. This reaction is the base-catalyzed combination of two molecules of an aldehyde to form a β-hydroxyaldehyde (6-52). If a ketone is used rather than an aldehyde, the product is a β-hydroxyketone (6-53).

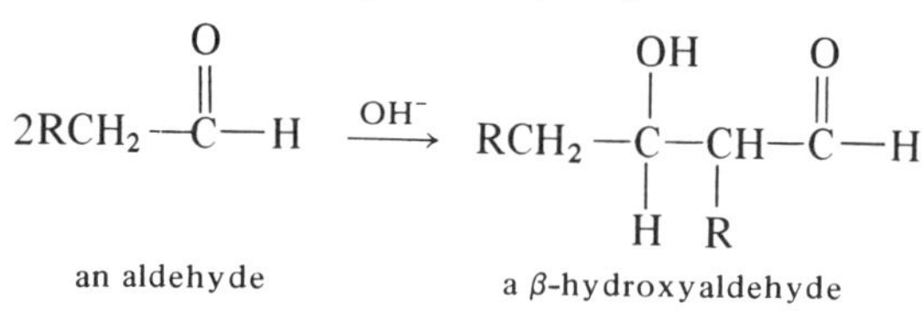

(6-52)

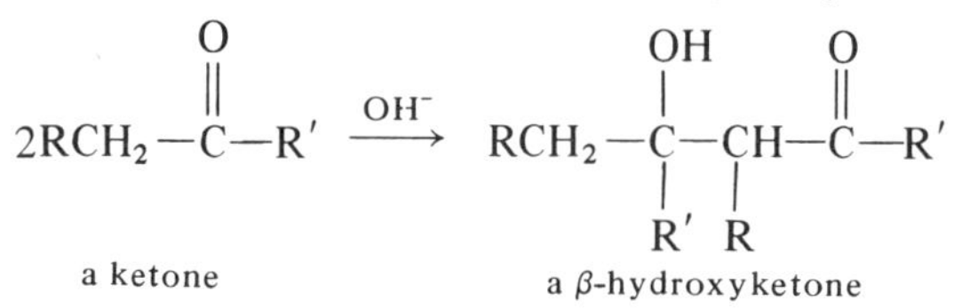

(6-53)

The mechanism of the reaction is shown in reactions (6-54), (6-55), and (6-56), in which acetaldehyde is the starting material. In the first step (6-54) an acidic α-hydrogen is abstracted by hydroxide ion, forming the anion CH_2CHO^-. This is a strongly basic species, which attacks the electron-deficient carbon in the carbonyl group of a second acetaldehyde molecule to form an alkoxide ion (6-55). In the final step (6-56) the alkoxide ion, also a base, abstracts a proton from water. This step yields the final product and also regenerates the hydroxide ion used in (6-54). There is thus no net consumption of OH^-; hydroxide is a catalyst, not a reagent.

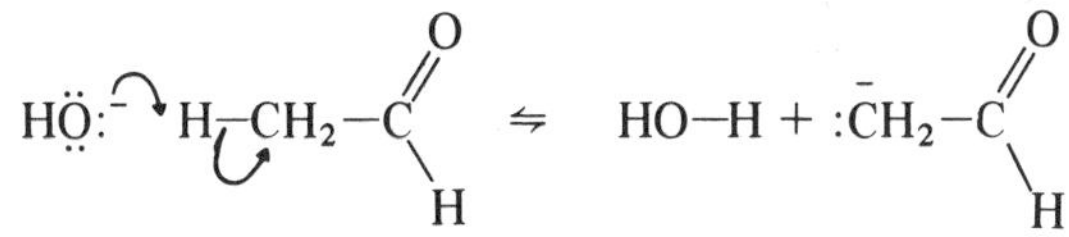

(6-54)

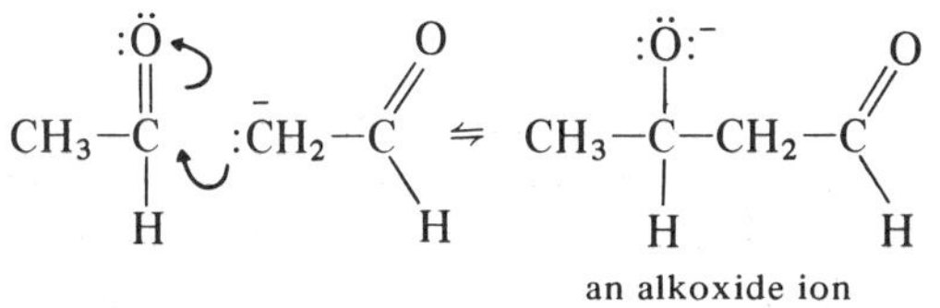

(6-55)

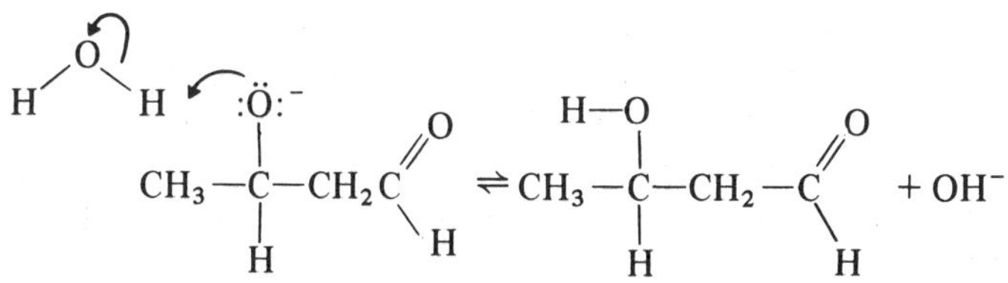

(6-56)

Reactions of this type are important in many organic syntheses. Look up the Claisen condensation in any introductory organic chemistry text and study the mechanism as we have done for the aldol condensation.

6.8 ALKENES: THEIR FORMATION AND BASIC CHARACTER

Alkenes are an important category of organic compound. The simplest member of this series, ethene, is produced in larger quantity than any other organic chemical; the annual U.S. production exceeds 28 billion pounds. In this section we discuss two general preparative methods (both acid-base reactions) and some aspects of the acid-base chemistry of alkenes.

A. Preparation of Alkenes

Alkenes are conveniently prepared in the laboratory by treating an alkyl halide with a strong base. Since in the reaction a hydrogen ion and a halide ion are eliminated from the alkyl halide, it is called a *dehydrohalogenation*. The mechanism (6-57) consists of attack

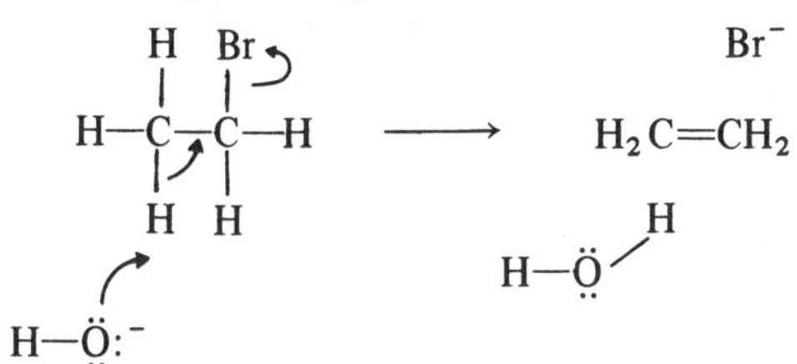

(6-57)

by base on a hydrogen atom (α to the halogen atom) and the loss of the halogen as an anion. Another method of alkene synthesis is the acid-catalyzed dehydration of an alcohol (6-58). In this reaction a proton attacks the electron-rich oxygen atom of an alcohol; following this a molecule of water is expelled, forming a carbocation. In the last step a proton is lost from the ion, yielding the alkene.

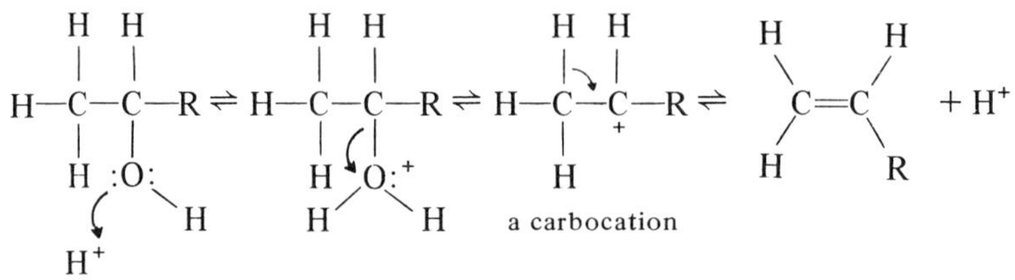

(6-58)

B. Basicity of Alkenes

Much of the chemistry of alkenes can be understood in terms of the electron-rich nature of the double bond. A double bond consists of a σ bond, in which the greatest electron density lies on the internuclear axis, and a π bond, in which the electron density does not lie on the internuclear axis. In a π bond there are two regions of large electron density; these are symmetrically located above and below the plane containing the carbon atoms and their substituents (see Figure 5.2). Compared to those of a σ bond, the electrons of a π bond are more readily available for reaction with electron-deficient species. When a Lewis acid or Brønsted acid reacts with an electron-rich double bond, it is the first step of a process known as an *electrophilic addition reaction*. In such reactions the alkene is a base.

One example is the reaction of an alkene with an inorganic acid, HCl. In the first step (6-59) the basic alkene donates π electrons to the electron-deficient H atom of the acid, forming a carbocation. In the next step (6-60) this positive ion bonds to the conjugate base, Cl^-, of the acid to form the addition product, an alkyl chloride. The general reaction, valid for most alkenes and most strong acids, is shown in (6-61).

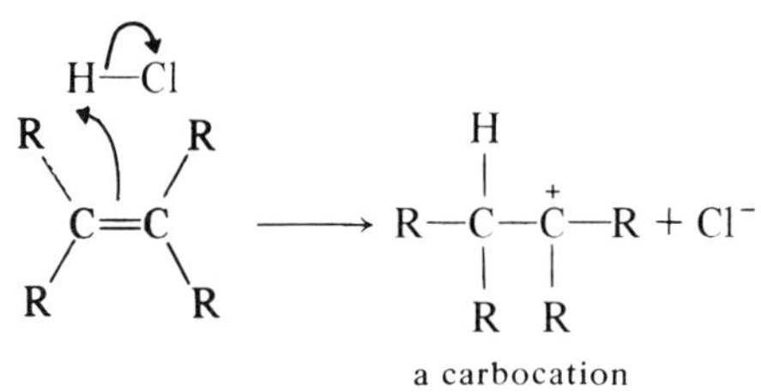

(6-59)

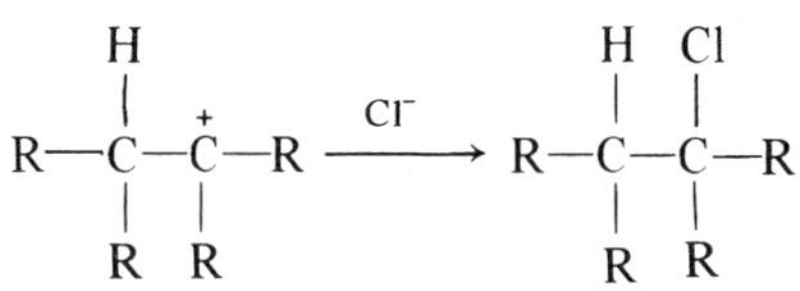

(6-60)

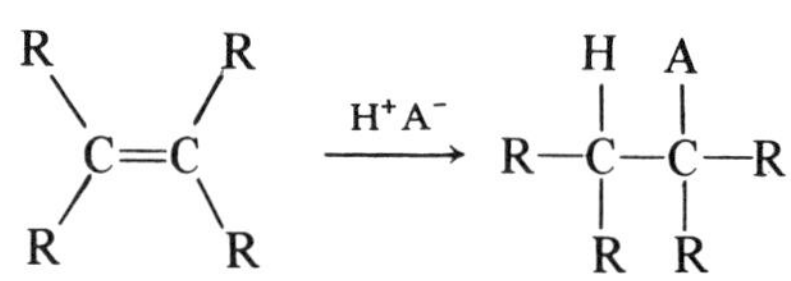

(6-61)

When an unsymmetrical alkene reacts, the acid can attack either one of the doubly bonded carbon atoms. Thus there are two possible products, as shown in reactions (6-62)

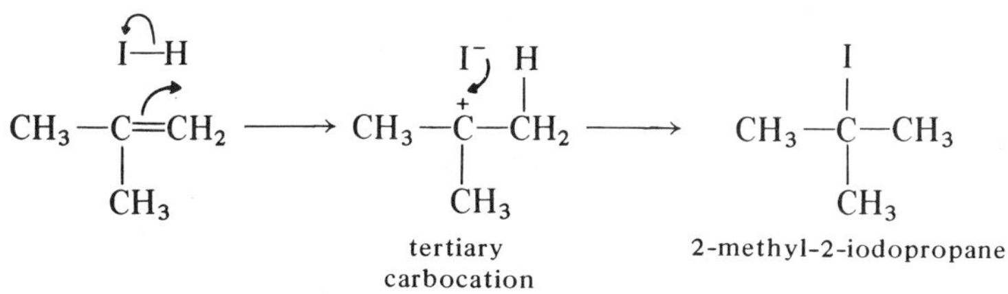

(6-62)

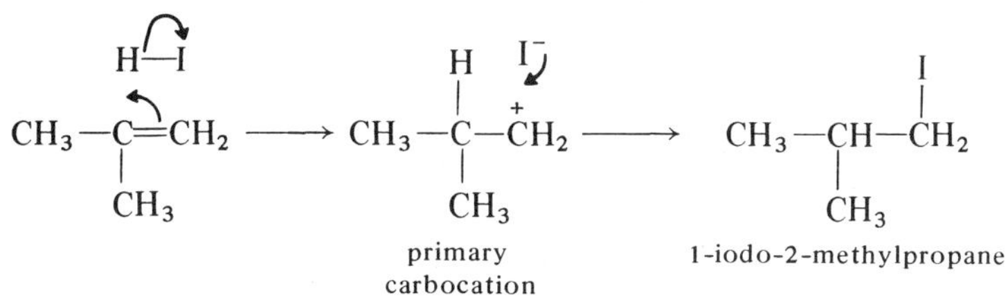

(6-63)

and (6-63) for the reaction of 2-methylpropane with hydrogen iodide. The major product is 2-methyl-2-iodopropane [reaction (6-62)]. This product distribution can be understood in terms of the stability of the carbocation intermediates. In general, reactions having more than one available path will proceed to the greatest extent via the most stable intermediate. Like all positive ions, carbocations are stabilized by electron-donating groups; stability increases as the number of electron-donating substituents increases. As we have already seen, alkyl groups are mild electron donors. The positive charge of the intermediate formed in (6-62) resides on a tertiary carbon atom (i.e., a carbon atom bonded to three other carbon atoms) and is subject to the influence of three electron-donating groups. This intermediate is therefore more stable than the primary carbocation formed in reaction (6-63), in which the charge is lessened by only one electron-donating substituent. Since the positive charge is weaker in the tertiary carbocation, its Lewis acid strength is less. The order of increasing carbocation stability and decreasing acid strength is shown below.

stability increases ⟶

$\overset{+}{C}H_3$	$R\overset{+}{C}H_2$	$R_2\overset{+}{C}H$	$R_3\overset{+}{C}$
methyl	primary	secondary	tertiary

Lewis acid strength decreases ⟶

In most electrophilic addition reactions the major product is the one that proceeds through the *weakest Lewis acid* intermediate (i.e., the most highly substituted carbocation).

6.9 BENZENE AS A BASE

The unusual stability of benzene is a consequence of resonance delocalization. The electronic structure is a resonance hybrid in which π-electron density is spread out over the entire molecule (see Section 5.3). The structure may be represented in several ways,

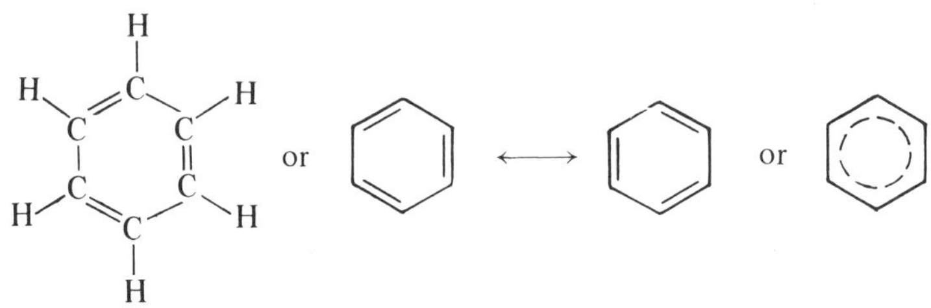

shown above. Even though some of these representations indicate the presence of double bonds, benzene is not an alkene. Addition reactions, discussed in Section 6.8 for alkenes, do not take place readily with benzene.

Nevertheless, the π-electron density in benzene is large; benzene is capable of acting as a Lewis base, in two important ways.

1. Benzene forms "π complexes" with metal cations such as Ag^+, similar to the π complexes formed by alkenes (see Section 4.6).
2. Benzene reacts with electrophilic reagents, although the mechanism involves *substitution* rather than addition. The product of a substitution reaction is a compound in which one of the hydrogen atoms of benzene has been replaced by a different atom or group. The resonance hybrid structure of the ring is maintained, and a substituted benzene has a stability comparable to that of benzene itself. It is commonly said that such reactions "maintain the integrity of the benzene ring."

In these electrophilic substitution reactions the ring system of benzene is attacked by an electrophile (usually a positive ion), resulting in an intermediate which decomposes to form the product. The reaction sequence is illustrated in reaction (6-64), in which E^+

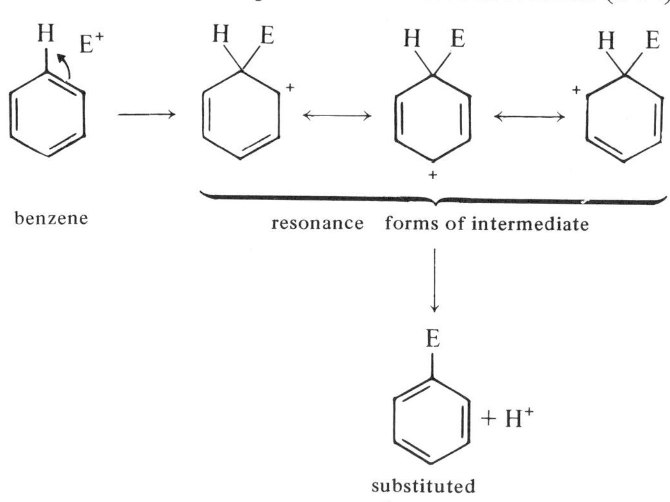

(6-64)

represents any electrophile (Lewis acid). The intermediate is stabilized by resonance delocalization of its positive charge. Elimination of a proton from the carbon atom which was attacked by E^+ results in a substituted benzene molecule.

In an electrophilic substitution the base, benzene, donates an electron pair to the acid, E^+. This acid-base interaction is strongly affected by the presence of substituents on the benzene ring. Electron-donating substituents, such as alkyl groups, enhance the basicity of the ring by increasing the availability of electrons to the attacking electrophile. Such compounds react faster than unsubstituted benzene. For example, methyl-substituted benzene (toluene) is more reactive than benzene. On the other hand, electron-withdrawing substituents, such as the nitro group, decrease the availability of electrons in the ring system. Nitrobenzene is a weaker base, and thus less reactive toward an electrophile than either benzene or toluene.

The Friedel-Crafts reaction is a specific example of electrophilic substitution in benzene. In this reaction benzene or a substituted benzene is treated with an alkyl halide (normally an alkyl chloride or bromide) in the presence of a Lewis acid (usually $AlCl_3$). The result is the substitution of an alkyl group for one of the hydrogen atoms of benzene, shown in reaction (6-65) for 2-chloropropane. In the first step (6-66) the Lewis acid abstracts a chloride ion from the alkyl halide, forming a carbocation and $AlCl_4^-$. The carbocation, an excellent electrophile, attacks the basic π-electron system of a benzene molecule. The result is another carbocation, which is stabilized to some degree by resonance (6-67). After the intermediate loses a proton (6-68), the resulting compound, isopropylbenzene, has regained the great stability of the benzene ring.

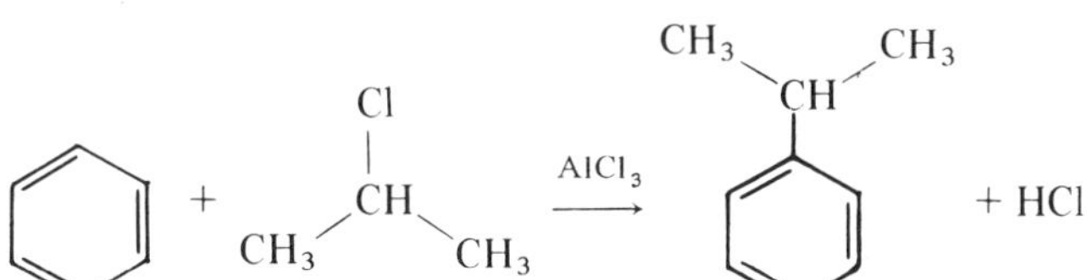

(6-65)

$$(CH_3)_2CHCl + AlCl_3 \longrightarrow (CH_3)_2\overset{+}{C}H + AlCl_4^- \qquad (6\text{-}66)$$

2-chloropropane

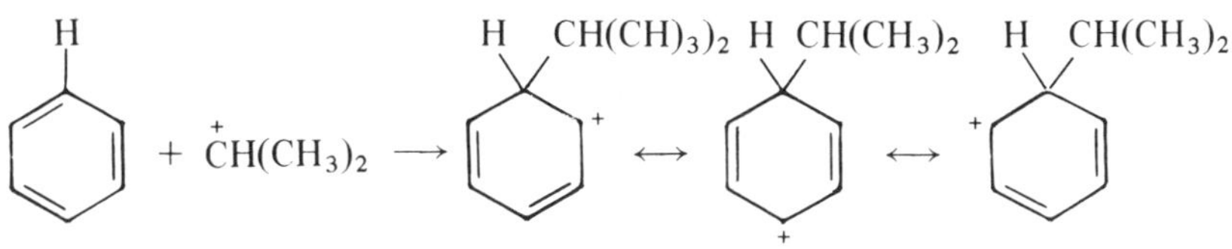

(6-67)

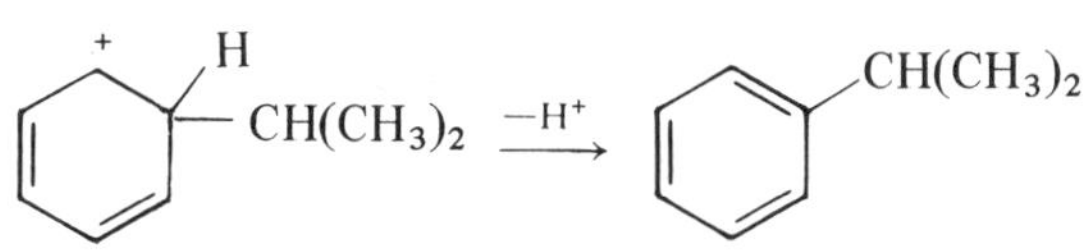

(6-68)

6.10 ACID-CATALYZED ESTER FORMATION

Many reactions take place too slowly in the absence of a catalyst to be useful in organic syntheses. We have discussed catalysis by base (the aldol condensation, Section 6.7B) and now we will consider acid catalysis. An example of this is the formation of an ester from a carboxylic acid and an alcohol; the overall reaction is given in (6-69).

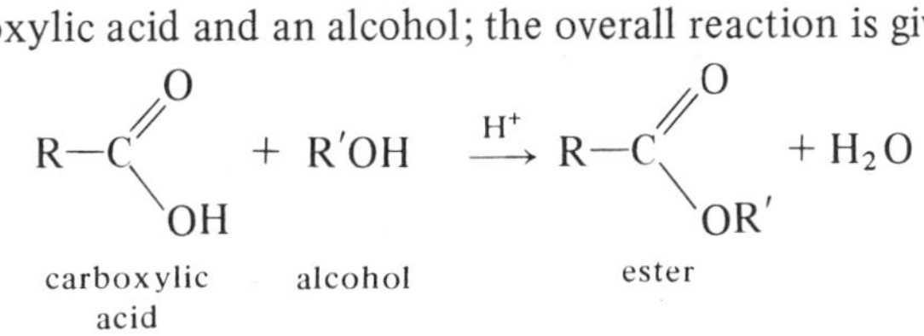

(6-69)

The mechanism of the reaction, given by (6-70) through (6-73), involves acid-base interactions in every step. The first step is the protonation of the carbonyl oxygen atom by the acid catalyst, shown for acetic acid in (6-70). Acetic acid functions as a

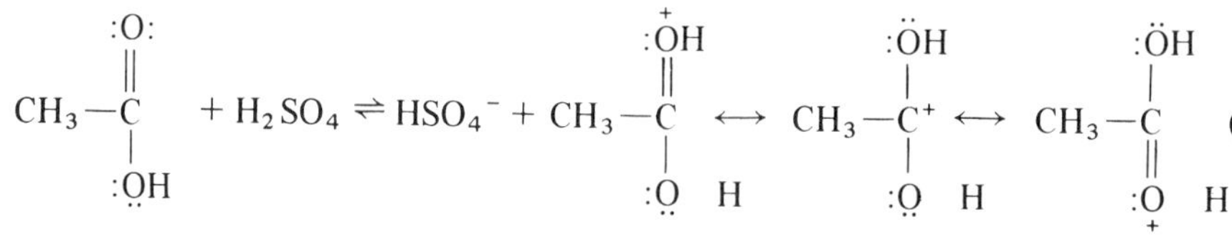

(6-70)

protonated form of acetic acid

Brϕnsted-Lowry base in this reaction, accepting a proton from the strong acid. In the protonated form, which is stabilized by resonance, the carbonyl carbon atom bears a partial positive charge; it has become more electron deficient and is therefore a stronger Lewis acid. In the next step (6-71) this Lewis acid site is attacked by the electron-rich oxygen atom of the alcohol, forming a Lewis acid-base adduct. The third step (6-72) involves *two* Brϕnsted-Lowry proton transfers, with the net result that the oxygen of the alcohol loses its proton and one of the carboxylic oxygen atoms becomes protonated.

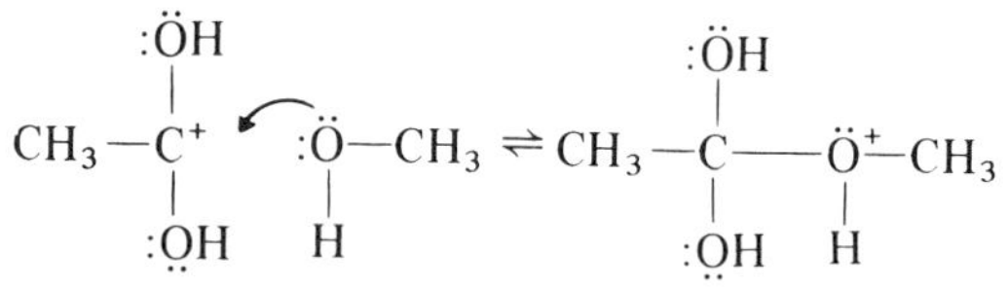

(6-71)

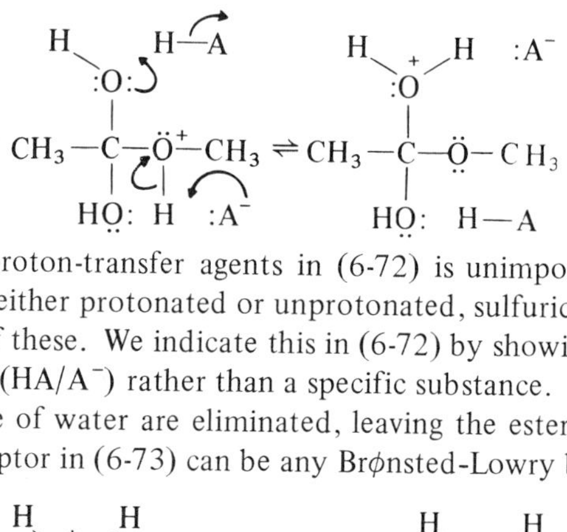

(6-72)

The identity of the proton-transfer agents in (6-72) is unimportant; they can be other acetic acid molecules, either protonated or unprotonated, sulfuric acid, solvent molecules, or any combination of these. We indicate this in (6-72) by showing the general Brϕnsted-Lowry conjugate pair (HA/A^-) rather than a specific substance. In the final step (6-73) a proton and a molecule of water are eliminated, leaving the ester (methyl acetate). As in (6-72) the proton acceptor in (6-73) can be any Brϕnsted-Lowry base.

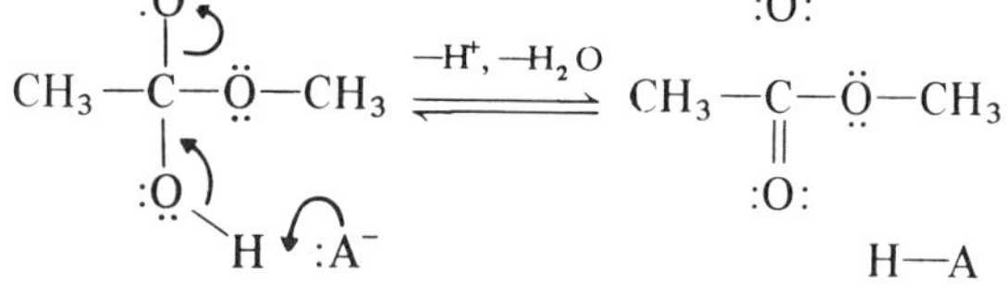

(6-73)

6.11 CONCLUSION

The selection of organic reactions in this chapter is necessarily limited, and no doubt other examples would have served as well. Our purpose has been to illustrate the importance of acid-base principles in understanding organic mechanisms. We hope that enough examples have been presented so that students can generalize the approach and begin to apply acid-base principles to the many other organic reactions that can and should be analyzed in this way.

PROBLEMS

6.1 Arrange the compounds in each of the following groups in order of decreasing acid strength (strongest acid first).

(a) 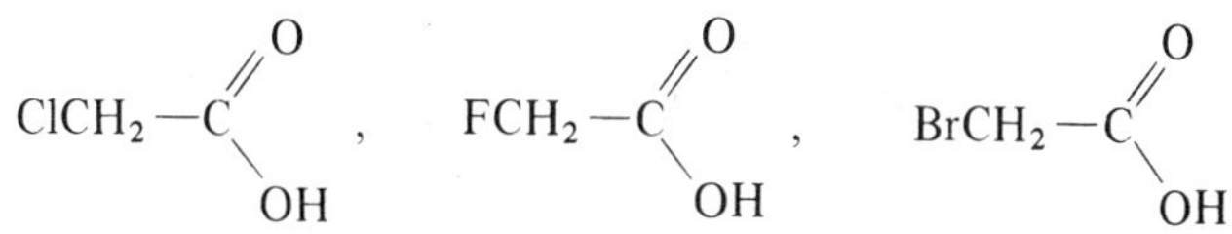

(b) 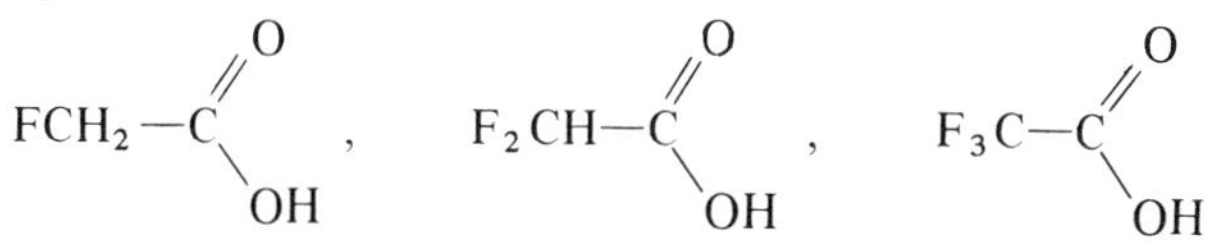

(c) 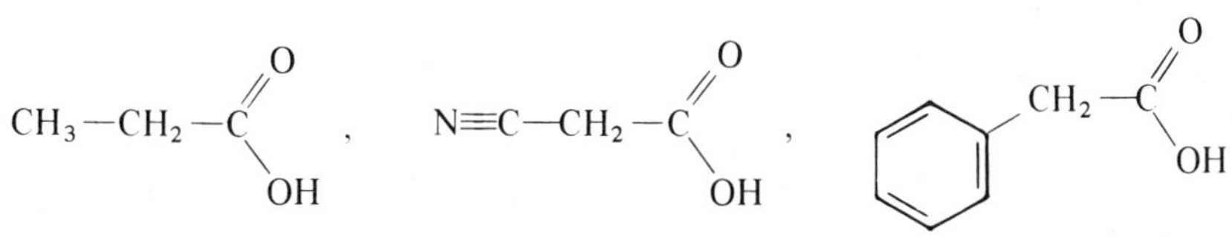

6.2 Arrange the compounds in each of the following groups in order of increasing base strength.

(a) 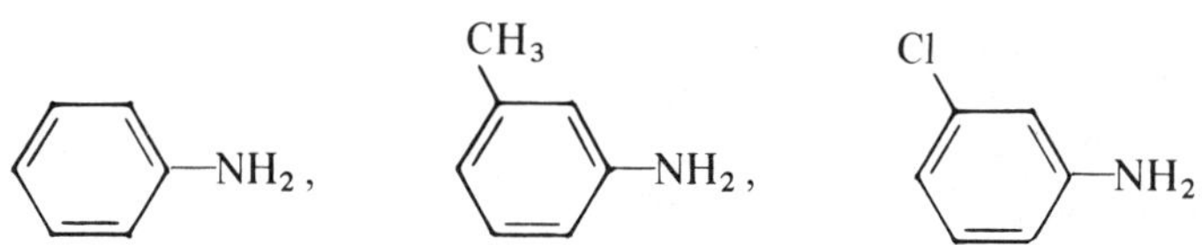

(b) 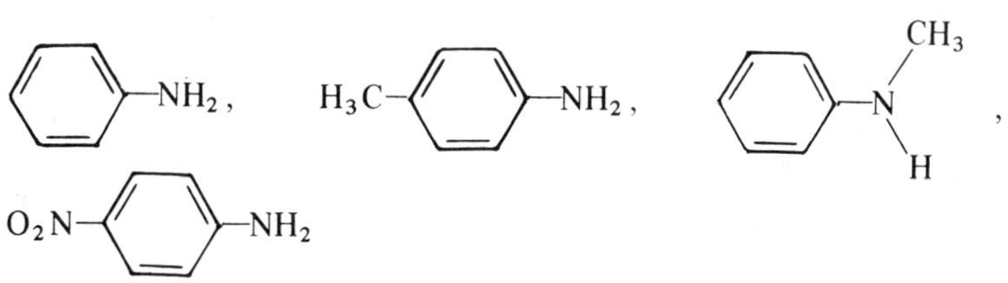

(c) 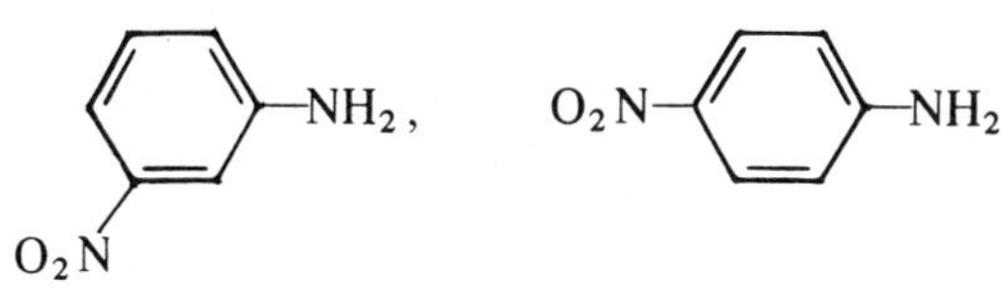

6.3 One synthesis of ethyl acetoacetate, the Claisen condensation, depends on the acidity of an α-hydrogen in the starting material, ethyl acetate. Show in detail the mechanism of the acid-base interaction which gives rise to the tetrahedral intermediate shown below. (*Hint:* See the aldol condensation, Section 6.7B.)

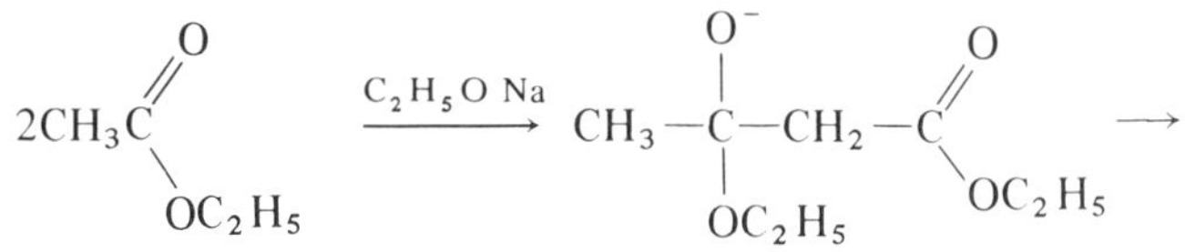

tetrahedral intermediate

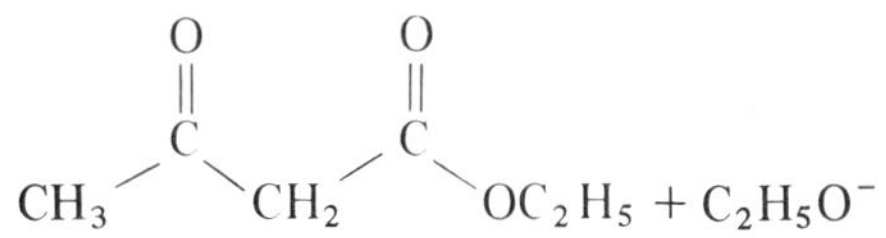

final product
(ethyl acetoacetate)

6.4 The first three compounds in the series of linear dicarboxylic acids are oxalic acid ($HO_2C{-}CO_2H$), $pK_1 = 1.27$; malonic acid ($HO_2C{-}CH_2{-}CO_2H$), $pK_1 = 2.9$; and succinic acid ($HO_2C{-}CH_2{-}CH_2{-}CO_2H$), $pK_1 = 4.2$. Explain the relative magnitudes of these pK_a values.

6.5 One of the strongest organic bases is guanidine. Suggest a reason why it is such a strong base. (*Hint:* Consider the stability of the conjugate acid.)

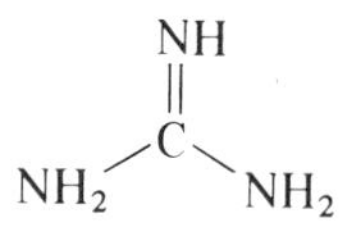

guanidine

6.6 Complete the following acid-base reactions.

(a) 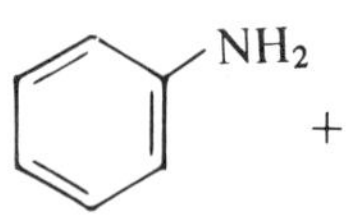 $+ HCl \longrightarrow$

(b) $(CH_3)_3N + H_2SO_4 \longrightarrow$

(c) $CH_3NH_2 + H_2O \longrightarrow$

(d) Product from part (c) + NaOH $\longrightarrow$

(e) $CH_3CH_2{-}MgBr + H_2O \longrightarrow$

(f) $CH_3CH_2{-}MgBr + CH_4 \longrightarrow$

(g) 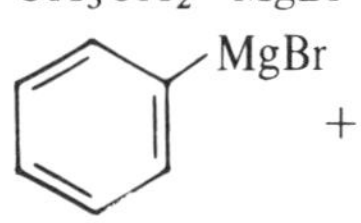 $+ NH_3 \longrightarrow$

(h) $H_3C{:}^- + CH_3C{\equiv}C{-}H \longrightarrow$

(i) Product from part (a) + NaOH $\longrightarrow$

(j) 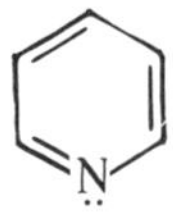 $+ HBr \longrightarrow$

(k) 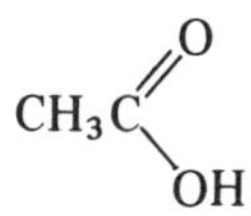 $+ CH_3NH_2 \longrightarrow$

(l) 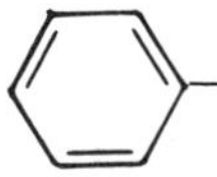 $—SO_3H + NH_2^- \longrightarrow$

(m) $CH_3CO_2H + H_2SO_4 \longrightarrow$

(n) Sodium acetate + 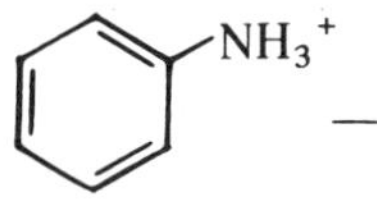 $\longrightarrow$

(o) Sodium acetate + HCl $\longrightarrow$

6.7 When semicarbazide is treated with dilute acid, one of its three nitrogen atoms is protonated. Which one? Draw the structure of this protonated form and explain your reasoning.

$$NH_2—\overset{\overset{\displaystyle O}{\|}}{C}—NH—NH_2 + H_3O^+ \longrightarrow H_2O + CH_6ON_3^+$$

6.8 Write a complete mechanism for the following esterification reaction.

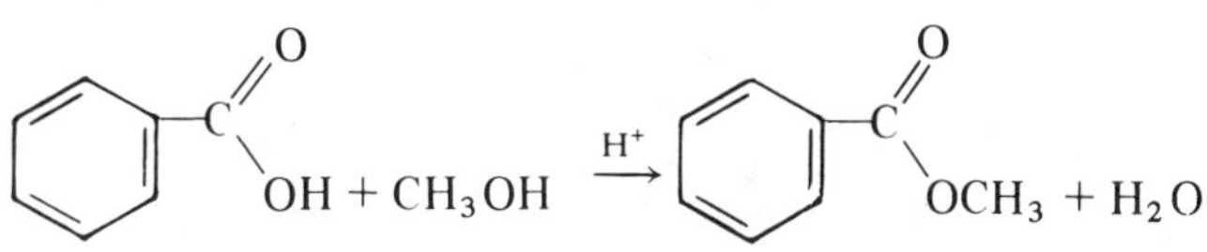

6.9 What problem might be encountered in an attempt at preparation of a Grignard reagent from *p*-chloroaniline (*p*-$ClC_6H_4NH_2$)?

6.10 *m*-Nitrobenzoic and *p*-nitrobenzoic acids are approximately equal in acid strength. The K_a of *p*-nitrophenol, however, is more than 15 times larger than the K_a of the *meta* isomer. Explain.

6.11 Explain the following observations. *p*-Toluidine is soluble in dilute HCl, whereas *p*-methylacetanilide is insoluble.

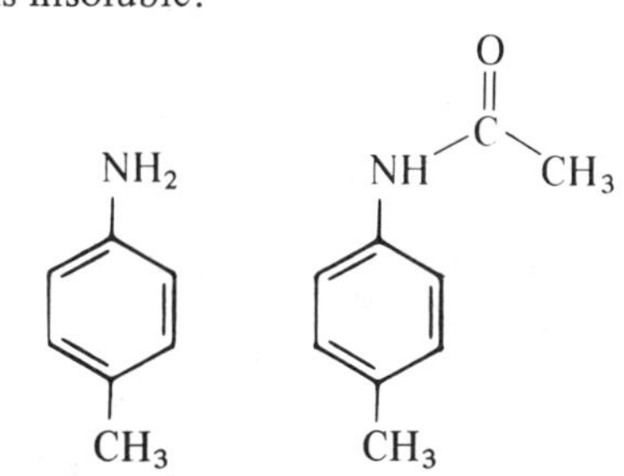

p-toluidine *p*-methylacetanilide

6.12 Grignard reagents react with CO_2, forming salts that yield carboxylic acids on treatment with dilute acid. Discuss a possible mechanism for this reaction in terms of acid-base principles.

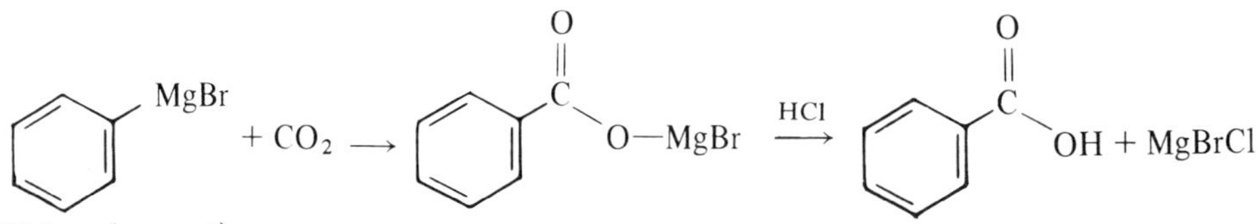

(Grignard reagent)

6.13 Choose the stronger acid in each of the following pairs and explain your reasoning.

(a) *m*-Fluorobenzoic acid or *m*-chlorobenzoic acid

(b) 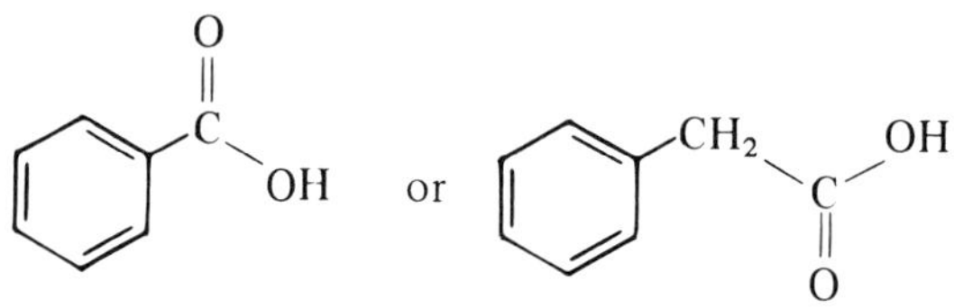

(c) 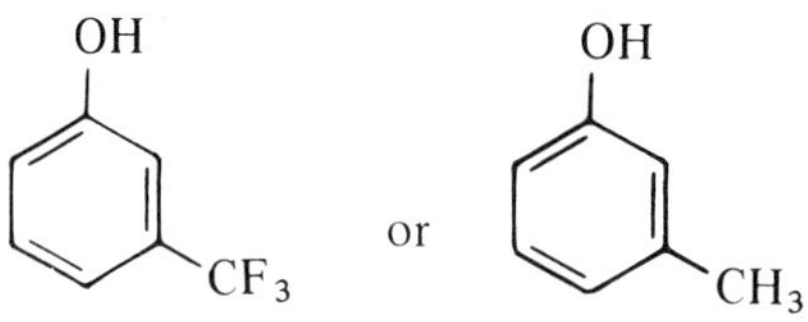

(d) 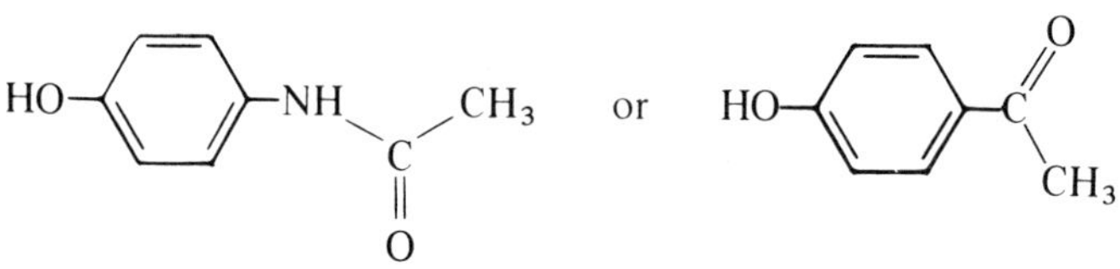

(e) 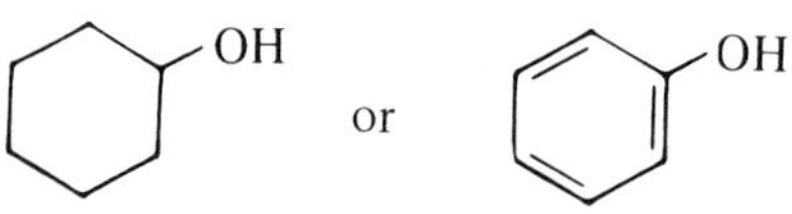

6.14 Explain why aniline is a weaker base than cyclohexylamine.

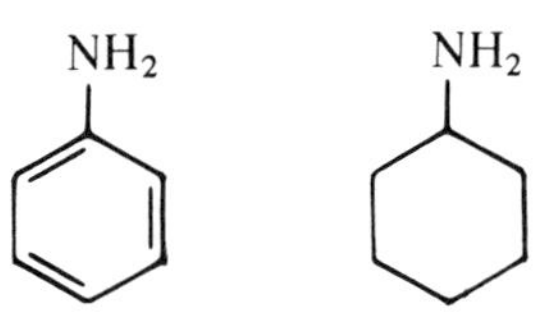

aniline cyclohexylamine

6.15 Arrange the labeled protons (A, B, etc.) in each molecule in order of decreasing acid strength.

(a) 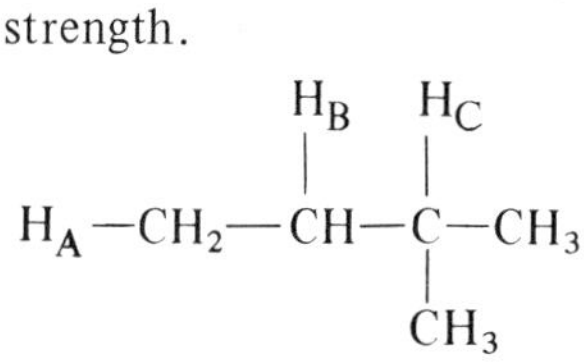

(b) 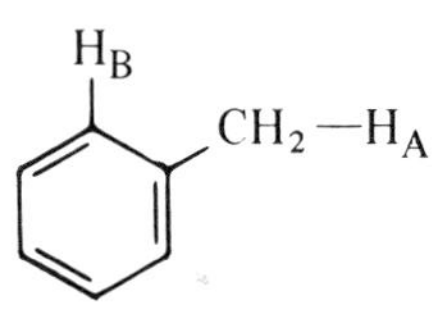

(c) 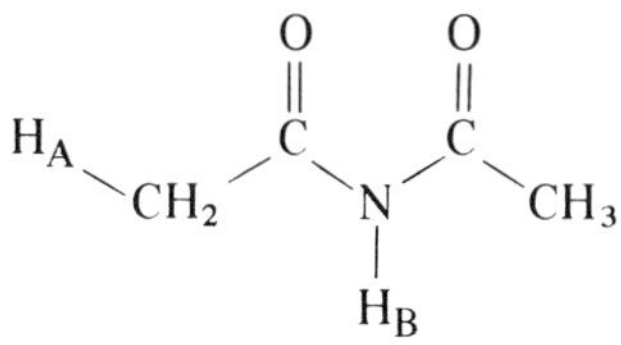

(d) 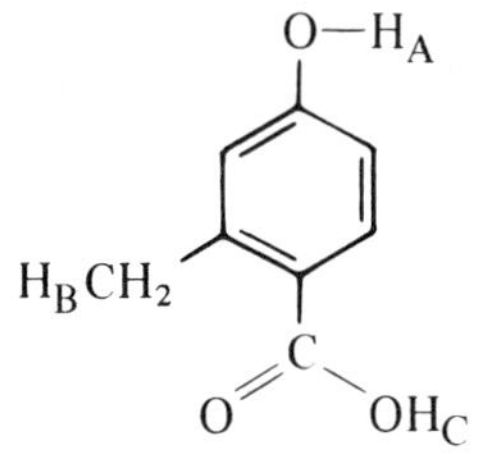

6.16 Explain the acidity order of the following species.

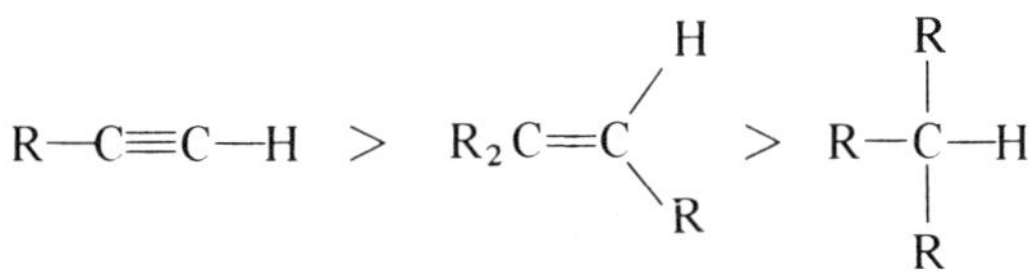

6.17 When a sulfonic acid (RSO_2OH) ionizes, the resulting anion is stabilized by resonance. Without consulting the text, draw three resonance structures.

6.18 Describe in detail how you might separate an ether solution of the following substances. Assume that the substances are all insoluble in water.

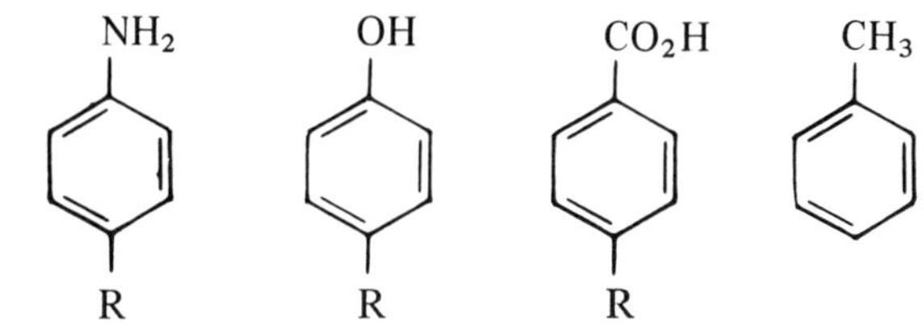

a substituted aniline — a substituted phenol — a substituted benzoic acid — toluene

(*Hint:* If the ether solution were thoroughly shaken with dilute HCl, two layers would form, since water and ether are mutually insoluble. What materials would remain in the ether layer? What materials would be extracted into the aqueous layer?)

6.19 Most low-molecular-weight organic compounds containing oxygen are soluble in concentrated H_2SO_4. Suggest a reason for this.

6.20 What product is formed when concentrated H_2SO_4 reacts with each of the following materials? (*Hint:* The reaction takes place at the most basic site.)

(a) CH_3CH_2OH

(b) CH_3OCH_3

(c) 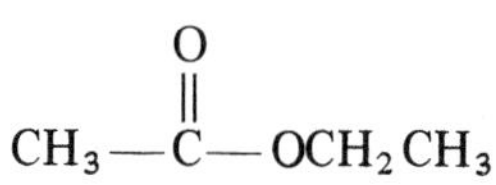

(d) 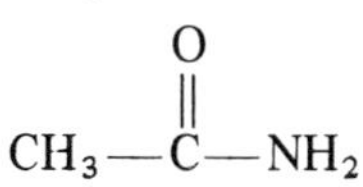

6.21 Which of the following compounds would be soluble in 10% HCl? Explain your reasoning, and show any reactions that take place.

(a) $CH_3CH_2CH_2CH_2NH_2$

(b) CH_3CH_2OH

(c) 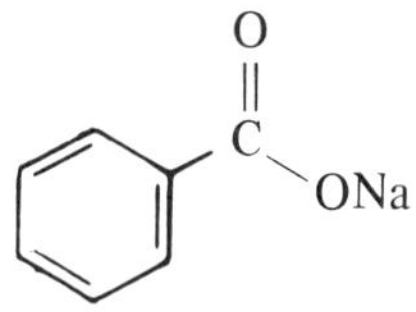

(d) 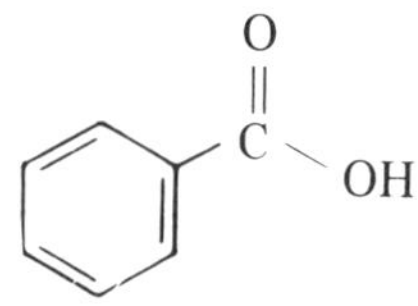

(e) 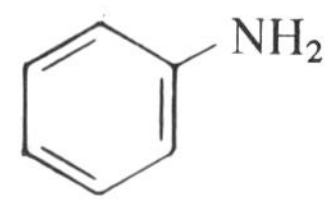

6.22 Which of the following compounds would be soluble in 10% NaOH? Explain your reasoning and show any reactions that take place.

(a) 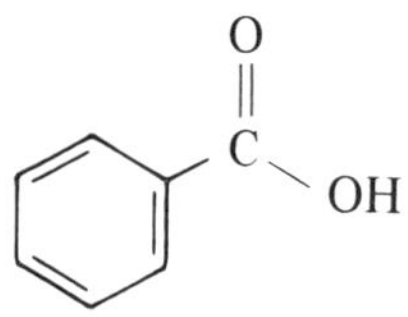

(b) $CH_3-\overset{O}{\overset{\|}{C}}-NH-\overset{O}{\overset{\|}{C}}-CH_3$

(c) CH_3CH_2OH

(d) 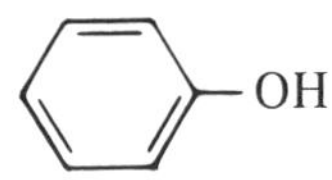

(e) 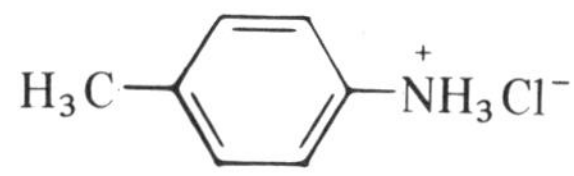

6.23 Explain fully how you would distinguish between a primary and a secondary amine using the Hinsberg test. Try to do this without referring to the text.

6.24 Alkenes react with acids such as HCl, HBr, and HI to yield alkyl halides according to the following reaction:

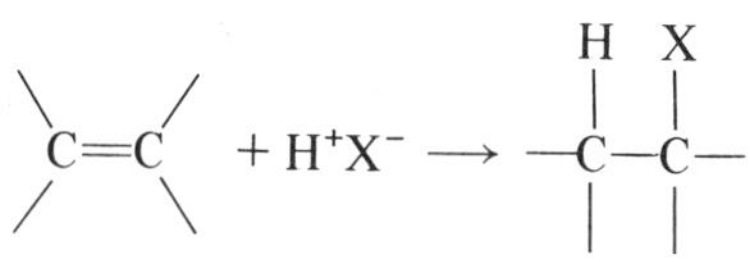

If the alkene is unsymmetrical, two possible halides may result. Predict the major and minor product that would be formed in the following reactions. Review Section 6.8B if necessary.

(a) $CH_3-CH=CH_2 \xrightarrow{HCl}$

(b) $CH_2=\underset{CH_3}{\underset{|}{C}}-CH_3 \xrightarrow{HBr}$

(c) $CH_3CH=CH-CH_3 \xrightarrow{HI}$

(d) $CH_3-\underset{CH_2}{\underset{\|}{C}}-CH_2CH_3 \xrightarrow{H_3O^+}$

6.25 Arrange the following series of weak acids (substituted phenols) in order of decreasing acid strength.

p-cresol

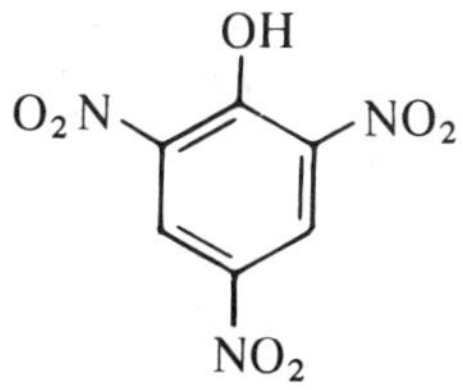

picric acid

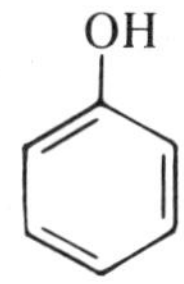

phenol

p-nitrophenol

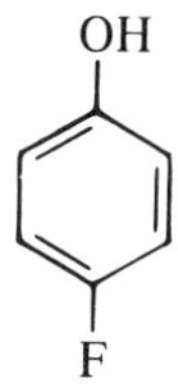

p-fluorophenol

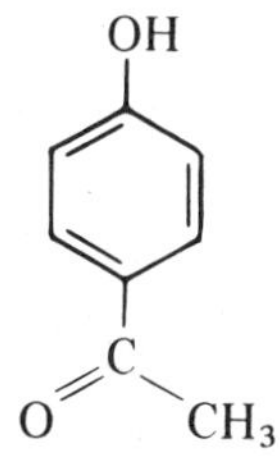

p-hydroxyacetophenone

6.26 When an amide is protonated by reaction with a Brønsted acid, it is the oxygen atom rather than the nitrogen atom which accepts the proton. Suggest a reason for this. (*Hint:* Consider the effect of resonance on stability.)

Chapter 7

Salt Hydrolysis and Buffer Action

7.1 Hydrolysis

When we accept the fundamental concept of the Brønsted-Lowry definition, namely that an acid-base reaction is a proton transfer from acid to base, we also accept some ideas that may seem a little strange at first. For example, the existence and stability of the ionic compound sodamide, $NaNH_2$, obliges us to regard ammonia as an *acid*[1] as well as a base. Similarly, the reaction of lime (CaO) with water is easily understood if we are willing to classify hydroxide ion as an acid whose conjugate base is O^{2-}. We are accustomed to thinking of acids and bases as neutral compounds that have certain properties; Brønsted-Lowry theory allows us to consider, as acids and bases, charged species as well. Thus HSO_4^-, HCl, and NH_4^+ are all acids; CO_3^{2-}, $(CH_3)_2NH$, and $H_2NCH_2CH_2NH_3^+$ are bases. Ionic acids such as NH_4^+ necessarily occur as part of a neutral ionic compound, for instance NH_4Cl. Such compounds are normally called *salts*, and Brønsted-Lowry theory invites the question (and also provides the answer): How is it that salts can also be acids or bases?

At this point we must make sure that we have a firm grasp on what is meant by the term "salt." Its meaning is simple, yet its definition presents some tricky problems. Prior to Brønsted-Lowry, the only compounds regarded as bases were those containing the hydroxide ion: NaOH, KOH, $Ca(OH)_2$, and so on. When hydroxide bases react with acids (neutralization reactions), the result is water and an ionic compound composed of the cationic part of the base and the anionic moiety of the acid. Such compounds were called *salts*, a name derived from the most common member of the class, sodium chloride.

$$NaOH + HCl \rightarrow H_2O + NaCl \tag{7-1}$$

$$Ca(OH)_2 + 2HNO_3 \rightarrow 2H_2O + Ca(NO_3)_2 \tag{7-2}$$

In Brønsted-Lowry theory, however, compounds other than hydroxides can be regarded as bases. If we maintain the definition of *neutralization* as "reaction of an acid with a base," two difficulties appear. According to Brønsted-Lowry, *any* substance that reacts by accepting a proton from an acid is a base, hence *all* reactions of acids are neutralizations. The word thus loses its specificity, and its usefulness as well. The second

[1]When sodamide dissolves in water to give sodium ion and amide ion, NH_2^-, the amide ion immediately reacts with water: $NH_2^- + H_2O \rightarrow NH_3 + OH^-$. This is a proton transfer reaction involving the base NH_2^- and the acid H_2O. The base NH_2^- of course has its conjugate acid, which is none other than ammonia, NH_3! Reactions of this sort are the main topic of this chapter.

difficulty impedes our effort to define "salt": the class of compounds defined as "products of neutralization reactions" contains some compounds which are not normally classified as salts. For example, addition of acid to the base sodium bicarbonate yields, in addition to water and salt, the very un-salt-like carbon dioxide [reaction (7-3)].

$$NaHCO_3 + HCl \rightarrow H_2O + NaCl + CO_2 \tag{7-3}$$

Although we could exclude compounds like CO_2 from our definition by specifying that salts are "products of the neutralization of *hydroxide* (Arrhenius) bases," we would then be excluding, for instance, NH_4Cl and other salts of ammonia. In spite of these deficiencies, some introductory textbooks continue to define salts as products of neutralization reactions.

Another definition that has been used is simply that salts are ionic compounds. This definition, we believe, is too broad to be of much use. For one thing, if "salt" is identical in meaning to "ionic compound," there is no need for the term "salt" at all! Also, although it may be true that not all salts can be formed in an acid-base neutralization, most of them can. Thus it seems appropriate to retain, in the definition of "salt," some relationship to acid-base chemistry.

In this text we define as *"salt" any (neutral) ionic compound derived from an (neutral) acid by replacing one or more of the acidic protons with one or more other cations.* It is useful to classify salts according to the strength of the acids and/or bases from which they are derived. For instance, the salt sodium benzoate ($C_6H_5CO_2Na$) is called a "salt of a weak acid" because it is derived from a weak acid, $C_6H_5CO_2H$, by replacing the one ionizable proton with a sodium cation. The salt ammonium chloride (NH_4Cl), on the other hand, is derived from the strong acid HCl. Since the ammonium cation is formed from the *weak* base NH_3, ammonium chloride is a "salt of a weak base." Ammonium benzoate is a salt of a weak acid *and* and weak base. In general, if the *anion* of a salt is conjugate to a weak acid, the salt is called a *salt of a weak acid*; if the *cation* of a salt is conjugate to a weak base, the salt is called a *salt of a weak base*. If the anion is conjugate to a weak acid and the cation is conjugate to a weak base, the salt is a salt of a weak acid and a weak base.

Let us consider the reaction with water of an unspecified Brϕnsted acid HA:

$$HA + H_2O \rightleftharpoons H_3O^+ + A^- \tag{7-4}$$

in which A^- stands for "the anionic part of HA." When we use the notation HA/A^-, it is clear that A^- also stands for "the base conjugate to the acid HA." Similarly, the base B reacts with water

$$B + H_2O \rightleftharpoons BH^+ + OH^- \tag{7-5}$$

and BH^+ stands for both "the protonated from of B" and "the acid conjugate to B." Now, since A^- is a *base* and BH^+ is an *acid*, they ought to react with water as does any other acid or base, that is, in ways analogous to those illustrated in reactions (7-4) and (7-5). In fact, they do:

$$BH^+ + H_2O \rightleftharpoons B + H_3O^+ \tag{7-6}$$

$$A^- + H_2O \rightleftharpoons HA + OH^- \tag{7-7}$$

Since BH^+ ions are available from salts (for instance, NH_4^+ from NH_4Cl), as are A^- ions ($C_6H_5CO_2^-$ from $C_6H_5CO_2Na$), reactions (7-6) and (7-7) occur as easily as reactions (7-4) and (7-5). Reaction (7-6) is essentially identical to (7-4): both are reactions of an acid with water and the sole difference is that whereas HA is a neutral acid, BH^+ is a cationic acid. Similarly, (7-5) and (7-7) are identical except that B is a neutral base whereas A^- is an anionic base. The similarities between these pairs of reactions are of considerably greater importance than the differences, as we will see throughout this chapter. Nonetheless, different names are used to describe them. Reactions (7-4) and (7-5) are acid and base *ionization* reactions, respectively, whereas (7-6) and (7-7) are *hydrolysis* reactions. In the context of acid-base chemistry, *hydrolysis* may be defined as the proton-transfer reaction of a salt with water. In hydrolysis, either the anion of the salt is a base (conjugate to a weak acid) and undergoes a reaction characteristic of a base, or the cation of the salt is an acid (conjugate to a weak base) and undergoes a reaction characteristic of an acid, or both.

We have established that salt hydrolysis is, except in electrical charge and in name, the same as the ionization of a weak acid or base. We should therefore expect the mathematical treatment of salt hydrolysis to be the same as was discussed in Chapter 3, and in fact this is so. Let us illustrate this by calculating the pH of a solution of ammonium chloride.

Salts are ionic compounds, and in dilute aqueous solution ionic compounds are completely dissociated into their constituent ions: that is, the species $NH_4Cl(aq)$ *does not exist*. In solution, only the ions NH_4^+ and Cl^- are present. Viewing it as a stepwise process, we write the dissociation of the salt (7-8) and then the hydrolysis of the ammonium ion (7-9). Next, we write the equilibrium expression (7-10) for the hydrolysis reaction

$$NH_4Cl(s) \rightarrow NH_4^+(aq) + Cl^-(aq) \tag{7-8}$$

$$NH_4^+(aq) + H_2O \rightleftharpoons NH_3 + H_3O^+ \tag{7-9}$$

$$K_h = \frac{[NH_3][H_3O^+]}{[NH_4^+]} \quad \text{or} \quad \frac{[NH_3][H^+]}{[NH_4^+]} \tag{7-10}$$

We have, as in Chapter 3, incorporated the constant H_2O concentration into the equilibrium constant. The only difference is of notation: we use K_h, the "hydrolysis constant," rather than K_a, the "acid ionization constant." As before, we define x, the extent of reaction, as the amount of NH_4^+ which has reacted, and write the initial and equilibrium concentration, using C for the original salt concentration.

	$NH_4^+(aq) + H_2O$	$\rightleftharpoons NH_3$	$+ H_3O^+$
initial:	C	0	0
at equilibrium:	$C - x$	x	x

The equilibrium expression becomes

$$K_h = \frac{(x)(x)}{C - x} = \frac{x^2}{C - x}$$

which is to be solved by any of the mathematical methods of Chapter 3, depending on the values of K_h and C. Since the value of K_h for ammonium ion is 5.37×10^{-10} at 25°C, the approximate solution is valid for any concentration greater than about 10^{-4} M (see Figure VI-1, Appendix VI). So for $C = 0.25\ M$,

$$x = [H_3O^+] = \sqrt{K_h C} = \sqrt{5.37 \times 10^{-10}(0.25)} = 1.16 \times 10^{-5}\ M$$

$$pH = -\log(1.16 \times 10^{-5}) = 4.94$$

In this example, and in those to follow, we have ignored any effects due to the hydrolysis of the chloride anion. The reason is that chloride anion is the base conjugate to a strong acid. It is therefore a very weak base and as such it undergoes little or no reaction with the acid H_2O. Hydrolysis can be safely neglected for salts of strong acids or bases. This point will be elaborated in the following section.

Example 7.1

Calculate the pH of a 0.035 M solution of sodium benzoate. (Assume that the salt ionizes completely in H_2O). The benzoate ion has K_h of 1.5×10^{-10} at 25°C.

Answer:

$$NaBz \rightarrow Na^+ + Bz^- \quad (100\%)$$

$$Bz^- + H_2O \rightleftharpoons HBz + OH^-$$

	Bz^-	HBz	OH^-
initially:	0.035	0	0
at equilibrium:	$0.035 - x$	x	x

$$K_h = 1.5 \times 10^{-10} = \frac{x^2}{0.035 - x} \approx \frac{x^2}{0.035}$$

$$x \approx \sqrt{1.51 \times 10^{-10}(0.035)}$$

$$\approx 2.29 \times 10^{-6}\ M$$

$x = [OH^-]$, hence pOH = 5.64 and pH = 8.36.

7.2 THE RELATIONSHIP BETWEEN K_h AND K_a

Consider the ionization of a weak acid:

$$HA + H_2O \rightleftharpoons H_3O^+ + A^- \qquad K_a = \frac{[H_3O^+][A^-]}{[HA]} \tag{7-11}$$

and the hydrolysis of its conjugate base:

$$A^- + H_2O \rightleftharpoons HA + OH^- \qquad K_h = \frac{[HA][OH^-]}{[A^-]} \tag{7-12}$$

If we multiply K_a by K_h and cancel the common terms, we get

$$K_aK_h = \left(\frac{[H_3O^+]\cancel{[A^-]}}{\cancel{[HA]}}\right)\left(\frac{\cancel{[HA]}[OH^-]}{\cancel{[A^-]}}\right) = [H_3O^+][OH^-] \quad (7\text{-}13)$$

Since the autoionization constant of water is defined by

$$K_w = [H_3O^+][OH^-]$$

it is true that

$$K_aK_h = K_w \quad (7\text{-}14)$$

or

$$K_h = \frac{K_w}{K_a} \quad (7\text{-}15)$$

Thus it is unnecessary to maintain separate listings of K_a for an acid and K_h for its conjugate base; any K_h may be quickly computed from a K_a value. An equally useful expression can be obtained by taking the log of both sides of equation (7-15) and multiplying by -1, giving

$$-\log K_h = -\log K_w - (-\log K_a)$$

$$pK_h = pK_w - pK_a \quad (7\text{-}16)$$

Example 7.2

Using the data in Appendix III, find K_h for sodium *p*-nitrobenzoate.

Answer: The listed value of pK_a, at 25°C, of *p*-nitrobenzoic acid is 3.42. Therefore,

$$pK_h = 14.00 - 3.42 = 10.58$$

$$K_h = \text{antilog}(-10.58) = 10^{-10.58} = 2.7 \times 10^{-11}$$

Example 7.3

At 20°C fluoroacetic acid has a K_a of 2.7×10^{-3}. Find K_h for the fluoroacetate ion.

Answer: At 20°C, $K_w = 6.81 \times 10^{-15}$ (see Appendix I). K_h is accordingly found from

$$K_h = \frac{K_w}{K_a} = \frac{6.81 \times 10^{-15}}{2.7 \times 10^{-3}} = 2.5 \times 10^{-12}$$

Although our derivation of the relationship between ionization constant and hydrolysis constant used a weak acid as an example, we could equally well have chosen a weak base. In that case we consider the base ionization and the hydrolysis of one of its salts.

$$B + H_2O \rightleftharpoons BH^+ + OH^- \qquad K_b = \frac{[BH^+][OH^-]}{[B]}$$

$$BH^+ + H_2O \rightleftharpoons B + H_3O^+ \qquad K_h = \frac{[B][H_3O^+]}{[BH^+]}$$

Equation (7-13) becomes

$$K_b K_h = \left(\frac{[BH^+][OH^-]}{[B]}\right)\left(\frac{[B][H_3O^+]}{[BH^+]}\right) = [H_3O^+][OH^-] \tag{7-17}$$

from which

$$K_h = \frac{K_w}{K_b} \tag{7-18}$$

or

$$pK_h = pK_w - pK_b \tag{7-19}$$

Note that the subscript h can refer to hydrolysis of a salt of either an acid or a base; it is not usual to distinguish between the two as we do for acid or base ionization.[2]

Example 7.4

Determine the hydrolysis constant for trimethylammonium chloride, $(CH_3)_3NHCl$.

Answer: The value of pK_b for trimethylamine is 4.20 at 25° C. K_h is therefore

$$pK_h = pK_w - pK_b = 14.00 - 4.20 = 9.80$$

$$K_h = \text{antilog}(-9.80) = 10^{-9.80} = 1.6 \times 10^{-10}$$

Our reasons for ignoring the hydrolysis of salts of *strong* acids or bases, above, can now be put on a firm mathematical footing. Equation (7-15) tells us that if K_a is large (strong acid), K_h is small; if K_h is small enough, the amount of hydrolysis is negligible. The ionization of an acid

$$HA \rightleftharpoons H^+ + A^-$$

$$(C - x) \quad x \quad x$$

will be regarded as complete if it proceeds to an extent greater than 99% in a 1 M solution. In this case

$$K_a = \frac{(0.99)(0.99)}{1.00 - 0.99} = 98$$

We will call this value approximately 100 for convenience and thus make the statement that strong acids have K_a values of *at least* 100. Salts of strong acids, therefore, have K_h values of *at most* $K_w/100$, or 10^{-16}.

[2] In some textbooks the ionization constant is given the symbol K_i, which can mean *either* K_a or K_b. This need not pose a problem to the careful student, because from the context it is always clear which is meant.

Example 7.5

What is the extent of hydrolysis in a 0.005 *M* solution of a salt with $K_h = 10^{-16}$?

Answer: Using the exact solution method of Appendix VI [equation (VI-15)] for the reaction

$$A^- + H_2O \rightleftharpoons HA + OH^-$$

we find the following equilibrium concentrations:

$$[OH^-] = 1.000025 \times 10^{-7}\ M$$

$$[H^+] = 0.999975 \times 10^{-7}\ M$$

$$[A^-] = 0.004999999995\ M$$

$$[HA] = 5 \times 10^{-12}\ M$$

The extent of hydrolysis is $[HA]/C = 5 \times 10^{-12}/0.005 = 10^{-9} = 10^{-7}\%$. We are therefore justified in our neglect of a reaction that occurs only once out of a billion times.

The situation is the same with respect to strong bases; we can show this mathematically by adapting the reasoning of Example 7.5 to base ionization. We define a strong base as one with $K_b > 100$, and then show that with $K_h < 10^{-16}$ the extent of hydrolysis is negiligibly small.

7.3 BUFFER SOLUTIONS

A solution that resists changes of pH is called a *buffer solution*. More precisely, a buffer solution is one to which a relatively large amount of strong acid or strong base may be added, without causing a large change in the pH of the solution. A buffer solution works because it contains substances that are capable of reacting with and neutralizing both acids and bases. As an example of this resistance to change of pH, let us consider 100 mL of a solution of pH 5, and calculate its pH after a small amount of acid is added. If the original solution is a *strong acid*, addition of 1 mL of 0.1 *M* HCl will cause the pH to jump from 5 to 3. (*Students*: Check this calculation to confirm that it is correct.) This solution is obviously not very resistant to changes in pH. On the other hand, a *buffer* solution of pH 5 can be prepared by mixing equal quantities of 0.100 *M* acetic acid and 0.174 *M* sodium acetate. If to 100 mL of this solution 1 mL of 0.1 *M* HCl is added, the resulting solution will have a pH of 4.994. Comparatively speaking, this buffer solution does indeed resist changes in pH. This calculation will be explained in detail in Section 7.5.

A buffer solution also maintains a constant pH when it is diluted. If 10 mL of 0.01 *M* HCl is diluted to 100 mL by adding water, its pH changes from 2 to 3. If 10 mL of an acetic acid (0.100 *M*)/sodium acetate (0.174 *M*) buffer is diluted to 100 mL, its pH remains constant at 5.00.

Buffer solutions consist of mixtures of a weak acid and one of its salts, as in the preceding example; or they can be mixtures of a weak base and one of its salts. An example of the latter type would be trimethylamine plus trimethylammonium chloride.

The crucial characteristic for buffer action is that *both the weak acid (or base) and its conjugate base (or acid) must be present, in comparable quantities.* To understand why buffer action takes place, it is necessary to review the common-ion effect and the principle of Le Châtelier.

Consider a solution of the weak acid acetic acid (HOAc). If to this solution is added a small amount of an acetate salt such as potassium acetate, the acetate ion in solution will have come from two sources: the partial ionization of the acetic acid, and the (complete) dissociation of the potassium acetate. The equilibrium set up in the original solution among H_3O^+, OAc^-, and HOAc will have been disturbed by the additional OAc^-. Le Châtelier's principle tells us that when an equilibrium system is disturbed, it will respond in a way that reduces the extent of the disturbance. In this case the disturbance is an increase in OAc^- and the response is to decrease the excess OAc^- in the only way possible, namely, by the combination of OAc^- with H_3O^+ to re-form HOAc. The extent of ionization of the original HOAc will decrease

$$\mathrm{HOAc} \longleftarrow \mathrm{H_3O^+} + \mathrm{OAc^-} \tag{7-20}$$

and it is usual to say that "the equilibrium shifts to the left." This is one example of the *common-ion effect*, so named because the acetate ion is common to both acetic acid and potassium acetate.

The *principle of Le Châtelier* is easily applied in acid-base chemistry, and it can be verified mathematically using the equilibrium expression. The equilibrium concentrations of H_3O^+ and OAc^- in a solution of acetic acid are found by solving the usual expression

$$K_a = \frac{(x)(x)}{C-x} \approx \frac{x^2}{C}$$

$$x \approx \sqrt{K_a C}$$

For acetic acid $K_a = 1.75 \times 10^{-5}$, and if we choose a concentration of 0.1 *M*, $x = [H_3O^+] = [OAc^-] = 1.32 \times 10^{-3}$ *M*. If to one liter of this solution we add even a relatively large amount of potassium acetate, say 0.200 mole, the resultant acetate ion concentration will *not* be the simple sum $0.200 + 1.32 \times 10^{-3}$ $M = 0.201$ *M* because some of the additional acetate will combine with H_3O^+ via (7-20). The amount of acetate ion that can react in this way is limited by the availability of H_3O^+; since initially $[H_3O^+] = 1.32 \times 10^{-3}$ *M*, no more than 1.32×10^{-3} *M* of the added acetate ion can be used up. The equilibrium concentration of acetate therefore lies between 0.200 and 0.201 *M*, and we will suffer no great loss in accuracy if we simply set $[OAc^-] = 0.2$ *M*. We now recalculate $[H_3O^+]$.

$$K_a = \frac{(x)(0.2)}{0.1-x} \approx \frac{(x)(0.2)}{0.1} = 2x$$

$$x \approx \frac{K_a}{2} \cong 8.8 \times 10^{-6}\ M$$

At this point, both because the common-ion effect is so important and in order to give some practice with base systems, we will repeat the foregoing argument using a weak base as an example. Trimethylamine (Me_3N, $K_b = 6.3 \times 10^{-5}$ at 25°C) ionizes according to

$$Me_3N + H_2O \rightleftharpoons Me_3NH^+ + OH^- \qquad K_b = \frac{[Me_3NH^+][OH^-]}{[Me_3N]} \tag{7-21}$$

and the equilibrium concentrations in a 0.02 *M* solution may be calculated by

$$K_b = \frac{(x)(x)}{0.02-x} \approx \frac{x^2}{0.02}$$

$$x = [Me_3NH^+] = [OH^-] = \sqrt{0.02K_b} = 1.1 \times 10^{-3}\ M \qquad \text{pOH} = 2.95$$

The extent of ionization is $1.1 \times 10^{-3}/0.02 = 0.055$ or 5.5%. If now some trimethylammonium chloride, Me_3NHCl, is added, the factor $[Me_3NH^+]$ in (7-21) increases. Since K_b is constant, this can be true only if there is a compensating decrease in $[OH^-]$ and/or an increase in $[Me_3N]$. In fact, both changes occur; the equilibrium shifts to the left as some of the extra Me_3NH^+ combines with OH^- to re-form Me_3N.

$$Me_3N \longleftarrow Me_3NH^+ + OH^-$$

Suppose that the amount of added salt is relatively large, say 0.5 *M*. The equilibrium concentration of Me_3NH^+ can be no greater than 0.500 + 0.001 = 0.501 *M* (if none of the OH^- reacts) and no less than 0.500 *M* (if all of the available OH^- reacts). We are justified in using the value $[Me_3NH^+] = 0.5\ M$, and the other equilibrium concentrations may be recalculated as follows:

$$K_b = \frac{(0.5)[OH^-]}{0.02-x} \approx \frac{0.5}{0.02}[OH^-]$$

$$[OH^-] = \frac{0.02}{0.5}K_b = 2.5 \times 10^{-6}\ M \qquad \text{pOH} = 5.60$$

In this calculation x is even smaller than in the pure base calculation, so it may safely be neglected. The extent of ionization is $2.5 \times 10^{-6}/0.02 = 0.0001$, or 0.01%. These calculations show that the presence of a large amount of salt having an ion in common can exert a considerable influence on acid or base ionization.

7.4 BUFFER CALCULATIONS

The equilibrium expression

$$K_a = \frac{[H^+][A^-]}{[HA]} \tag{7-22}$$

is true in *any* solution in which the species HA and A^- are present. Solutions of weak acids *or* their salts, or solutions prepared from both acid and salt, all obey equation (7-22). It is very important that this fact be understood and remembered. Rearrangement of (7-22) leads to an expression by which $[H_3O^+]$ and hence pH may be calculated.

$$[H_3O^+] = K_a \frac{[HA]}{[A^-]} \tag{7-23}$$

We learned in Chapter 3 that throughout much of the range of K_a values and concentrations normally encountered in laboratory work the extent of dissociation of weak acids is small enough so that the amount of nonionized acid present in solution at equilibrium is approximately the same as the amount originally present (page 48). In terms of the present discussion, in the acid ionization we represent by C_a the amount of acid originally added to the solution. If $HA \rightarrow H_3O^+ + A^-$, and x is the extent of ionization, then at equilibrium

$$[HA] = C_a - x \approx C_a \tag{7-24}$$

If in addition to the acid some salt is dissolved, the common-ion effect decreases x even further, and the approximation (7-24) is even more valid. The same arguments apply to the salt: A^- undergoes a basic hydrolysis reaction

$$A^- + H_2O \longrightarrow HA + OH^- \tag{7-12}$$

But unless HA is an exceedingly weak acid ($K_a < 10^{-13}$), A^- is a weak base: the extent of its hydrolysis, y, is small even if no additional acid is in solution. The presence of HA from another source (*added* acid) will, by the principle of Le Châtelier, shift the equilibrium to the left and decrease y even further. The concentration of A^- at equilibrium is therefore approximately equal to the amount of salt (C_s) originally added to the solution.

$$[A^-] = C_s - y \approx C_s \tag{7-25}$$

To improve our approximation, we should say that in a buffer at equilibrium $[A^-]$ equals the amount originally added, C_s, *minus* y, the amount lost in hydrolysis, *plus* x, the amount contributed from the ionization of the acid component.

$$[A^-] = C_s - y + x \approx C_s \tag{7-26}$$

Since both x and y are negligibly small, it is still true that $[A^-] \approx C_s$. Similar considerations apply to [HA], which is decreased by ionization but in a buffer is also increased by hydrolysis of the salt constituent.

$$[HA] = C_a - x + y \approx C_a \tag{7-27}$$

Equation (7-23) thus becomes

$$\begin{aligned} [H_3O^+] &= K_a \frac{C_a - x + y}{C_s - y + x} \\ [H_3O^+] &\approx K_a \frac{C_a}{C_s} \end{aligned} \tag{7-28}$$

Taking logarithms of both sides and multiplying by -1, we obtain

$$-\log [H_3O^+] = -\log K_a - \log \frac{C_a}{C_s}$$

or

$$\mathrm{pH} = \mathrm{p}K_a - \log\left(\frac{C_a}{C_s}\right) \tag{7-29}$$

This equation is usually presented in its mathematically equivalent form

$$\mathrm{pH} = \mathrm{p}K_a + \log\left(\frac{C_s}{C_a}\right) \tag{7-30}$$

which is known as the *Henderson-Hasselbalch equation*.

Equation (7-30) is an approximate form, based on the two assumptions that the extent of ionization (of HA) and hydrolysis (of A^-) are negligibly small and the autoionization of water can be neglected. As we will see when we develop the exact form, these conditions are satisfied whenever C_a and C_s are considerably larger than both $[H_3O^+]$ and $[OH^-]$. This in turn means that equation (7-30) may be used with good results except in very dilute solutions or when the pH is either very low (<3) or very high (>11).

Buffer calculations using the Henderson-Hasselbalch equation are, provided that one is familiar with logarithms, among the simplest that are encountered in ionic equilibria. Some examples follow.

Example 7.6

Calculate the pH of a solution which is 0.05 *M* in benzoic acid and 0.02 *M* in sodium benzoate. The $\mathrm{p}K_a$ of benzoic acid is 4.18.

Answer: The given values are substituted into the Henderson-Hasselbalch equation.

$$\mathrm{pH} = \mathrm{p}K_a + \log\left(\frac{C_s}{C_a}\right)$$

$$= 4.18 + \log\left(\frac{0.02}{0.05}\right)$$

$$= 4.18 + (-0.40)$$

$$= 3.78$$

Example 7.7

A buffer solution was analyzed and found to be 0.183 *M* in acid and 0.625 *M* in salt. Its pH was measured and found to be 5.82. Calculate K_a for the acid.

Answer: The given values are substituted into the Henderson-Hasselbalch equation.

$$\mathrm{pH} = \mathrm{p}K_a + \log\left(\frac{C_s}{C_a}\right)$$

$$5.82 = \mathrm{p}K_a + \log\left(\frac{0.625}{0.183}\right)$$

$$5.82 = \mathrm{p}K_a + 0.53$$

$$\mathrm{p}K_a = 5.29$$

$$K_a = \text{antilog}(-5.29) = 10^{-5.29} = 5.13 \times 10^{-6}$$

Example 7.8

(a) What should the ratio of sodium acetate concentration to acetic acid concentration be to produce a buffer of pH 5.00?
(b) If the acid concentration is 0.100 *M*, what should the sodium acetate concentration be? For acetic acid, $K_a = 1.75 \times 10^{-5}$.

Answer: (a) The given values are substituted into the Henderson-Hasselbalch equation, which is solved for the ratio C_s/C_a.

$$\mathrm{pH} = \mathrm{p}K_a + \log\left(\frac{C_s}{C_a}\right)$$

$$5.00 = -\log(1.75 \times 10^{-5}) + \log\left(\frac{C_s}{C_a}\right)$$

$$= 4.76 + \log\left(\frac{C_s}{C_a}\right)$$

$$\left(\frac{C_s}{C_a}\right) = \text{antilog } 0.24 = 10^{0.24} = 1.74$$

(b) The concentration of sodium acetate must be 1.74 times the concentration of acetic acid: since the acid concentration is 0.100 *M*, the sodium acetate concentration is (0.100 *M*) (1.74) = 0.174 *M*.

Example 7.9

The $\mathrm{p}K_a$ of *m*-nitrophenol is 8.40. How many moles of the salt sodium *m*-nitrophenolate must be added to 500 mL of a 0.5 *M* solution in *m*-nitrophenol to bring its pH to 7.60? (Assume the solution volume to be unchanged by the addition of the salt.)

Answer:

$$\mathrm{pH} = \mathrm{p}K_a + \log\left(\frac{C_s}{C_a}\right)$$

$$7.60 = 8.40 + \log\left(\frac{C_s}{0.5}\right)$$

$$-0.80 = \log\left(\frac{C_s}{0.5}\right)$$

$$\frac{C_s}{0.5} = 10^{-0.80} = 0.16$$

$$C_s = (0.16)(0.5) = 0.080 \text{ mol/liter}$$

Only half of this amount, 0.040 mol, is required for this solution of 500 mL.

An equation analogous to the Henderson-Hasselbalch equation, for buffer systems involving weak bases and their salts, may be derived by inspection. The reasoning behind the assumptions is the same for both acids and bases, so that we need only modify the Henderson-Hasselbalch equation by substituting pOH for pH, pK_b for pK_a, and C_b for C_a. The result is

$$\text{pOH} = pK_b + \log\left(\frac{C_s}{C_b}\right) \tag{7-31}$$

The use of this equation is precisely analogous to the use of equation (7-30). Often the result is expressed as pH rather than pOH, in which case equation (7-31) must be modified by substituting the relationship pK_w = pH + pOH.

$$\text{pH} = pK_w - pK_b - \log\left(\frac{C_s}{C_b}\right) \tag{7-32}$$

Example 7.10

The base cyanomethylamine ($NCCH_2NH_2$) has a pK_b of 8.66. What must be the ratio of concentration of cyanomethylammonium chloride to cyanomethylamine in order to produce a buffer of pH 5.50?

Answer: First, convert pH to pOH (assume that $t = 25°\text{C}$ so that $pK_w = 14.0$).

$$\text{pOH} = 14 - \text{pH} = 14 - 5.50 = 8.50$$

Then substitute the given values into equation (7-31) and solve for the ratio C_s/C_b.

$$\text{pOH} = pK_b + \log\left(\frac{C_s}{C_b}\right)$$

$$8.50 = 8.66 + \log\left(\frac{C_s}{C_b}\right)$$

$$-0.16 = \log\left(\frac{C_s}{C_b}\right)$$

$$\frac{C_s}{C_b} = \text{antilog}(-0.16) = 10^{-0.16} = 0.69$$

7.5 BUFFER CAPACITY

A buffer solution does very well in reducing the change in pH when a small amount of either a strong acid or strong base is added to the solution. However, it is not reasonable to suppose that a buffer can absorb a large amount of acid or base without significant pH change. Suppose that a buffer contains 1 mol each of a weak acid and its salt;

since it absorbs OH^- by reaction with the weak acid component, if 1 mol of OH^- is added, the capacity of the buffer to respond will have been exhausted. Moreover, we might expect that its ability to respond would have been seriously impaired well before the equivalent amount of base had been added.

The term *buffer capacity* refers to the amount of base (or acid) the solution can absorb without significant change in pH. Precise, quantitative discussions of buffer capacity must of course include a specification of what is meant by "significant change of pH," and this specification differs according to the requirements of the experimental situation. The quantitative aspects of buffer capacity will not be treated in this book.[3] Some qualitative remarks, however, can be made as a guide to intelligent choice of buffers. Since a buffer responds by reaction of one of its components with the added acid (or base), it is clear that the buffer capacity is dependent on the concentration of weak acid and/or its salt in the buffer. Usually, we refer to the *total buffer concentration* $C = [HA] + [A^-]$, and note that buffer capacity is proportional to C.

As an illustration, let us calculate the effect of the addition of 5.0 mL of 1.0 *M* NaOH to 1.00 liter of the buffer solution of Example 7.8 (0.100 *M* acetic acid, 0.174 *M* sodium acetate). The molar amount of NaOH added is (1.0 mol/liter) (0.0050 liter) = 0.0050 mol. This amount reacts, neutralizing 0.0050 mol of the 0.100 mol of acetic acid originally present, and producing 0.0050 mol of sodium acetate in addition to the 0.174 mol already there. After addition of the base, the concentrations are: acid, 0.100 − 0.0050 = 0.095 *M*; and salt, 0.174 + 0.0050 = 0.179 *M*. The pH is calculated from the Henderson-Hasselbalch equation as

$$\begin{aligned} \text{pH} &= \text{p}K_a + \log\left(\frac{C_s}{C_a}\right) \\ &= 4.76 + \log\left(\frac{0.179}{0.095}\right) \\ &= 5.035 \quad \text{or an increase of only 0.035 pH unit} \end{aligned}$$

Suppose that the same amount of base (0.0050 mol) were added to a buffer of the same pH, but only 0.1 as concentrated (0.0100 *M* acetic acid, 0.0174 *M* sodium acetate). In this case the amount of acid neutralized, 0.0050 mol, is a larger fraction of

[3] As a guide to those who would like some basis for our conclusions about buffer capacity, there follows a terse outline. We define buffer capacity β (sometimes called "buffer index") in differential terms as $\beta = d[B]/d\text{pH}$, the increment of strong base B needed to change the pH of the buffer by one unit. The larger the value of β, the greater is the capacity of the buffer to resist changes in pH. Each mole of added base increases $[A^-]$ by 1 mol, so that $d[B]/d\text{pH} = d[A^-]/d\text{pH}$. Furthermore, $d[A^-]/d\text{pH} = (d[H^+]/d\text{pH})\,(d[A^-]/d[H^+])$. The first of these factors is equal to $-2.303[H^+]$ (since $d \ln x = (1/x)dx$); the second is found by differentiation of $[A^-] = CK_a/(K_a + [H^+])$, which in turn comes from the equilibrium expression and the relationship $C = [HA] + [A^-]$.

$$\beta = \frac{d[B]}{d\text{pH}} = 2.303\,\frac{K_a C[H^+]}{(K_a + [H^+])^2}$$

When β is plotted against pH, it shows a maximum at pH = pK_a and drops to low values at 1½ pH units away from the maximum. When comparing or selecting buffers, one should choose so that pH ≈ pK_a and C is not too small. At large values of C activity coefficients of ions differ markedly from 1, causing errors in pH calculations (see Appendix IV).

the amount originally present. After addition of base, the concentrations are: acid, $0.0100 - 0.0050 = 0.0050$; and salt, $0.0174 + 0.0050 = 0.0224\ M$. The pH is

$$\text{pH} = 4.76 + \log\left(\frac{0.0224}{0.0050}\right) = 5.41$$

The more dilute buffer suffers a greater change in pH for the same amount of added base; its buffer capacity is less.

A second criterion for choice of buffers is that the pK_a value should be as close as possible to the desired pH. There are two reasons for this rule. The Henderson-Hasselbalch equation tells us that when $\text{pH} = pK_a$, $C_a = C_s$. A buffer in which $C_a = C_s$ is desirable because it has equal ability to respond to added acid or base. Secondly, the buffer capacity is largest when $pK_a = \text{pH}$. A buffer in which the pH differs from the pK_a by 1 unit undergoes, for a given amount of added acid or base, more than twice the change of pH compared to a buffer of $\text{pH} \approx pK_a$. The choice of $pK_a \approx \text{pH}$ not only gives even-handed treatment to either added acid or added base, but also maximizes the ability of the buffer to maintain constant pH.

7.6 PREPARATION OF BUFFERS

Buffer solutions are normally prepared in either of two ways. The most straightforward method is the one that has been implicit in our discussion up to now, namely, the dissolving of comparable amounts of both a weak acid and one of its salts in the same solution. The second method, equally convenient, is to *partly neutralize* a weak acid by adding a strong base such as NaOH. Since OH^- is a strong base, it is completely consumed in the reaction

$$\text{HA} + \text{OH}^- \longrightarrow \text{H}_2\text{O} + \text{A}^-$$

and each OH^- ion added to the solution results in the conversion of one HA molecule to one A^- ion (provided, of course, that no more OH^- is added than the amount of HA originally present).

Suppose that we begin with a solution containing 1 mol of HA, and add to it 1/3 mol of NaOH. After the partial neutralization 1/3 mol of A^- will be present, and 2/3 mol of HA will remain. The resulting solution, and this is the important point, *is identical in all respects to one prepared from 2/3 mol HA and 1/3 mol of its sodium salt*. Once in solution, an individual A^- ion has no memory of whether it was originally part of an acid or of a salt.

Example 7.11

The pK_a of phenylboric acid is 8.86. Calculate the pH of the solution that results when 10 g of KOH is added to 1 liter of a 0.450 M solution of phenylboric acid.

Answer: 0.450 mol of acid is present. When 10 g of KOH (0.178 mol) is added, neutralization of 0.178 mol of phenylboric acid takes place. The amount of phenylborate ion formed is 0.178 mol; the amount of phenylboric acid remaining is 0.450 − 0.178 = 0.272. Therefore, $C_a = 0.272\,M$, $C_s = 0.178\,M$.

$$\begin{aligned} \text{pH} &= \text{p}K + \log\left(\frac{C_s}{C_a}\right) \\ &= 8.86 + \log\left(\frac{0.178}{0.272}\right) \\ &= 8.86 - 0.18 \\ &= 8.68 \end{aligned}$$

Example 7.12

Calculate the pH of a solution prepared by mixing equal volumes of 0.625 M phenylboric acid and 0.250 M KOH.

Answer: Since the problem says only "equal volumes," we are free to choose any amount of solution we please. Of course, the pH will be the same, whether we choose one drop or one bucket. For convenience, choose 1 liter of each solution.

One liter of 0.250 M KOH contains 0.250 mol of OH^-, which reacts with 0.250 mol of the phenylboric acid. Then 0.250 mol of phenylborate ion is formed, and the amount of remaining acid is 0.625 − 0.250 = 0.375 mol. Since 1 liter of each solution was used, the volume is now 2 liters, and C_a = 0.375 mol/2 liters = 0.1875 M, and C_s = 0.250 mol/2 liters = 0.125 M.

$$\begin{aligned} \text{pH} &= \text{p}K_a + \log\left(\frac{C_s}{C_a}\right) \\ &= 8.86 + \log\left(\frac{0.125}{0.1875}\right) \\ &= 8.86 - 0.18 \\ &= 8.68 \end{aligned}$$

7.7 EXACT TREATMENT

The Henderson-Hasselbalch equation is based on the assumption that the autoionization of water is negligible. This assumption is valid throughout much of the concentration and pH range in which buffers are commonly used, but becomes a source of error at low concentration of total buffer or outside the approximate pH range 3.0 to 11.0. It is not very difficult to develop an exact expression for buffer pH, and much of the work has already been done. In Appendix VI, we develop an exact equation for the pH of a solution of a weak acid. One of the assumptions of that treatment is that the only source of A^- ions is dissociation of the original C_a moles per liter of HA, so that

$$[HA] + [A^-] = C_a \text{ (acid only)} \tag{VI-8}$$

In buffer systems there is an additional source, the salt of A^-, so that

$$[HA] + [A^-] = C_a + C_s \tag{7-33}$$

A second assumption was that the solution is electrically neutral, that is, that the concentration of cations must be equal to the concentration of anions

$$[H_3O^+] = [A^-] + [OH^-] \text{ (acid only)} \tag{VI-9}$$

In buffer solutions we need to account for the fact that some additional positive charge will be present (from the cationic part of the salt), so that, taking a sodium salt as an example,

$$[H_3O^+] + [Na^+] = [A^-] + [OH^-] \tag{7-34}$$

Sodium ion does not participate in the buffer equilibrium, so its concentration is constant and equal to the initial concentration of salt:

$$[Na^+] = C_s \tag{7-35}$$

Equation (7-34) becomes

$$[H_3O^+] + C_s = [A^-] + [OH^-] \tag{7-36}$$

We have now completed the only changes necessary to include buffer systems in the exact expression, by assembling four equations

$$[HA] + [A^-] = C_a + C_s \tag{7-33}$$

$$[H_3O^+] + C_s = [A^-] + [OH^-] \tag{7-36}$$

$$K_a = \frac{[H_3O^+][A^-]}{[HA]} \tag{7-22}$$

$$K_w = [H_3O^+][OH^-] \tag{7-37}$$

from which the four unknowns $[H_3O^+]$, $[OH^-]$, $[HA]$, and $[A^-]$ may be determined. The following steps summarize the derivation in Appendix VI. Equation (7-22) is substituted into (7-33) to eliminate $[HA]$, resulting in

$$[A^-]\left(1 + \frac{[H_3O^+]}{K_a}\right) = C_a + C_s \tag{7-38}$$

Equation (7-36) is substituted into (7-38) to eliminate $[A^-]$, giving

$$([H_3O^+] - [OH^-] + C_s)\left(1 + \frac{[H_3O^+]}{K_a}\right) = C_a + C_s \tag{7-39}$$

Equation (7-37) is substituted into (7-39) to eliminate $[OH^-]$; the result is

$$\left([H_3O^+] - \frac{K_w}{[H_3O^+]} + C_s\right)\left(1 + \frac{[H_3O^+]}{K_a}\right) = C_a + C_s \tag{7-40}$$

The terms in parentheses are multiplied out, giving

$$[H_3O^+] + \frac{[H_3O^+]^2}{K_a} - \frac{K_w}{[H_3O^+]} - \frac{K_w}{K_a} + C_s + \frac{[H_3O^+]C_s}{K_a} = C_a + C_s \tag{7-41}$$

When both sides of equation (7-41) are multiplied by $[H_3O^+]$ and the resulting terms grouped by powers of $[H_3O^+]$, the final equation (7-42) is obtained.[4]

$$\frac{1}{K_a}[H_3O^+]^3 + \left(1 + \frac{C_s}{K_a}\right)[H_3O^+]^2 - \left(C_a + \frac{K_w}{K_a}\right)[H_3O^+] - K_w = 0 \tag{7-42}$$

This equation, like its companion equation (VI-15), can easily be solved on a computer. A BASIC program for the pH of buffer systems is given in Section II.3 of Appendix II.

7.8 VALIDITY OF THE HENDERSON-HASSELBALCH EQUATION

Although equation (7-42) can easily be solved, it is not very informative; it is so complicated that looking at it gives us no insight into its meaning and validity. Fortunately, it is possible to put the equation into a more instructive form. Although the result [equation (7-46)] can be derived in a few steps by algebraic manipulation of (7-42) using $K_w = [H_3O^+][OH^-]$, it is useful to see how it can be derived from more fundamental considerations. We begin with the equilibrium expression

$$[H_3O^+] = K_a \frac{[HA]}{[A^-]} \tag{7-23}$$

take logarithms of both sides and multiply by -1; the result is

$$pH = pK_a + \log \frac{[A^-]}{[HA]} \tag{7-43}$$

[Note that this is *not* the same as equation (7-30), which deals with the *initial* concentrations of salt and acid.] Equation (7-36) can be rewritten as

$$[A^-] = C_s + [H_3O^+] - [OH^-] \tag{7-44}$$

[4] If the salt of a divalent cation [e.g., $Mg(NO_2)_2$] is used rather than a sodium salt, the factor C_s in equation (7-42) must be multiplied by 2.

and (7-33) can be rewritten as

$$[\mathrm{HA}] = C_a + C_s - [\mathrm{A}^-] \tag{7-45}$$

When we combine (7-43) and (7-44) with (7-45), the result is

$$\mathrm{pH} = \mathrm{p}K_a + \log\left(\frac{C_s + [\mathrm{H_3O^+}] - [\mathrm{OH^-}]}{C_a - [\mathrm{H_3O^+}] + [\mathrm{OH^-}]}\right) \tag{7-46}$$

If the initial concentrations of salt and acid are significantly greater than both $[H_3O^+]$ and $[OH^-]$, then

$$C_s + [\mathrm{H_3O^+}] - [\mathrm{OH^-}] \approx C_s$$

$$C_a - [\mathrm{H_3O^+}] + [\mathrm{OH^-}] \approx C_a$$

and equation (7-46) becomes identical with the approximate form, the Henderson-Hasselbalch equation. Both $[H_3O^+]$ and $[OH^-]$ will be less than 0.001 *M* when the pH is in the range 3 to 11; C_a and C_s are rarely less than 0.1 *M* in practical buffers. Therefore, the Henderson-Hasselbalch equation may be used with confidence when the pH is between 3 and 11.

7.9 ERRORS AND VALIDITY RANGES

The sources of error in buffer solutions can be placed in three categories: imprecisely known $\mathrm{p}K_a$ values, the use of approximate mathematical methods, and errors arising from solute activity coefficients. As mentioned earlier, the solute activity problem is beyond the scope of this book.

The accuracy of most acid-strength determinations is such that the $\mathrm{p}K_a$ is known to ±0.01 unit. In exceptional cases, such as acids that are used as standards for pH measurements, pK values are sometimes measured to ±0.001 unit. This is difficult to do, however, and requires specialized equipment and painstaking laboratory technique. Many pK values, in fact, are known to no better than ±0.1 unit. From the form of the Henderson-Hasselbalch equation, or its more exact counterpart [equation (7-46)], it can be seen that the uncertainty in pH is directly related to that in pK. If $\mathrm{p}K_a$ is in error by ±0.01 unit, the error in pH must be at least this large. For most buffer systems, then, there is an accuracy limit of ±0.01 pH unit regardless of which equation is used.

In the preceding section it was remarked that the Henderson-Hasselbalch equation may be safely used if the pH lies between 3 and 11. A more detailed consideration shows that this range of validity depends on the total buffer (acid plus salt) concentration, and is somewhat narrower for dilute buffers. The results of some exact and approximate calculations are presented in Figure 7.1. Taking a series of buffers of differing $\mathrm{p}K_a$ (but always with the ratio $C_s/C_a = 2$) we have calculated and plotted the pH at two different total buffer concentrations. The Henderson-Hasselbalch equation becomes $\mathrm{pH} = \mathrm{p}K_a + \log 2$, or $\mathrm{pH} = \mathrm{p}K_a + 0.301$. This is the straight line of unit slope in Figure 7.1 and it is

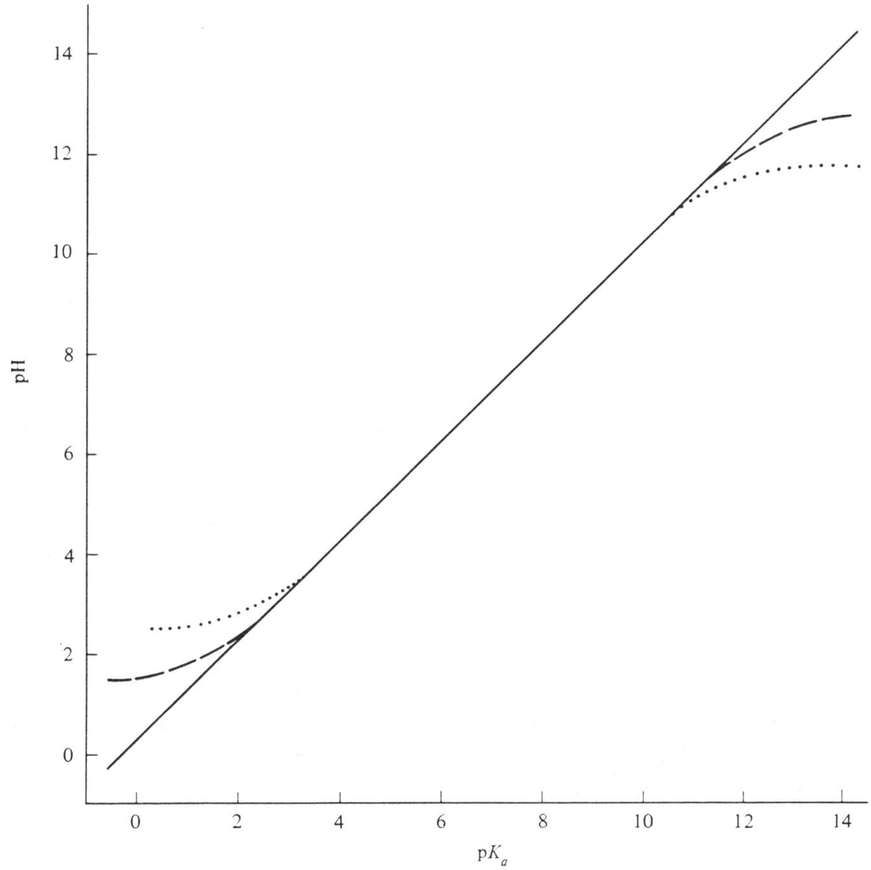

Figure 7.1 pH of a 2:1 (C_s:C_a) buffer as a function of acid strength (pK_a). Solid line, Henderson-Hasselbalch equation; dashed line, exact calculation for $C_s + C_a = 0.1\ M$; dotted line, exact calculation for $C_s + C_a = 0.01\ M$.

independent of total buffer concentration. The exact calculation yields the two curved lines. If the total buffer is 0.1 *M*, the Henderson-Hasselbalch line approaches within 0.02 pH unit when the pH = 3.0 and within 0.01 unit when pH = 3.3. When the total buffer concentration is 0.01 *M*, however, the Henderson-Hasselbalch line does not approach the exact line within 0.01 unit until pH = 4.3. Thus the "3 to 11 pH range" of validity applies only when the concentration is at least 0.1 *M*.

Although buffers more dilute than 0.1 *M* are rarely used, since their buffer capacity is small, it is instructive to see what happens to the pH as the buffer concentration is decreased to zero. This information is presented in Figure 7.2, in which buffer pH is plotted against total buffer concentration. As expected, as the solution becomes more dilute, the pH of *any* buffer approaches the pH of pure water. The horizontal straight lines in Figure 7.2 are the results of Henderson-Hasselbalch calculations, and they are seen to be grossly in error in very dilute solutions. As can be seen in Figure 7.2, this is especially true for solutions more dilute than 10^{-4} *M*.

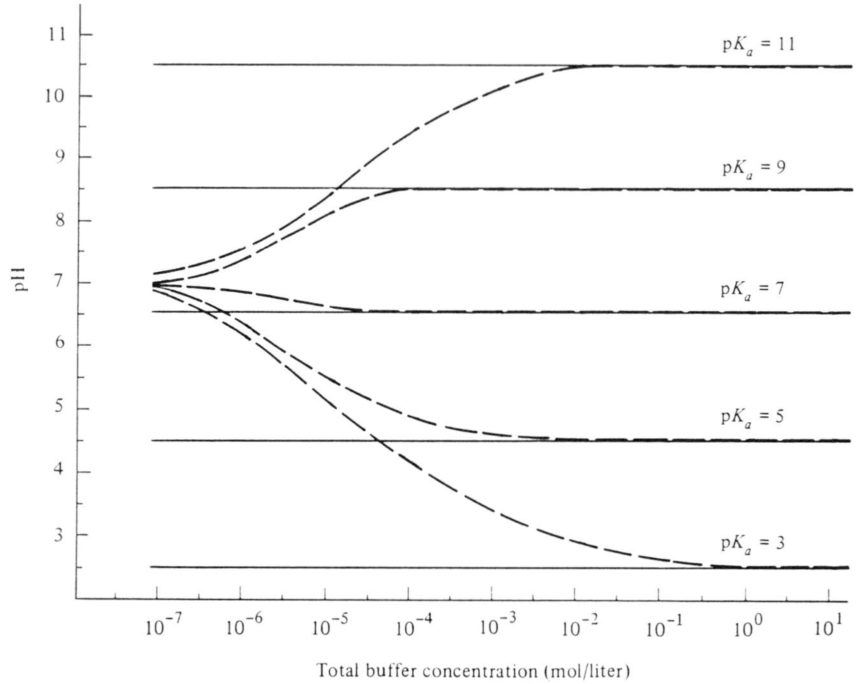

Figure 7.2 pH of very dilute buffer solutions. All calculations are for a buffer in which $C_s/C_a = 1/3$. Solid lines, Henderson-Hasselbalch equations; dashed lines, exact calculations. "Total buffer concentration" is $C_s + C_a$.

PROBLEMS

7.1 Identify the acid that is conjugate to each of the following bases.

(a) $NH(CH_3)_2$ (b) S^{2-} (c) O^{2-} (d) $H_2NCH_2CH_2\overset{+}{N}H_3$ (e) $H_2PO_4^-$ (f) CN^- (g) C_5H_5N (h) $H_3\overset{+}{N}CH(CH_3)CO_2^-$ (i) NH_3 (j) NH_2^- (k) H^-

7.2 Give the name and formula of the acid from which each of the following salts is derived.

(a) KCl (b) NH_4Cl (c) $(CH_3)_3NHCl$ (d) K_2SO_4 (e) ammonium benzoate (f) KOH

7.3 Identify each of the following reactions as "hydrolysis" or "ionization."

(a) $CH_3CO_2H + H_2O \rightleftharpoons CH_3CO_2^- + H_3O^+$

(b) $(CH_3)_3N + H_2O \rightleftharpoons (CH_3)_3NH + OH^-$

(c) $NH_2^- + H_2O \rightleftharpoons NH_3 + OH^-$

(d) $O^{2-} + H_2O \rightleftharpoons 2OH^-$

(e) $NH_3 + H_2O \rightleftharpoons NH_2^- + H_3O^+$

(f) 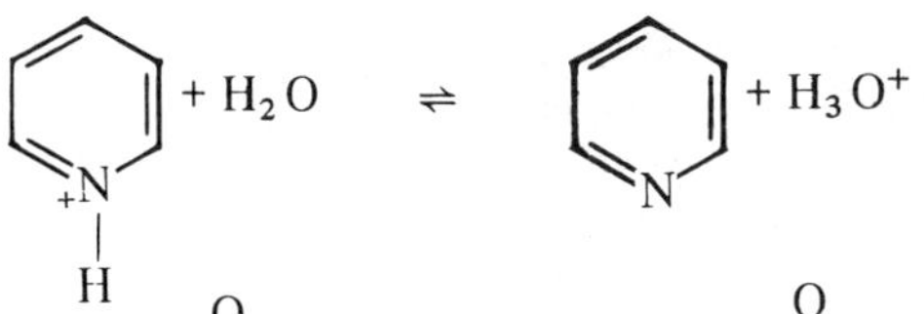

(g) 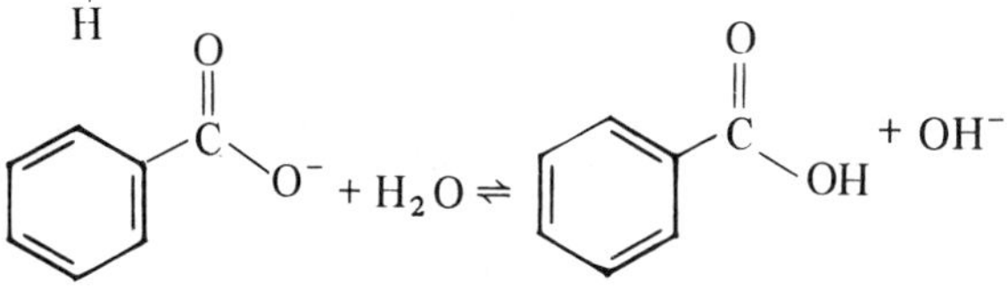

(h) 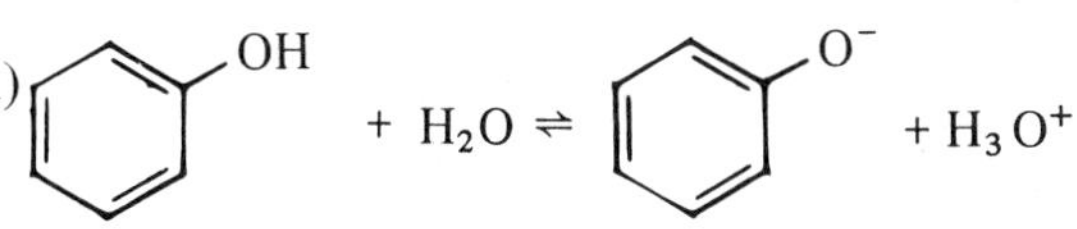

(i) $CH_2(CO_2H)_2 + 2H_2O \rightleftharpoons CH_2(CO_2)^{2-} + 2H_3O^+$

7.4 Using the values in Appendix III, calculate K_h and pK_h for the following salts:
(a) potassium chlorite
(b) sodium fluoroacetate
(c) potassium *o*-nitrophenoxide
(d) sodium oxalate
(e) sodium hydrogen oxalate
(f) anilinium chloride
(g) ethylammonium chloride
(h) 2-methylpyridinium chloride

7.5 For each of the following solutions, tell whether the approximate method will yield an acceptable value for pH or whether a more exact method must be used.
(a) 0.05 *M* potassium chlorite
(b) 0.003 *M* sodium fluoroacetate
(c) 0.2 *M* potassium *o*-nitrophenoxide
(d) 0.001 *M* sodium oxalate
(e) 1.3×10^{-5} *M* sodium phenylacetate
(f) 0.00015 *M* anilinium chloride
(g) 0.35 *M* ethylammonium chloride
(h) 7.2×10^{-6} *M* 2-methylpyridinium chloride

7.6 For each of the solutions in Problem 7.5, calculate the pH by an appropriate method.

7.7 For each of the solutions in Problem 7.6, estimate the percent error incurred by the use of the approximate method (see Appendix VI.1).

7.8 Calculate the pH of a solution which is 0.05 *M* in formic acid and 0.05 *M* in sodium formate.

7.9 Calculate the pH of a formic acid/sodium formate solution which is 0.3 *M* in total buffer and has a 2:1 acid/salt ratio. Repeat the calculation for a 1:2 ratio.

7.10 What ratio of salt to acid concentration must be used to prepare a phenylacetic acid buffer of pH 5.00?

7.11 What is the pH of a solution of 0.025 *M* dimethylamine and 0.075 *M* dimethylammonium chloride?

7.12 Choose an appropriate acid/salt or base/salt pair to produce a buffer for each of the pH values 3.0, 4.0, 5.0, 6.0, 7.0, 8.0, 9.0, 10.0, and 11.0.

7.13 Calculate the acid/salt or base/salt ratio necessary for each buffer solution in Problem 7.12.

7.14 Derive an equation, analogous to the Henderson-Hasselbalch equation but involving pK_b, that is appropriate to buffers consisting of base plus salt.

7.15 Calculate the pH of a solution prepared by mixing 10.0 g each of phenylacetic acid and potassium phenylacetate together with enough water to bring the final volume to 1.00 liter.

7.16 An equimolar mixture of an unknown acid and its sodium salt, when dissolved in water, yields a solution of pH 8.3; is this enough information from which K_a can be calculated? If so, calculate K_a. If not, what additional information is required?

7.17 If to 500 mL of a 0.5 *M* solution of acetic acid is added 500 mL of 0.2 *M* KOH, what will be the pH?

7.18 Given a buffer solution that is 0.250 *M* in formic acid and 0.150 *M* in sodium formate, what will the pH be if to 50 mL of this solution is added 50 mL of 0.100 *N* HCl?

7.19 How many grams of potassium *o*-nitrophenoxide and how many milliliters of 0.100 *M* *o*-nitrophenol (plus water as needed) are required to prepare 1.00 liter of a buffer of pH 7.00?

7.20 Calculate the pH of a solution prepared by dissolving 2.50 g each of pyridinium chloride and potassium hydroxide in enough water to make the solution volume equal to 2.00 liter.

7.21 What is the pH of a solution that is 1.0 *M* in nitrous acid and 1.0 *M* in sodium nitrite?

(a) If to 100 mL of this buffer is added 10 mL of 1.00 *N* HCl, by how many units will the pH change? Repeat the calculation for the following buffers:

(b) 0.1 *M* HNO_2, 0.1 *M* $NaNO_2$

(c) 0.05 *M* HNO_2, 0.15 *M* $NaNO_2$

(d) 0.15 *M* HNO_2, 0.05 *M* $NaNO_2$

Which of the solutions (b), (c), or (d) has the greatest buffer capacity?

7.22 Concentrated HCl is about 12 *M*, and 1 drop is about 0.05 mL. Calculate the pH before and after 1 drop of concentrated HCl is added to 50 mL of each of the following solutions.

(a) 1.00×10^{-4} *M* KOH

(b) a solution that is 0.0100 *M* in methylamine and 0.0500 *M* in methylammonium bromide

(c) pure water

7.23 How many grams each of ethylamine and ethylammonium iodide must be used to prepare 1 liter of a buffer of pH 10.25? The total buffer concentration is to be 0.080 M.

7.24 Calculate the pH of a solution if 1 liter contains:
(a) 3.00 g of 2-naphthoic acid
(b) 3.00 g of 2-naphthoic acid and 0.50 g of KOH
(c) 3.00 g of 2-naphthoic acid and 1.50 g of KOH

7.25 Use the BASIC Program II.3 of Appendix II to calculate pH values of various buffers and compare with the results of the Henderson-Hasselbalch equation. Vary the ratio of salt to acid and also their total concentration. Try to develop a feeling for when the simpler method is sufficiently accurate.

Chapter 8

Titrations

8.1 INTRODUCTION AND DEFINITIONS

According to its usual definition, a *titration* is a determination of the volume of a reagent solution which is stoichiometrically equivalent to a given sample of another reagent. For example, one might measure how many milliliters of 0.100 N KOH is required to neutralize exactly a sample containing an acid, and from this information determine how much acid was in the sample. Normally, this would be done by slowly dispensing the KOH solution from a 50-mL *buret*, which allows the experimenter to use any desired amount of reagent and to measure its volume to an accuracy of ±0.02 mL. The *equivalence point* of a titration is reached when an amount of base has been added which is stoichiometrically equivalent to the amount of acid in the sample.

As the KOH solution drains from the buret into the beaker containing the sample, the experimenter must make frequent observations of some physical property of the solution. The physical property, which may be color, pH, rate of heat release, and so on, is one that undergoes a more or less abrupt and easily noticeable change at or near the equivalence point. The point at which this change takes place is called the *end point* of the titration. It differs from the equivalence point because it occurs near but not necessarily *at* the point of true stoichiometric equivalence. The most commonly observed physical property is color; if the reagents themselves undergo no color change as the pH of the solution changes, a small amount of highly colored *indicator* substance is added to signal the end point.

The quantitative calculations are based on the normality of the solution, that is, on concentration expressed as *equivalents per liter* (see Chapter 3). It follows from this definition that the number of equivalents in a sample is equal to its volume times its normality.

$$\text{equiv. solute} = (\text{volume})(\text{normality}) \tag{8-1}$$

At the equivalence point of a titration of an acid with a base the number of equivalents of acid originally present in the sample is equal to the number of equivalents of base that has been added:

$$\text{equiv. acid} = \text{equiv. base} \tag{8-2}$$

$$V_a N_a = V_b N_b \tag{8-3}$$

where V and N are the volume in liters and the normality, and the subscripts a and b refer to acid and base. The object of a titration may be to determine concentration, molecular weight, total amount, weight percent acid, and so on. Furthermore, the technique is not limited to acid-base reactions, but is used also with redox reactions, complex-forming reactions, precipitation reactions, exothermic reactions, and others.

Example 8.1

If 37.85 mL of 0.0997 N base is required to titrate 25.0 mL of acid, what is the normality of the acid?

Answer: First we note that normality is defined as equivalents per *liter*, so we convert the given data to liters by the appropriate conversion factor

$$37.85 \text{ mL} = (37.85 \text{ mL})(0.001 \text{ liter/mL})$$

$$25.0 \text{ mL} = (25.0 \text{ mL})(0.001 \text{ liter/mL})$$

Next the given values are substituted into equation (8-2):

$$(25.0 \text{ mL})(0.001 \text{ liter/mL})N_a = (37.85 \text{ mL})(0.001 \text{ liter/mL})(0.0997 \text{ equiv./liter})$$

Dividing by (25.0)(0.001) yields

$$N_a = (0.0997 \text{ equiv./liter}) \frac{(37.85 \text{ mL})(0.001 \text{ liter/mL})}{(25.0 \text{ mL})(0.001 \text{ liter/mL})}$$

$$= 0.151 \text{ equiv./liter}$$

When the problem is set up in this way, it is obvious that the preliminary conversion from milliliters to liters was unnecessary; the conversion factors cancel out in the final step. This would have been true even if the volumes had been given in some peculiar unit, such as fluid ounces or cubic feet. When solving the equation $V_aN_a = V_bN_b$, *any* volume units may be used *without* conversion to liters, provided that V_a and V_b are expressed in the same units.

One very important use of titration is in determining the equivalent and/or molecular weight of an unknown acid or base. This is illustrated in Example 8.2.

Example 8.2

A sample of acid weighing 0.617 g is titrated and found to require 18.35 mL of 0.178 N base. Calculate the equivalent weight of the acid.

Answer: In the example the volume and normality of the acid are not given because the acid sample exists as a pure substance (either solid or liquid) rather than as a solution.

Equation (8-3) does not apply, and we must use the fundamental equation (8-2) of acid-base titrations.

$$\text{equiv. acid} = \text{equiv. base}$$

$$\frac{\text{wt acid}}{\text{equiv. wt}} = V_b N_b$$

We insert the data given in the problem:

$$\frac{(0.617\text{ g})}{\text{equiv. wt}} = (18.35\text{ mL})(0.001\text{ liter/mL})(0.178\text{ equiv./liter})$$

$$\frac{0.617\text{ g}}{(18.35\text{ mL})(0.001\text{ liter/mL})(0.178\text{ equiv./liter})} = \text{equiv. wt}$$

$$189\text{ g/equiv.} = \text{equiv. wt}$$

Note that we cannot ignore the volume units in this example. Had we omitted the factor (0.001 liter/mL), the answer would have been 0.189 g/equiv., which is of course an impossible value for a molecular or equivalent weight. *Units may be ignored only when the equation $V_aN_a = V_bN_b$ is used, for only then does the conversion factor cancel out.* According to the definition (Chapter 3), the equivalent weight of an acid is the weight, in grams, of the quantity which can supply 1 mol of protons in acid-base reactions. If the acid is monoprotic, 1 mol is the same as one equivalent, and the equivalent and molecular weights are the same. If the acid is diprotic, however, only ½ mol is required in order to supply 1 mol of protons, and 1 equivalent equals ½ mol. In this case the equivalent weight is only half the molecular weight, or the molecular weight is twice the equivalent weight. If the acid is triprotic, the molecular weight is thrice the equivalent weight. Thus in order to determine molecular weight from titration data it is necessary to know whether the unknown acid is mono-, di-, or triprotic. This can be determined from the shape of the titration curve, as will be shown later.

When an acid solution is titrated, its pH rises steadily as the base is added. At the equivalence point, when complete neutralization has been achieved, the pH is close to 7. (We shall see shortly that the pH at the equivalence point is exactly 7.00 only for *strong acids* titrated with *strong bases*.) This information can conveniently be summarized in a *titration curve*, which is a graph of pH against volume of added base. Figure 8.1 shows several such titration curves, calculated for the titration of 50 mL of a strong acid such as HNO_3 with a strong base such as KOH. In the solid curve the acid is 0.1 N, and so is the base. In the dashed curve both acid and base are 0.001 N.

Several aspects of Figure 8.1 should be noted. First let us consider the effect of concentration. Since HNO_3 is a strong acid it is completely ionized. The pH at the beginning of the titration is given by the negative logarithm of the concentration, and is 1.0 for 10^{-1} N acid and 3.0 for 10^{-3} N acid. The pH values of the two curves differ by two units at the high end of the titration also. In this region all the acid in the sample has been neutralized; the solution contains only the strong base (and the salt, which does not

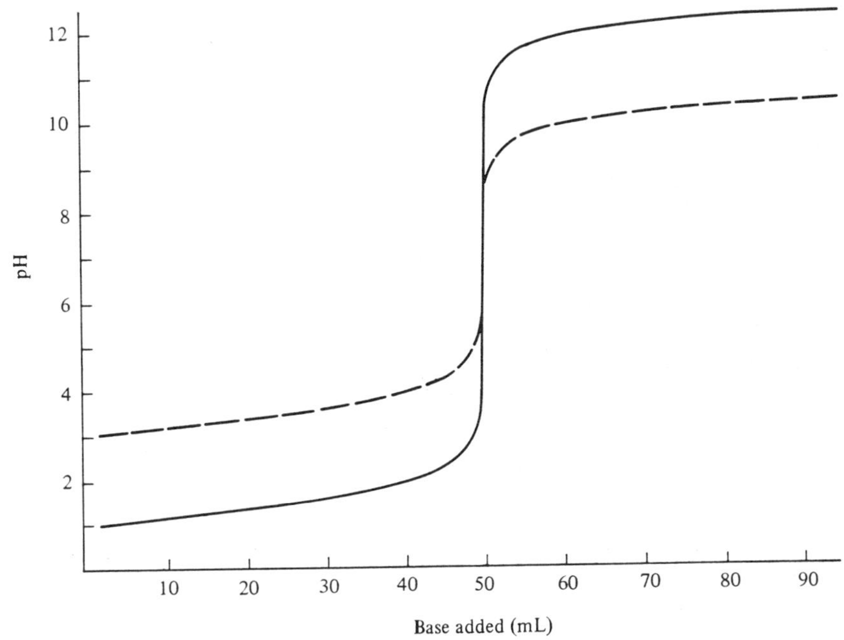

Figure 8.1 Titration of a strong acid with a strong base, showing pH as a function of volume of added titrant. Solid line, 50 mL of 0.1 *N* acid titrated with 0.1 *N* base; dashed line, 50 mL of 0.001 *N* acid titrated with 0.001 *N* base.

affect pH[1]). For one titration the titrant is 0.1 *N* KOH, pH = 13, and for the other it is 0.001 *N*, pH = 11. The titration curves show that the final pH is slightly less than the pH of the titrant. After 50 mL of titrant has been added, the solution is neutral and contains no excess OH^-. Addition of another 50 mL of titrant results in a solution that contains as much OH^- as 50 mL of titrant but whose volume is 150 mL. The *concentration* of OH^- is only one-third that which exists in the titrant, and the pH accordingly is less by about half a pH unit. Second, the shape of both curves changes drastically as the volume of added base approaches the equivalence point of 50.0 mL. Whereas during the first 40 mL the pH changes by less than 1 unit, the addition of the 49th and 50th mL causes the pH to rise from less than 3 all the way to 7. An additional 2 mL causes a further rise, to almost 12. This change in shape near the equivalence point is characteristic of all titrations and is referred to as the *break* in the curve. Titrations can be loosely categorized by the *sharpness* of their breaks. By "sharpness" we mean the steepness of the rise as the equivalence point is passed.

In the region near the end point, both curves look identical in Figure 8.1. In fact, the titration of the more concentrated acid has the sharper break. This is shown in Figure 8.2, in which the end-point region of Figure 8.1 is expanded. Here it is clear that the two curves are not the same.

[1] Hydrolysis of the salt of a strong acid does not take place to a significant extent. See Chapter 7, especially Example 7.5.

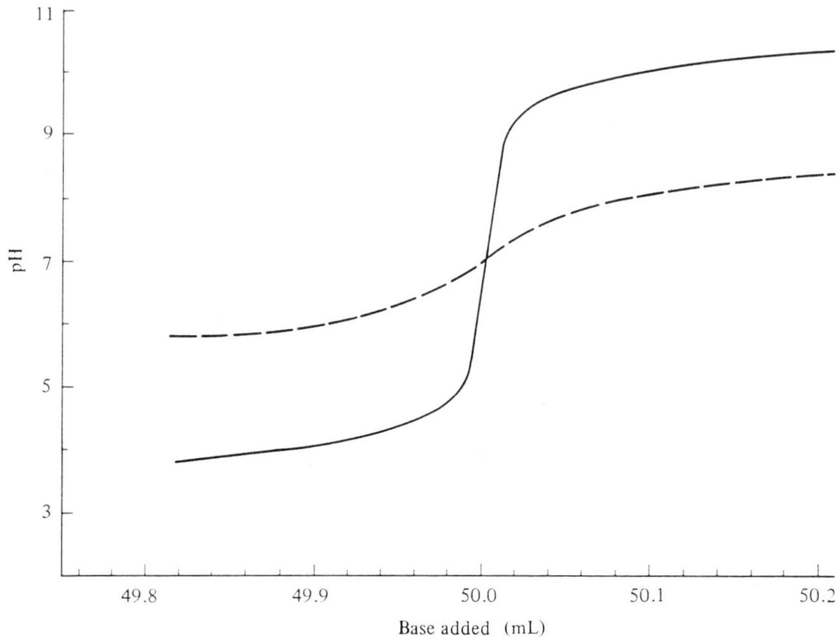

Figure 8.2 Titration of a strong acid with a strong base, showing pH as a function of added titrant. Solid line, 50 mL of 0.1 *N* acid titrated with 0.1 *N* base; dashed line, 50 mL of 0.001 *N* acid titrated with 0.001 *N* base. The region near the end point is expanded; otherwise, these are the same titrations as in Figure 8.1.

The break of titration curves can be illustrated in another way. The volume of a single drop of base from the type of buret most frequently used in titrations is about 0.05 mL. We can calculate, at any point in the titration, the *change* in pH (ΔpH) caused by the addition of a single additional drop of titrant; for sharply breaking curves this change in pH is larger. For instance, we calculate the quantity ΔpH = pH (after 40.05 mL of added base) − pH (after 40.00 mL of added base). The results of such calculations for the two titrations of Figure 8.1 are presented in Figure 8.3. Note that there is only one line in Figure 8.3. This is because the curves for the two concentrations, 0.1 *N* and 0.001 *N*, are superimposable except at the equivalence point. Between 50.00 and 50.05 mL of added titrant, the 0.001 *N* acid change is from 7.00 to 7.72, whereas for the 0.1 *N* acid the change is from 7.00 to 8.69. The arrows in Figure 8.3 indicate the maximum to which ΔpH rises in the titrations. The more concentrated the acid sample and the base titrant, the sharper is the break at the equivalence point.

8.2 CALCULATIONS

Before proceeding with the discussion of titration curves, we should give an account of how they can be calculated. It is convenient to consider separately the three cases (1) *undertitration*, or the addition of less than the equivalent amount of base; (2) equivalence point; and (3) *overtitration*, the addition of more than the equivalent amount of base.

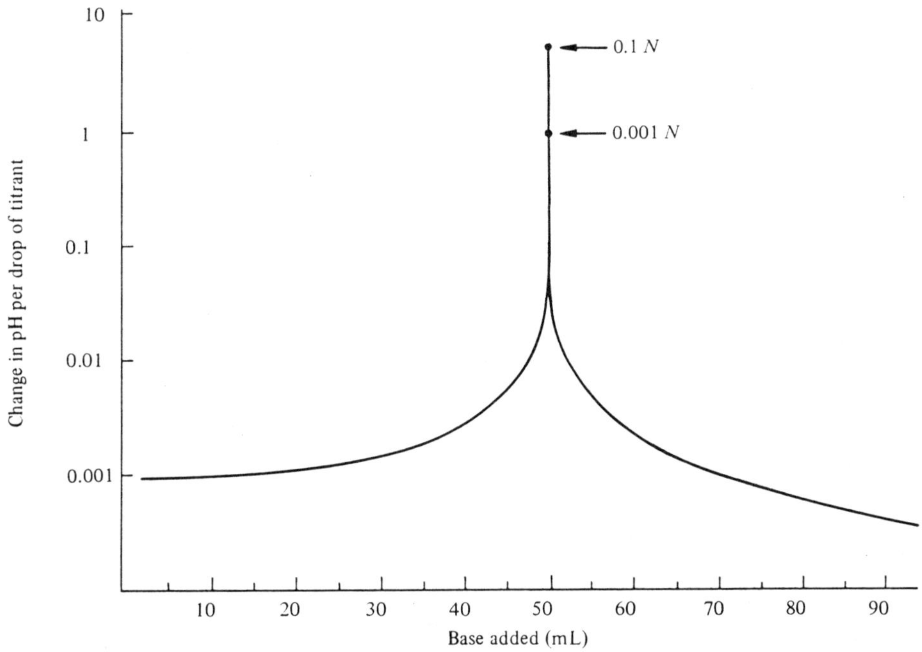

Figure 8.3 Titration of a strong acid with a strong base. pH change per drop (0.05 mL) of titrant, plotted as a function of total volume of added titrant; 50 mL 0.1 *N* acid titrated with 0.1 *N* base. The curve for 50 mL of 0.001 *N* acid titrated with 0.001 *N* base is superimposable except between 49 and 51 mL; there at the cusp ΔpH is about 1 unit less. The maximum points of the two curves are indicated by arrows.

1. If a strong acid is undertitrated (with a strong base the titrant is *always* strong in routine laboratory procedures), it is only partially neutralized. The solution contains both the unneutralized acid and the salt which is a product of the neutralization. Since the salt of a strong acid and a strong base does not hydrolyze to any significant extent, it can be ignored in pH calculations. All of the added base has reacted with the acid in the sample, so that the unneutralized acid alone determines the pH.

Example 8.3

Calculate the pH after 26.9 mL of 0.1 *N* KOH has been added to 40.0 mL of 0.2 *N* HCl.

Answer: First, convert the volumes to equivalents. It is very common and very convenient to use milliequivalents (1 meq = 0.001 equivalent) rather than equivalents, since in titrations, volumes are almost always expressed in milliliters rather than liters.

By definition:

$$\text{normality} = \frac{\text{equiv.}}{\text{liter}} = \frac{\text{equiv.}/1000}{\text{liter}/1000} = \frac{\text{meq}}{\text{mL}}$$

That is, a 0.1 N solution has 0.1 meq/mL as well as 0.1 equivalent/liter, so that

$$\text{meq} = \text{volume (mL)} \times \text{normality (meq/mL)}$$

In the example,

$$\begin{array}{lll} \text{meq acid in sample} & = (40.0\text{ mL})(0.2\ N) = & 8.00\text{ meq} \\ -\text{meq added base} & = (26.9\text{ mL})(0.1\ N) = & 2.69\text{ meq} \\ \hline \text{meq acid remaining} = & & 5.31\text{ meq} \end{array}$$

(2.69 meq KCl has been produced, but as mentioned above, it does not affect the pH.)

The concentration of remaining acid is calculated next; *note that the volume of the solution has increased.* This must not be overlooked in the calculation of concentration.

$$\frac{5.31\text{ meq}}{40.0\text{ mL} + 26.9\text{ mL}} = 0.0794\text{ meq/mL} = 0.0794\ N$$

Since HCl is a strong acid, $[H_3O^+]$ is equal to the normality, and

$$\text{pH} = -\log[H_3O^+] = -\log(0.0794) = 1.10$$

The calculation of the undertitration of a weak acid is similar but more complicated. Following the addition of a less-than-equivalent amount of base to a solution of a weak acid, the solution contains unneutralized acid and salt of the acid. As before, all of the added base has disappeared by reaction with the acid. However, in this case the salt is subject to hydrolysis because it is the salt of a weak acid. The solution contains the weak acid and its salt; it is a buffer solution, and its pH is calculated by the methods discussed in Chapter 7.

Example 8.4

Calculate the pH of a sample of 50.0 mL of 0.1 N *p*-iodobenzoic acid ($pK_a = 4.00$) following the addition of 32.0 mL of 0.1 N KOH.

Answer: Convert the volumes to milliequivalents and calculate the amount of salt produced and acid remaining.

$$\begin{array}{lll} \text{meq acid in sample} & = (50.0\text{ mL})(0.1\ N) = & 5.00\text{ meq} \\ -\text{meq base added} & = (32.0\text{ mL})(0.1\ N) = & 3.20\text{ meq} \\ \hline \text{meq acid remaining} = & & 1.80\text{ meq} \\ \text{meq salt produced} & = \text{meq based added} = & 3.20\text{ meq} \\ \text{total volume} & = 82.0\text{ mL} & \end{array}$$

Calculate the pH from the Henderson-Hasselbalch equation.

$$pH = pK_a + \log \frac{[\text{salt}]}{[\text{acid}]}$$

$$= 4.00 + \log \frac{3.20 \text{ meq}/82.0 \text{ mL}}{1.80 \text{ meq}/82.0 \text{ mL}} = 4.0 + \log \frac{3.20}{1.80}$$

$$= 4.00 + 0.25 = 4.25$$

(Note that the solution volumes cancel out and need not have been considered. This will always occur in the Henderson-Hasselbalch equation, but not if the exact method [equation (VI-15)] is used.)

If the acid is very dilute or if the pH is outside the range 3.0 to 11.0, it may be necessary to use the BASIC program (Appendix Section II.3) to solve the exact buffer calculation. This is rarely the case for an actual laboratory operation. As a matter of interest, the exact calculation of the problem in Example 8.4 gives pH = 4.2516. (Calculation of a pH value to four decimal places is unjustifiable, except to illustrate a point, when pK_a is known only to 0.01 unit.)

2. The pH at the equivalence point of the titration of a strong acid with a strong base is easily calculated: neutralization is complete and exact, and the solution contains only a salt which does not hydrolyze and whose presence does not affect the pH. The pH is therefore the same as that of neutral water, or 7.00 (at 25°C, at other temperatures it will not be exactly 7.00). The calculation is more complicated in the case of a weak acid. At the equivalence point the solution contains only the neutralization product, which is a salt of a weak acid and which hydrolyzes to a greater or lesser extent. The pH is calculated following the method of Example 7.1 as well as that of Example 8.5.

Example 8.5

Calculate the pH at the equivalence point of a titration of 30.0 mL of 0.0500 *N* *o*-methoxybenzoic acid ($K_a = 8.1 \times 10^{-5}$) with 0.100 *N* KOH.

Answer: At the equivalence point the solution contains only the salt of a weak acid, we first need to calculate the concentration of the salt, then its extent of hydrolysis. To calculate the concentration we find the number of equivalents and the volume.

$$\text{meq of acid in sample} = (30.0 \text{ mL})(0.0500 \text{ meq/mL}) = 1.50 \text{ meq}$$

Since at the equivalence point meq acid = meq base, the required volume of titrant is

$$\text{volume base} = \frac{1.50 \text{ meq}}{0.100 \text{ meq/mL}} = 15.0 \text{ mL}$$

The total volume is 30.0 mL + 15.0 mL = 45.0 mL and the concentration of the salt, potassium *o*-methoxybenzoate, is

$$[o\text{-methoxybenzoate}] = \frac{1.50 \text{ meq}}{45.0 \text{ mL}} = 0.0333\ N$$

Next we calculate the extent of hydrolysis, using the hydrolysis constant K_h. K_a for this acid is 8.1×10^{-5}, so that K_h for its salt is $K_w/K_a = 1.0 \times 10^{-14}/8.1 \times 10^{-5} = 1.2 \times 10^{-10}$.

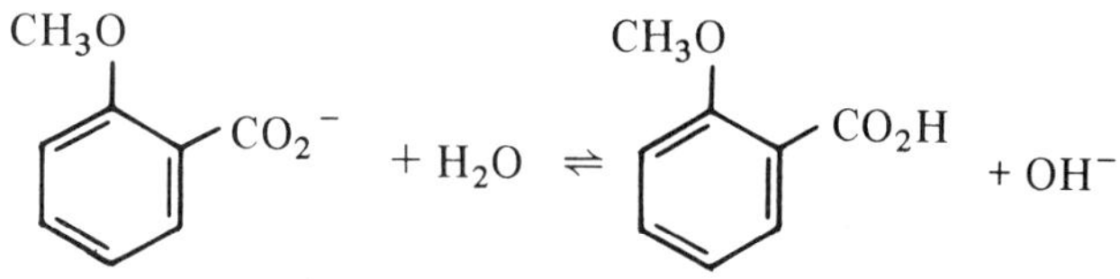

initially:	0.0333	0	0
at equilibrium:	$0.0333 - x$	x	x

$$K_h = \frac{x^2}{0.0333 - x} \approx \frac{x^2}{0.0333}$$

$$x^2 = (K_h)(0.0333)$$

$$x = [OH^-] \approx \sqrt{(1.2 \times 10^{-10})(0.0333)} = 2.00 \times 10^{-6}$$

Finally, the pH is calculated from the extent of hydrolysis:

$$pOH = -\log(2 \times 10^{-6}) = 5.70$$

$$pH = 14 - pOH = 8.30$$

Although it may seem a little strange to follow a calculational procedure that first assumes complete conversion of the weak acid to its salt and then considers the re-formation of a small amount (about 0.006% in Example 8.5) of the acid by hydrolysis of the salt, we are simply making use of the fact that there are *no* differences between a salt solution prepared by dissolving the salt and one prepared by neutralizing the acid. Since the calculations are easier in the former instance, that is the way it should be done.

3. When the sample is overtitrated, all the acid has reacted; the solution contains excess strong base and the salt produced in the neutralization. If the original sample was a strong acid, its salt does not hydrolyze and does not affect the pH. We need only determine the concentration of excess base to calculate pH.

Example 8.6

Calculate the pH after 37.2 mL of 0.100 N KOH has been added to a sample of 25.0 mL of 0.125 N HCl.

Answer:

meq added base	= (37.2 mL)(0.100 meq/mL)	= 3.72 meq
−meq original acid	= (25.0 mL)(0.125 meq/mL)	= 3.13 meq
meq excess base		= 0.59 meq

total volume = 62.2 mL

$$[OH^-] = \frac{0.59 \text{ meq}}{62.2 \text{ mL}} = 0.0095\ N$$

$$pOH = -\log[OH^-] = 2.02$$

$$pH = 14.0 - pOH = 11.98$$

The calculation does not change if the sample being titrated is a weak acid. It is true that the salt produced is itself a weak base and thus subject to hydrolysis.

$$A^- + H_2O \rightleftharpoons HA + OH^- \tag{8-4}$$

However, the presence, beyond the equivalence point, of excess OH^- from the titrant represses the hydrolysis by driving the equilibruim (8-4) to the left. The amount of OH^- produced by reaction (8-4) is negligibly small in comparison to the amount supplied by the excess titrant. The system behaves as if there were no hydrolysis and the calculation is the same as for a strong acid. Only if the acid being titrated is very weak, with pK_a greater than about 10, does this method give inaccurate results. Consider the titration of 50 mL of each of a series of 0.1 N acid samples having pK_a values of -2 (strong acid), 2, 4, 6, 8, and 10. If the titrant is 0.100 N KOH, the equivalence point occurs at the addition of 50 mL of base. Calculation of the pH after the addition of, say, 56 mL of base by the method of Example 8.6 gives pH = 11.75. The result is the same for all pK_a values, because we are treating the overtitration calculation as if no salt hydroylsis occurs. The exact calculation also gives pH = 11.75 for all acids except the one with $pK_a = 10$, for which the result is pH = 11.81. The discrepancy increases as pK_a increases above 10.

The exact calculation of the pH of an overtitrated weak acid is similar to the exact treatment of buffer systems (see Chapter 7) and is outlined very briefly here. As in buffer systems, the calculation is based on conservation of mass and charge. The amount of acid, HA, plus the amount of salt, A^-, is unchanged by the addition of titrant. The concentration of acid plus salt is given by the amount of acid originally present in the sample, $M_a V_a$, divided by the total volume of sample (V_a) plus the volume of added base (V_b). Conservation of mass, expressed in terms of concentration, is

$$[HA] + [A^-] = \frac{M_a V_a}{V_a + V_b} \tag{8-5}$$

Conservation of charge requires that

$$[H_3O^+] + [Na^+] = [A^-] + [OH^-] \tag{8-6}$$

Sodium ion comes from the titrant, which we have assumed to be NaOH. The concentration of sodium ion is given by

$$[Na^+] = \frac{M_b V_b}{V_a + V_b} \tag{8-7}$$

The two equilibrium expressions (8-8) and (8-9),

$$K_w = [H_3O^+][OH^-] \tag{8-8}$$

$$K_a = \frac{[H_3O^+][A^-]}{[HA]} \tag{8-9}$$

are combined with (8-5), (8-6), and (8-7) to give

$$[H_3O^+]^3 + [H_3O^+]^2\left[\frac{M_a V_b}{V_a + V_b} + K_a\right] - [H_3O^+]\left[\frac{K_a(M_a V_a - M_b V_b)}{V_a + V_b} + K_w\right] - K_w K_a = 0 \tag{8-10}$$

This equation may be solved by the methods of Appendix II.

Equation (8-10) is appropriate for the undertitration and equivalence point calculations as well, provided that nothing other than the original acid sample and the titrant is present in the solution. (The derivation of the exact solution has been outlined for the

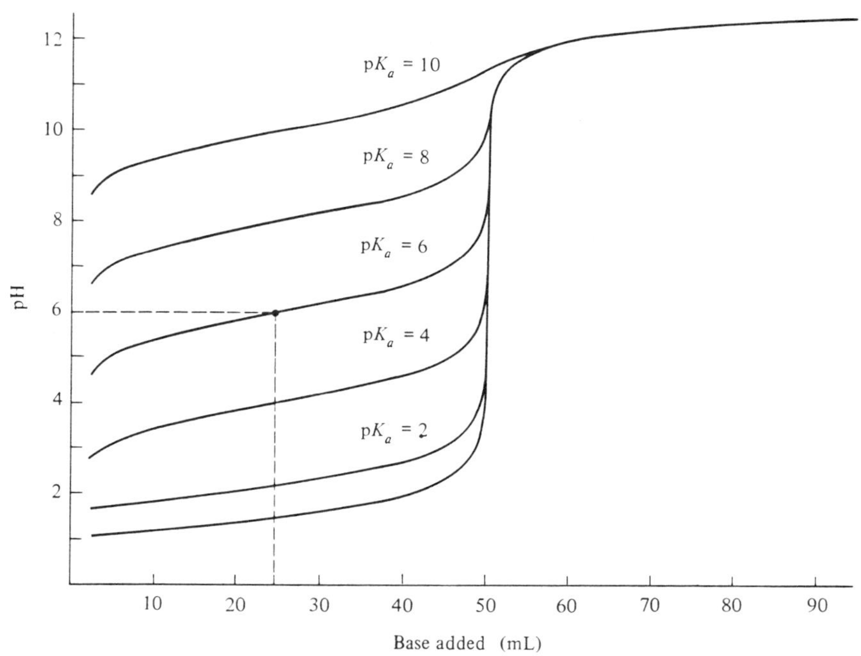

Figure 8.4. Titration of a weak acid with a strong base: pH as a function of added base. The 50 mL acid sample is 0.100 *N*; the titrant is 0.100 *N* base. The number above each curve gives the pK_a of the acid. The lowest curve is for a strong acid. The half-titration point is marked for pK_a = 6. At this point [HA] = [A$^-$] and pH = pK_a.

sake of completeness. The simpler methods given earlier in this chapter are much easier to use and sufficiently accurate for the calculation of titration curves.)

8.3 WEAK ACID TITRATION CURVES

Following the scheme outlined in the preceding section we can calculate titration curves for weak acids of various concentrations and pK_a values. Figure 8.4 shows some examples. Since the sample volume and titrant concentration are the same in all cases, the curves become superimposable very soon after the equivalence point is reached. Several points can be noted from Figure 8.4. First, as the acid gets weaker, the break becomes less sharp and moves to higher pH. Second, at the half-titration point, when exactly half of the acid has been neutralized, the concentrations of acid and salt are equal; [HA] = [A^-]. It follows from the Henderson-Hasselbalch equation that at this point pH = pK_a. This point is marked on the curve for pK_a = 6. The pK_a of an unknown acid can be estimated at a glance from the titration curve.

The relative sharpness of the break for acids of differing strengths is shown in Figure 8.5, where pH change per drop (0.05 mL) of titrant near the equivalence point is plotted. For an acid of moderate strength, with $K_a = 10^{-4}$, the pH change per drop at the equivalence point is about 2 units. This is adequate to produce a striking color change in most indicators (see Section 8.4). However, for an acid of $K_a = 10^{-8}$ the maximum pH change per drop is only about 0.05 pH unit, which is not enough to produce a large color change

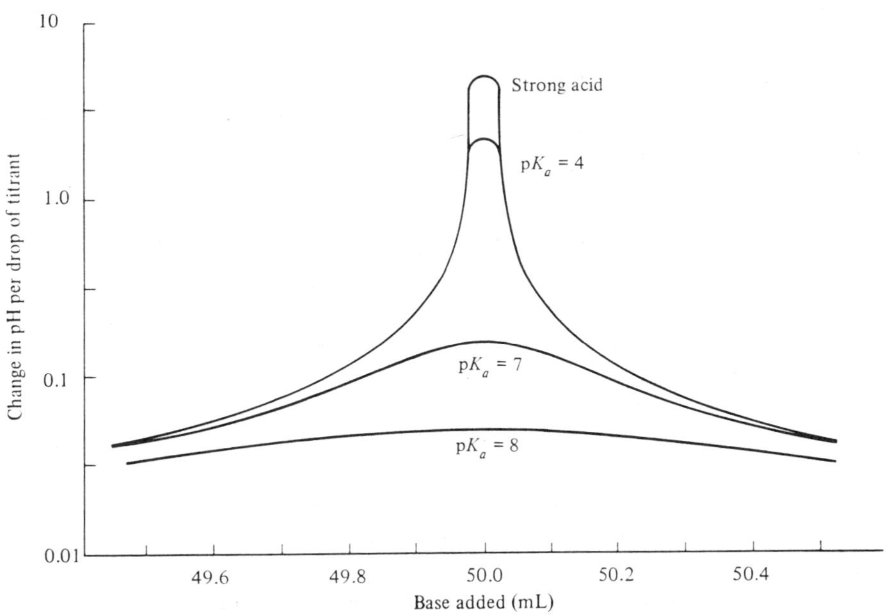

Figure 8.5 Titration of acids of various strength with a strong base: pH change per drop (0.05 mL) of 0.1 *N* base, plotted as a function of total volume of added titrant. The top curve is for a strong acid; the others are identified by the value of pK_a. The curves are shown in expanded scale near the equivalence point.

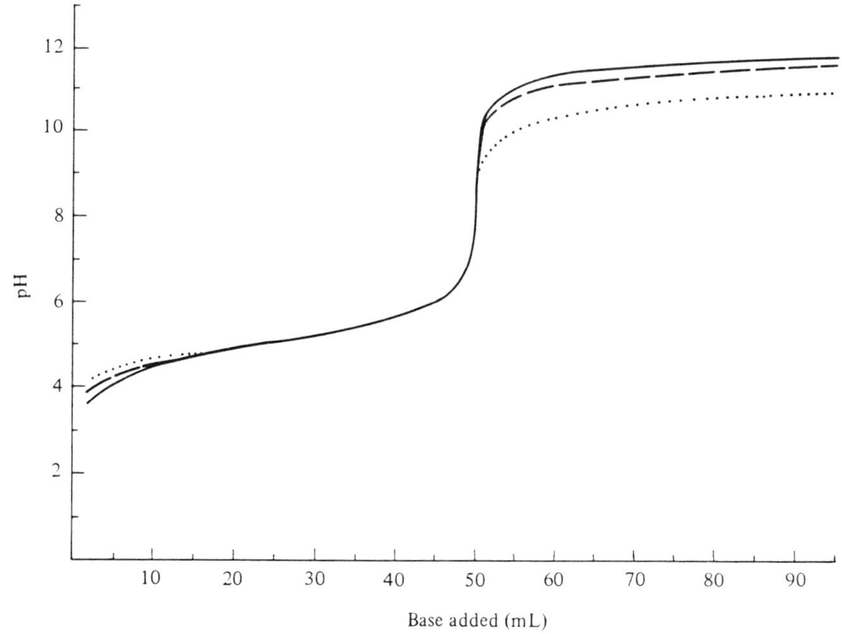

Figure 8.6 Titration of a weak acid ($pK_a = 5$) at several concentrations. The titrant is 0.01 *N* strong base in each case. Solid line, 0.1 *N*, 5 mL; dashed line, 0.01 *N*, 50 mL; dotted line, 0.001 *N*, 500 mL.

at the equivalence point. Furthermore, the pH change per drop at the equivalence point is not greatly different from that several milliliters earlier or later in the titration. Such titrations require excellent equipment and lab technique, and cannot be done by visual observation of color change.

The acid concentration has less effect on the form of the titration curve in the case of weak acids than it has for strong acids. Figure 8.6 shows a weak acid, $pK_a = 5$, being titrated with 0.01 *N* strong base. Each of the three curves refers to the same *amount* (0.5 meq) of acid, but the concentrations are 0.1 *N*, 0.01 *N*, and 0.001 *N*. The early portions of the curves are essentially the same, because an undertitrated weak acid is a buffer whose pH depends on the relative amounts of HA and A^-, and not on their concentrations. The curves diverge beyond the end point because the solution is no longer a buffer (it contains salt and base, not salt and acid), and the excess titrant is diluted to different extents in the three cases.

Our discussion so far has been limited to the titration of acid samples with strong bases. Although the titration of weak acids with weak bases is of no practical importance and will not be discussed, the titration of moderately to weakly basic samples with strong acids is a common laboratory practice. Such titrations are analogous to the titration of a weak acid with a strong base. Figures 8.1 to 8.5 remain unchanged if we substitute pOH for pH, K_b for K_a, and interchange the words "acid" and "base" in the figure axes and their captions. All calculations are likewise unchanged, after the foregoing substitutions have been made.

Example 8.7

A 100.0-mL sample of 0.01 N benzamidine,[2] whose pK_b is 2.6 at 20°C, is titrated with 0.1 N HCl. Calculate the pH after the addition of (a) 5 mL, (b) 8 mL, (c) 10 mL, and (d) 15 mL of titrant.

Answer:

(a)

$$\begin{array}{lll} \text{amount of original base} & = (100 \text{ mL})(0.01 \text{ meq/mL}) & = 1.00 \text{ meq} \\ -\text{amount of added acid} & = (5.00 \text{ mL})(0.1 \text{ meq/mL}) & = 0.50 \text{ meq} \\ \hline \end{array}$$

The calculation need not go any further; we notice that the titration is exactly half complete, and according to the Henderson-Hasselbalch equation, as modified for base buffers (7-31),

$$pOH = pK_b = 2.6$$

(This corresponds to the relationship for acid buffers; $pH = pK_a$ when $[HA] = [A^-]$.)

We are asked for pH, and we note that the temperature is not 25°C; therefore, $pK_w \neq 14.0$. From Appendix I we find $K_w = 6.81 \times 10^{-15}$, $pK_w = 14.2$, and

$$pH = pK_w - pOH = 14.2 - 2.6 = 11.6$$

(b) Calculate the pH after the addition of 8 mL of titrant.

$$\begin{array}{lll} \text{amount of original base} & = (100 \text{ mL})(0.01 \text{ meq/mL}) & = 1.00 \text{ meq} \\ -\text{amount of added acid} & = (8.00 \text{ mL})(0.1 \text{ meq/mL}) & = 0.80 \text{ meq} \\ \hline \text{amount of base remaining} & = & = 0.20 \text{ meq} \\ \text{amount of salt formed} & = & = 0.80 \text{ meq} \\ \text{total volume} & = 108 \text{ mL} & \end{array}$$

Since the pK_b is known only to 0.1 unit, there is no need for a precise calculation; the base buffer equation is adequate.

$$pOH = pK_b + \log \frac{[\text{salt}]}{[\text{base}]}$$

$$= 2.6 + \log \frac{[0.80]/\cancel{108 \text{mL}}}{[0.20]/\cancel{108 \text{mL}}} = 2.6 + 0.6 = 3.2$$

[2] Benzamidine is an unusually strong organic base because the cation is stabilized by resonance. The extra stability imparted by resonance pulls the ionization equilibrium to the right.

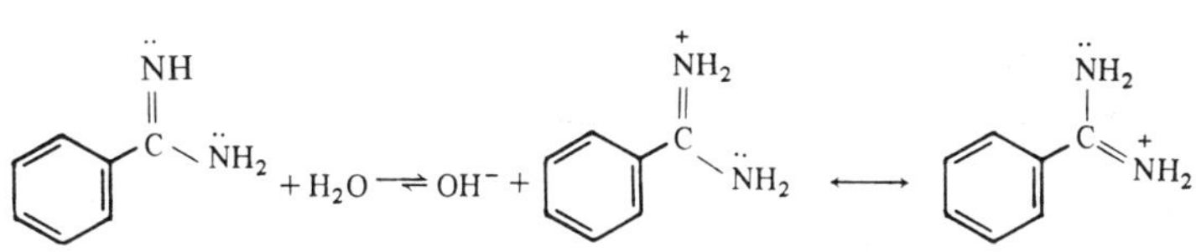

$$\mathrm{pH} = 14.2 - \mathrm{pOH} = 11.0$$

Note that this longer method could also be used to solve for pOH in part (a), since [salt] = [base] and log(1) = 0.

(c) Calculate the pH after the addition of 10 mL of titrant.

$$\begin{array}{lll} \text{amount of original base} & = (100\ \text{mL})(0.01\ \text{meq/mL}) & = 1.00\ \text{meq} \\ -\text{amount of acid added} & = (10.0\ \text{mL})(0.1\ \text{meq/mL}) & = 1.00\ \text{meq} \\ \hline \end{array}$$

This is the equivalence point; the solution consists of 1.00 meq of the salt (benzamidinium chloride) dissolved in 110 mL of water, or a concentration of 0.00909 N.

$$K_h = \frac{K_w}{K_b} = \frac{6.81 \times 10^{-15}}{\text{antilog}(-2.6)}$$

$$= \frac{6.81 \times 10^{-15}}{2.51 \times 10^{-3}} = 2.71 \times 10^{-12}$$

Now solve the salt hydrolysis problem.

$$\begin{array}{lcccc} & C_7H_8N_2H^+ + H_2O & \rightleftharpoons & C_7H_8N_2 & + \ H_3O^+ \\ \text{initially:} & 0.00909 & & 0 & 0 \\ \text{at equilibrium:} & 0.00909 - x & & x & x \end{array}$$

$$K_h = \frac{x^2}{0.00909 - x} \approx \frac{x^2}{0.00909}$$

$$x^2 = 0.00909 K_h = (0.00909)(2.71 \times 10^{-12})$$

$$x = 1.57 \times 10^{-7}\ M = [H_3O^+]$$

$$\mathrm{pH} = -\log(1.57 \times 10^{-7}) = 6.8$$

(d) Finally, we calculate the pH after 15 mL of acid has been added.

$$\begin{array}{lll} \text{amount of added acid} & = (15.0\ \text{mL})(0.1\ \text{meq/mL}) & = 1.50\ \text{meq} \\ -\text{amount of original base} & = (100\ \text{mL})(0.01\ \text{meq/mL}) & = 1.00\ \text{meq} \\ \hline \text{amount of excess strong acid} & & = 0.50\ \text{meq} \end{array}$$

The salt benzamidinium chloride is a very weak acid, which is apparent from the fact that its K_a (i.e., its K_h)[3] is about 10^{-12}; its contribution to acidity is negligible in com-

[3] There are some possibilities for confusion by the two different symbols commonly used for this equilibrium constant. If the reaction of the salt $BH^+ + H_2O \rightleftharpoons B + H_3O^+$ is discussed in the context of hydrolysis reactions, its equilibrium constant is denoted K_h. If it is desired to emphasize that the reaction is in fact an acid reaction, it is usual to denote the same equilibrium constant by K_a.

parison to the 0.50 meq of HCl present. The acidity is therefore

$$[H^+] = \frac{0.50 \text{ meq}}{115 \text{ ml}} = 4.35 \times 10^{-3}\ M$$

$$pH = -\log(4.35 \times 10^{-3}) = 2.4$$

8.4 INDICATORS

An acid-base titration indicator is a very strongly colored substance, whose color when dissolved in acidic solutions is strikingly different from that when dissolved in basic solutions. A familiar example is litmus, which is blue in basic solutions and pink in acid solutions. Another example is phenolphthalein, which is colorless in acid solutions and red in basic solutions.

Indicators are themselves weak acids, or, less commonly, weak bases. As such, they exist in protonated and nonprotonated forms, which are in equilibrium in solution. If the neutral form of the indicator loses a proton, the equilibrium may be represented as

$$HIn + H_2O \rightleftharpoons In^- + H_3O^+ \tag{8-11}$$

which is similar to all other weak acid ionizations.

The indicator methyl red is an acid of this type.

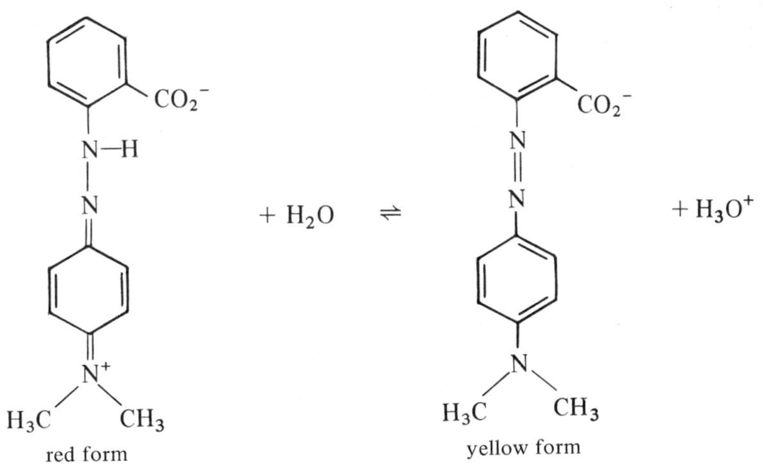

Addition of base to a solution of methyl red causes the conversion of the red form to the yellow, by removing the acidic proton from the nitrogen atom.

$$\underset{\text{red}}{HIn} + OH^- \rightarrow \underset{\text{yellow}}{In^-} + H_2O \tag{8-12}$$

If, on the other hand, the neutral form is a proton acceptor, the equilibrium may be represented as

$$In + H_2O \rightleftharpoons HIn^+ + OH^- \tag{8-13}$$

Here the indicator is a base; an example of this type of indicator is *N*, *N*-dimethyl-2,4,6-trinitroaniline.

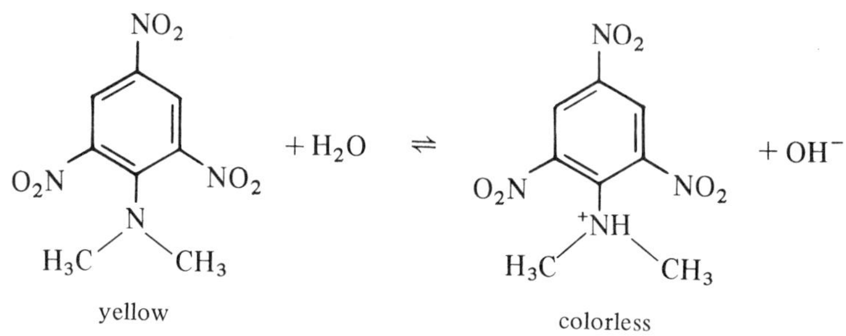

Addition of acid to a solution of *N*, *N*-dimethyl-2,4,6-trinitroaniline causes a conversion of the yellow form to the colorless form.

$$\underset{\text{yellow}}{In} + H_3O^+ \rightleftharpoons \underset{\text{colorless}}{HIn^+} + H_2O \tag{8-14}$$

Most of the commonly used indicators are acids, so our general discussion will be based on the equilibrium (8-11). As always, however, all aspects of the discussion may be transferred to bases by making the substitutions OH^- for H^+, K_b for K_a, and so on.

In the treatment of buffer solutions we were concerned with the way in which the relative concentrations of the protonated and deprotonated forms affect the pH. In titrations, however, indicators are present in such small concentrations (typically 10^{-10} *M* or less) that they have little effect on pH, and the focus becomes: How does pH affect the relative concentration of the protonated (HIn) and deprotonated (In^-) forms? It is, of course, the relative amount of the two differently colored forms which determines the overall color of the solution.

The anionic form of the indicator methyl red is yellow in color. We will assume that if a solution contains as much as 90% of one form, it will have the color characteristic of that form. (This is introduced as an assumption because color perception varies from one observer to the next. Also, indicators vary in the strength of their colors, and indicator concentration affects color. The 90% figure is arbitrary, and can differ among different situations.) The pK_a of methyl red is about 4.8, from which we can calculate the pH at which the concentration ratio will be $90{:}10 = [In^-]:[HIn]$.

$$pH = pK_a + \log \frac{[\text{salt}]}{[\text{acid}]}$$

$$= 4.8 + \log \frac{90}{10} = 5.8$$

Thus at a pH of 5.8, *or any larger value*, as the solution becomes increasingly basic, the solution contains at least 90% In^-; it has the color characteristic of In^-, in this case yellow. Similarly, we calculate the pH at which the solution contains 90% HIn:

$$pH = pK_a + \log \frac{[\text{salt}]}{[\text{acid}]}$$

$$= 4.8 + \log \frac{10}{90} = 3.8$$

At a pH of 3.8, or any lesser value, at least 90% of the indicator is in the HIn form, and the color is red. If the pH of the solution steadily increases, as in the titration of an acid, then during the early part of the titration the solution is red. As the pH goes through the range 3.8 to 5.8, the color changes from red to yellow. The pH region in which the color undergoes the change is called the *pH range* of the indicator, or sometimes the *indicator range*. Since indicators and observers vary, the pH range of all indicators is not necessarily 2 units wide as is the case in our example, nor is it necessarily centered on the pK of the indicator. For these reasons, when indicators are listed in handbooks and textbooks, they are arranged by pH range than by pK_a. Furthermore, different tables may give slightly different ranges for the same indicator. The pH range of methyl red, for example, is variously given as 4.0 to 6.0, 4.2 to 6.3, and 4.2 to 6.2. This small uncertainty in range fortunately has no effect on the usefulness of indicators in acid-base titrations.

The titration curve for a strong acid (Figure 8.1) shows that the pH changes from about 2 to 11 very abruptly near the equivalence point. In fact, the final drop of titrant, which brings the total added base from 49.95 to 50.00 mL,[4] causes a pH change from 4.3 to 7.0. Thus a single drop of titrant causes a complete color change of the methyl red, and the end point is established as closely as can be done without trying to manipulate fractions of a drop. Note that the end point (the point of color change) and the equivalence point (added base equals originally present acid) are reached *in the same drop* in this example.

The situation can be quite different if the sample is a weak acid with, for instance, $K_a = 10^{-5}$. At the point of *half-titration*, pH = pK_a = 5.0, and the color of a methyl red indicator would be about half changed. At the equivalence point the pH is 8.85, and the color change has been completed well before the equivalence point has been reached. Not only does the end point fail to coincide with the equivalence point, but the color change (which occurs approximately between 10 and 95% titration) is so gradual that end "*point*" is scarcely a useful description. Suppose, however, that instead of methyl red the indicator phenolphthalein is used. Phenolphthalein is colorless in acid media, red-purple in base, and has a pH range of 8.0 to 9.8. If the acid being titrated is strong, a single drop (50.00 to 50.05 mL) changes the pH from 7.00 to 9.70. The pH change lacks 0.1 unit of completely encompassing the indicator's pH range, and the color change is essentially complete. The end point and equivalence point again coincide within an accuracy of 1 drop. In the weak acid example the pH reaches 8.00 after 49.95 mL of added base; after 50.00 mL, pH = 8.85 (note from the methyl red discussion that this is the *equivalence point*); after 50.05 mL, pH = 9.7. So an acid of pK_a = 5 takes about twice as much (2 drops) titrant to effect a phenolphthalein color change as does a strong acid.

[4] In this calculation, and all others in the section on indicators, we assume an acid sample of 0.100 N, 50.0 mL, and a titrant concentration of 0.100 N.

Although the end point may no longer be said to *coincide* with the equivalence point, it is certainly no more than 1 drop away. One should bear in mind that, with a total titrant volume of 50 mL, an error of 1 drop is a percentage error of only 0.1%, well within normal laboratory requirements.

The preceding arguments are much easier to appreciate and remember if they are presented in pictorial form. In Figure 8.7 we show weak-acid titration curves (similar to those in Figure 8.4), together with an overlay of the indicator pH ranges of methyl red and phenolphthalein. If we follow the pH of the acid with $pK_a = 5$ as it changes during the titration, we see that the pH enters the methyl red range shortly after the titration begins. The pH leaves this range, with the solution color completely changed, several milliliters before the equivalence point. However, since (except for exceedingly weak acids) the pH rises sharply immediately after the equivalence point, an indicator with a high pH range, such as phenolphthalein, will give a "good end point." A good end point is one that occurs very close to the equivalence point. As can be see from Figure 8.7, this will occur whenever the pH range of the indicator lies well above the pK_a of the acid being titrated. Many indicators are available, with pH ranges 1.5 to 2 units wide, covering the span from pH = 0 to 14. With proper attention to the choice of indicator, weak acids can be titrated with considerable accuracy. For strong acid titration, indicator choice is not a problem; any one of many indicators will serve.

Some indicators in common use are polyprotic acids, that is, acids having two or more ionizable hydrogens. Such indicators have more than two forms, whose relative concen-

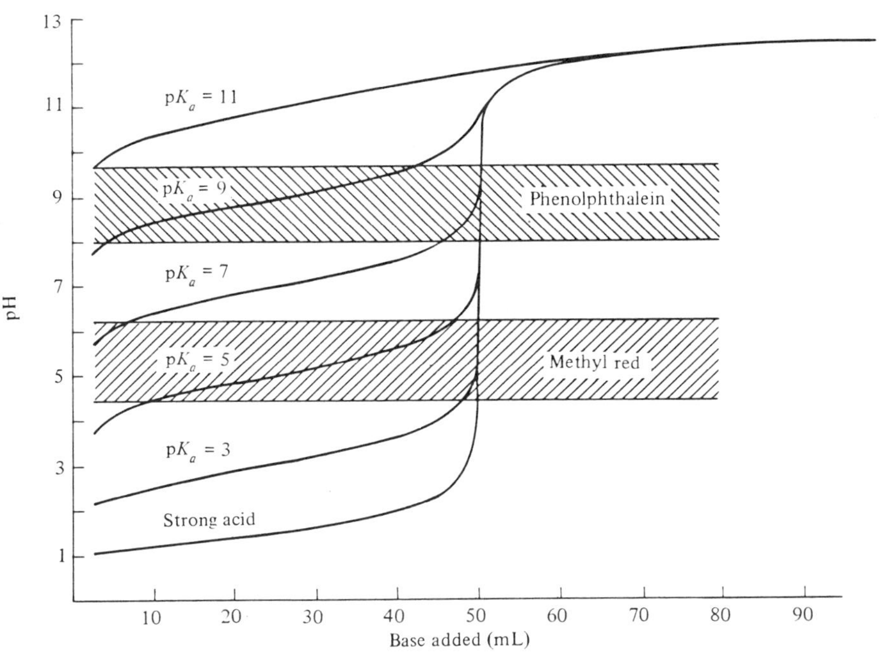

Figure 8.7 Weak acid titration curves compared to indicator pH ranges. All acid samples are 0.100 *N*, 50 mL, and the titrant is 0.100 *N* strong base. The number above each curve gives the acid pK_a, and the shaded regions show the pH range in which the color change of the designated indicator occurs.

trations change according to the pH of the solution. Although the mathematical treatment is different, the result is the same: there is one color in acid media and another color in basic media, regardless of how many protons are required for the transition. The pH range of polyprotic indicators is in general no wider than that of monoprotics. From a practical laboratory point of view it makes no difference whatsoever.

8.5 VERY WEAK ACIDS

It is apparent from Figures 8.4, 8.5, and 8.7 that it is difficult to measure the end point of a titration of a very weak acid. If the pK_a is much larger than 9 or 10, it may be difficult even to notice the end point. However, the conjugate-pair relationship between an acid and its salt suggests an alternate approach to pK_a determination. In this modification the weak acid is treated with a strong base and converted to its salt. For instance, phenol (pK_a = 10.0) is converted to sodium phenoxide:

$$C_6H_5OH + NaOH \longrightarrow C_6H_5ONa + H_2O$$

The phenoxide ion, as the salt of a very weak acid, is a base of moderate strength ($K_b = K_w/K_a = 10^{-14}/10^{-10} = 10^{-4}$, $pK_b = 4$). The titration of sodium phenoxide with a strong acid shows a relatively sharp break, exactly like the titration of an acid with $pK_a =$ 4. The conjugate acid-base pair relationship is such that if an acid has a pK_a greater than 7, a sharper end point will be observed in the titration of the conjugate base.

PROBLEMS

8.1 Calculate the normality of the following solutions.
(a) 3.72 g of NaOH in 75 mL of water
(b) 0.0537 g of HNO_3 in 8.52 mL of water
(c) 1.46 g of H_2SO_4 in 25.00 mL of water
(d) 5.72×10^{-1} g of $Ba(OH)_2$ in 230 mL of water

8.2 Distinguish between "equivalence point" and "end point" in an acid-base titration.

8.3 Titration of a 10.0 mL sample of base required 26.45 mL of 0.0987 *N* acid. Calculate the normality of the base.

8.4 How much 0.503 *N* acid is required to titrate 20.0 mL of 0.0783 *N* base?

8.5 An unknown monoprotic acid weighing 0.912 g was titrated with 0.107 *N* base and found to require exactly 62.7 mL to reach the end point. Calculate the number of equivalents. What is the molecular weight? The equivalent weight?

8.6 Repeat Problem 8.5, assuming that the acid is:
(a) diprotic
(b) triprotic

8.7 Discuss the statement: "In a titration of a weak acid with a strong base the end point is reached before the equivalence point."

8.8 When a weak acid is titrated with a strong base, does the equivalence point occur at a pH greater than, less than or equal to 7? Does your answer depend on temperature? Suppose that the titration is of a weak base with a strong acid; does this affect the answer?

8.9 Verify the "last-drop" calculation of Section 8.1. That is, in the titration of 50.0 mL of 0.001 N strong acid with 0.001 N strong base, calculate the pH after the addition of 49.5, 50.00 and 50.05 mL of base.

8.10 Show that in the titration of a weak acid with a strong base of the same normality, the pH at the equivalence point is given by the expression

$$\text{pH} = \frac{1}{2}\left(\log \frac{C}{2} + \text{p}K_a + \text{p}K_w\right)$$

where C is the normality of both acid and titrant.

8.11 To "standardize" a titrant solution means to make an accurate determination of its normality by using it to titrate an accurately known number of moles of acid (or base). A convenient "standard substance" for this purpose is potassium hydrogen phthalate (KHP), a monoprotic acid of molecular weight 204.22 g/mol. A student prepares a NaOH solution of approximately 0.5 M, and finds that exactly 32.15 mL is required to titrate 3.237 g of KHP. What is the exact molarity of the base?

8.12 Consider the titration of a 50.0 mL sample of a weak acid ($\text{p}K_a = 4.0$) of normality C_a with 0.1 N strong base. Calculate the pH under each of the following conditions.
(a) $C_a = 0.1\ N$; volume of added base = 30, 50, and 70 mL.
(b) $C_a = 0.01\ N$; volume of added base = 3, 5, and 7 mL.

8.13 Repeat Problem 8.12 for an acid of $\text{p}K_a = 8.0$.

8.14 Repeat Problem 8.12 for a strong acid.

8.15 Sketch (without detailed calculations) a curve for the titration of 100 mL of 0.1 N weak base ($K_b = 10^{-5}$) with a strong acid. On the same sketch include the titration curve for a strong base.

Chapter 9

Polyprotic Acids and Amphiprotic Substances

9.1 POLYPROTIC ACIDS

For those acids having more than one ionizable hydrogen (i.e., polyprotic acids), our thinking and our calculations will necessarily be more complicated than for monoprotic acids. In Chapter 3 we learned to regard, for calculational purposes, the process of weak acid ionization as a separable two-step sequence; first the acid dissolves, then it ionizes and comes to equilibrium. We learned further that this model leads to numerically correct answers because of an important principle of thermodynamics, namely, that the parameters of an equilibrium system do not depend on the path by which the system reaches equilibrium. In discussing polyprotics we will find it useful to adopt the same approach, modified to recognize that ionization does in fact take place in sequential steps, one corresponding to each of the ionizable protons. Thus when phosphoric acid ionizes, it does so in three chemically distinct steps:

$$H_3PO_4 + H_2O \rightleftharpoons H_2PO_4^- + H_3O^+ \qquad K_1 \qquad (9\text{-}1)$$

$$H_2PO_4^- + H_2O \rightleftharpoons HPO_4^{2-} + H_3O^+ \qquad K_2 \qquad (9\text{-}2)$$

$$HPO_4^{2-} + H_2O \rightleftharpoons PO_4^{3-} + H_3O^+ \qquad K_3 \qquad (9\text{-}3)$$

If the polyprotic acid is weak, as is phosphoric acid, there is an acid ionization constant associated with each of the ionization steps. These are denoted by K_1, K_2, and K_3. Similarly, for the diprotic carbonic acid, ionization is a two-step sequence and the separate acid ionization constants are denoted by K_1 and K_2.

$$H_2CO_3 + H_2O \rightleftharpoons HCO_3^- + H_3O^+ \qquad K_1 \qquad (9\text{-}4)$$

$$HCO_3^- + H_2O \rightleftharpoons CO_3^{2-} + H_3O^+ \qquad K_2 \qquad (9\text{-}5)$$

It is sometimes desirable to regard polyprotic ionization as a unitary process, for instance, if we wish to consider the total amount of hydronium ion an acid can deliver. In such a situation we would write the equilibrium as

$$H_3PO_4 + 3H_2O \rightleftharpoons PO_4^{3-} + 3H_3O^+ \qquad (9\text{-}6)$$

It is important to realize that equations (9-1) through (9-3) are *not* in conflict with equation (9-6); they merely emphasize different aspects of the equilibrium system. Since reaction (9-6) describes an equilibrium system, there must be an equilibrium constant associated with it. It is denoted by K, called the overall ionization constant, and is defined in the usual way by

$$K = \frac{[H_3O^+]^3 [PO_4^{3-}]}{[H_3PO_4]} \tag{9-7}$$

It is not hard to show what must be the relationship between K and the individual ionization constants K_1, K_2, and K_3. First we write the equilibrium constant expressions for each of reactions (9-1) through (9-3).

$$K_1 = \frac{[H_3O^+][H_2PO_4^-]}{[H_3PO_4]} \tag{9-8}$$

$$K_2 = \frac{[H_3O^+][HPO_4^{2-}]}{[H_2PO_4^-]} \tag{9-9}$$

$$K_3 = \frac{[H_3O^+][PO_4^{3-}]}{[HPO_4^{2-}]} \tag{9-10}$$

Then, by multiplying together K_1, K_2, and K_3 and canceling out those concentration terms that appear in both numerator and denominator,

$$K_1 \cdot K_2 \cdot K_3 = \frac{[H_3O^+]\cancel{[H_2PO_4^-]}}{[H_3PO_4]} \cdot \frac{[H_3O^+]\cancel{[HPO_4^{2-}]}}{\cancel{[H_2PO_4^-]}} \cdot \frac{[H_3O^+][PO_4^{3-}]}{\cancel{[HPO_4^{2-}]}} \tag{9-11}$$

we achieve the desired result,

$$K_1K_2K_3 = \frac{[H_2O^+]^3 [PO_4^{3-}]}{[H_3PO_4]}$$

or

$$K_1K_2K_3 = K \tag{9-12}$$

This result is quite general; the equilibrium constant of any overall process is equal to the product of the equilibrium constants of the substeps that make up the overall process. A related feature of such systems can be stated in another, nonmathematical way: if an overall reaction system, such as (9-6), is in equilibrium, so are the individual steps. This cannot be otherwise, for it is the *system* that is in equilibrium; equations (9-1) through (9-3) and equation (9-6) are alternative ways of describing the system.

The several equilibrium constants of a polyprotic acid system have in general different numerical values, a fact that can be understood in terms of the electrical forces involved. The acid H_2S, for instance, has $K_1 = 1.0 \times 10^{-7}$ and $K_2 = 1.2 \times 10^{-13}$, corresponding to the first and second ionization steps

$$H_2S + H_2O \rightleftharpoons HS^- + H_3O^+ \tag{9-13}$$

$$HS^- + H_2O \rightleftharpoons S^{2-} + H_3O^+ \tag{9-14}$$

Even though H_2S is a weak acid, the first step (9-13) is comparatively easy. The removal of a proton from H_2S is opposed by the electrostatic attraction between the proton and the newly formed HS^-. In the second ionization the proton removal is opposed by the electrostatic force between the proton and the newly formed S^{2-} ion. Since sulfide ion carries a double negative charge, the force is considerably greater. For this reason, the second ionization step is more difficult, the equilibrium lies farther to the left, and K_2 is smaller than K_1.

This reasoning holds for all polyprotic acids and it is quite generally true that each ionization step has a smaller ionization constant than the preceding step. This electrostatic line of thinking also tells us that the effect is less pronounced in larger species, because the negative charge is distributed over a larger region. In hydrogen sulfide the difference between pK_1 and pK_2 is about 6 units, meaning that the acid HS^- is about 10^6 times weaker than the acid H_2S. By comparison, this difference (pK_2 - pK_1) for carbonic acid is 4, while that for terephthalic acid (another diprotic acid: see below) is only 0.9.

Although size is a factor in the decrease of pK difference between subsequent steps in polyprotic acid ionization, it is not the only one. A quick demonstration that other factors are important is provided by comparison of the acidities of the three positional isomers of phthalic acid, which are all essentially the same size.

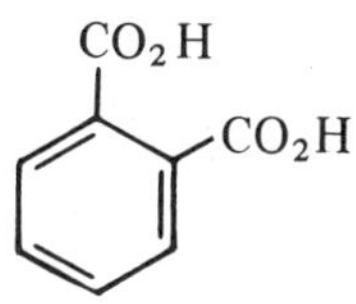

o-phthalic acid
(phthalic acid)
$pK_1 = 2.95$
$pK_2 = 5.41$

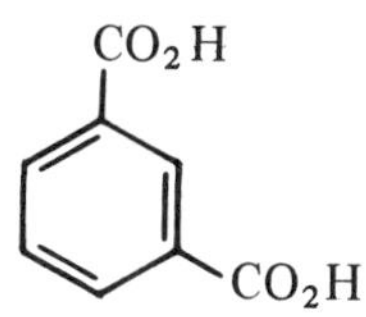

m-phthalic acid
(isophthalic acid)
$pK_1 = 3.62$
$pK_2 = 4.60$

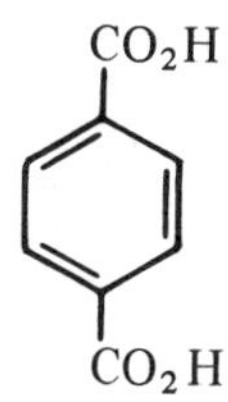

p-phthalic acid
(terephthalic acid)
$pK_1 = 3.54$
$pK_2 = 4.46$

In this series the second ionization step becomes easier (lower pK) as the distance between the two acidic sites increases. This is really just a modification of the "size" argument, however, and although previously we discussed it in terms of the increased electrostatic force opposing the removal of the second proton, most chemists would prefer to put it in terms of *stability*. A molecule or ion becomes more stable if centers of like charge can be accommodated in parts of the molecule remote from one another. In the dianions of the phthalic acids, the negative charges are confined to the carboxyl groups and are most widely separated in the *para* isomer. Since the *para* dianion is the most stable, the second ionization step proceeds farthest to the right, with the largest K_2 and the smallest pK_2.

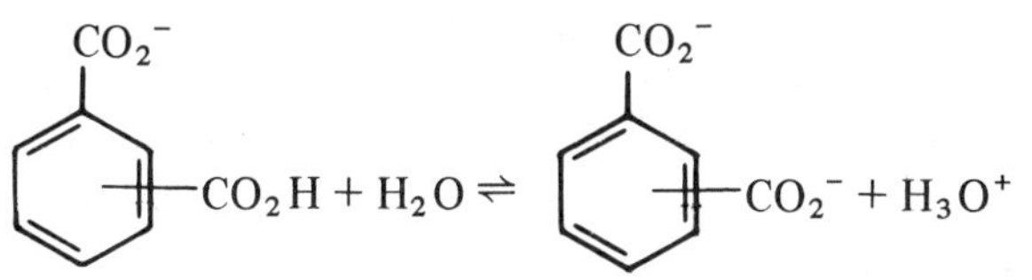

K_2 (9-15)

Although it cannot be relied on as the sole factor, the general correlation between acid size and the difference between pK_2 and pK_1 remains useful. Since most organic polyprotic acids are large and most inorganic polyprotics (H_2CO_3, H_3PO_4, H_3AsO_4, H_2SiO_3, etc.) are small, it is helpful to remember that for inorganic acids the successive ionization steps have large differences in pK, while in general the differences are smaller for organic polyprotics.

9.2 POLYHYDRIC BASES

As their chemical formulas strongly suggest, the hydroxides of group II elements are dihydric bases. The group II hydroxides such as $Ba(OH)_2$ are ionic, that is, they dissociate (ionize) completely in solution:

$$Ba(OH)_2 \rightarrow Ba^{2+} + 2OH^- \qquad (9\text{-}16)$$

The undissociated $Ba(OH)_2$ species does not exist in solution. These hydroxides [although they may have limited solubility, as does $Mg(OH)_2$] are *strong* bases.

Group III hydroxides exhibit more complicated behavior. Not only are they amphoteric, but they tend to form coordination complexes and for this reason they should be discussed together with the transition metals. Coordination complexes are adducts of a Lewis acid (a metal cation, often a *transition metal*) and several Lewis bases, called *ligands*. Their chemistry is extensive and fascinating, and we will discuss some aspects of it in the following chapter.

There are many possibilities for polyhydric base behavior among organic compounds; the majority of organic bases are amines or other derivatives of ammonia, and show basic behavior for the same reason that ammonia itself does. As an example, consider 1,2-diaminoethane (ethylenediamine), whose base ionization constants are $K_1 = 1.2 \times 10^{-4}$ and $K_2 = 1 \times 10^{-7}$.

$$H_2NCH_2CH_2NH_2 + H_2O \rightleftharpoons H_2NCH_2CH_2\overset{+}{N}H_3 + OH^- \quad K_1 \qquad (9\text{-}17)$$

$$H_2NCH_2CH_2\overset{+}{N}H_3 + H_2O \rightleftharpoons H_3\overset{+}{N}CH_2CH_2\overset{+}{N}H_3 + OH^- \quad K_2 \qquad (9\text{-}18)$$

As with polyprotic acids, and for the same reasons, each succeeding step of polyhydric base ionization is weaker, with a smaller K_b. The size effect is operative here as well as in acids. In 1,2-diaminoethane only two CH_2 groups separate the basic groups, and the $(pK_2 - pK_1)$ difference is about 3.1; we would expect this difference to decrease if there were more CH_2 groups, and it does. 1,4-Diaminobutane $[H_2N{-}(CH_2)_4{-}NH_2]$ has $pK_2 - pK_1 = 1.5$, and 1,8-diaminooctane $[H_2N{-}(CH_2)_8{-}NH_2]$ has $pK_2 - pK_1 = 0.9$.

9.3 EXACT CALCULATIONS

It is not difficult to adapt the exact methods of Chapter 3 and Appendix VI to polyprotic acids. The approach is a general one, so that the result is applicable to buffer solutions and salt solutions as well as to solutions containing only the polyprotic acid itself. We develop this in detail for phosphoric acid, and then simply give the result for the case of a diprotic acid.

The equilibrium constant expressions (9-8) through (9-10) for the three ionization steps are first rearranged to give

$$[H_2PO_4^-] = \frac{K_1 [H_3PO_4]}{[H_3O^+]} \qquad (9\text{-}19)$$

$$[HPO_4^{2-}] = \frac{K_2 [H_2PO_4^-]}{[H_3O^+]} \qquad (9\text{-}20)$$

$$[PO_4^{3-}] = \frac{K_3[HPO_4^{2-}]}{[H_3O^+]} \qquad (9\text{-}21)$$

Conservation of mass requires that the total amount of phosphorus be unaffected by the ionization processes.

$$\text{total P} = [H_3PO_4] + [H_2PO_4^-] + [HPO_4^{2-}] + [PO_4^{3-}] = T = \text{constant} \qquad (9\text{-}22)$$

The total phosphorus is always equal to the total amount initially present. Define C_1, C_2, C_3, and C_4 such that[1]

$$\begin{aligned} C_1 &= \text{initial molarity of acid } H_3PO_4 \\ C_2 &= \text{initial molarity of salt } NaH_2PO_4 \\ C_3 &= \text{initial molarity of salt } Na_2HPO_4 \\ C_4 &= \text{initial molarity of salt } Na_3PO_4 \end{aligned} \qquad (9\text{-}23)$$

It follows that

$$T = C_1 + C_2 + C_3 + C_4 \qquad (9\text{-}24)$$

The solution must be electrically neutral, so that the total positive charge equals the total negative charge. The doubly charged species HPO_4^{2-} contributes twice as much charge as a singly charged ion, so its concentration must be multiplied by 2; similarly, the concentration of the triply charged PO_4^{3-} ion must be multiplied by 3.

$$[H_3O^+] + [Na^+] = [OH^-] + [H_2PO_4^-] + 2[HPO_4^{2-}] + 3[PO_4^{3-}] \qquad (9\text{-}25)$$

Finally, we recognize the autoionization of the solvent

$$[H_3O^+][OH^-] = K_w \qquad (9\text{-}26)$$

There are thus seven equations; the four equilibria, (9-19) through (9-21) and (9-26); conservation of mass and charge, (9-22) and (9-25); and the initial conditions, (9-24). There are seven unknowns: $[H_3O^+]$, $[OH^-]$, $[H_3PO_4]$, $[H_2PO_4^-]$, $[HPO_4^{2-}]$, $[PO_4^{3-}]$, and $[Na^+]$. Since the number of unknowns equals the number of independent equations, we are assured that a solution exists. From (9-25) we eliminate $[H_2PO_4^-]$, $[HPO_4^{2-}]$, and $[PO_4^{3-}]$ by using (9-19), (9-20), and (9-21); we also eliminate $[OH^-]$ using (9-26), and obtain

$$[H_3O^+] + [Na^+] = \frac{K_w}{[H_3O^+]} + \frac{K_1[H_3PO_4]}{[H_3O^+]} + \frac{2K_1K_2[H_3PO_4]}{[H_3O^+]^2} + \frac{3K_1K_2K_3[H_3PO_4]}{[H_3O^+]^3}$$

[1] Any of these may be set equal to zero; to calculate the pH of a solution of the intermediate salt Na_2HPO_4, for example, C_1, C_2 and C_4 are set equal to zero. To calculate the pH of a solution containing only the acid from H_3PO_4, the initial salt concentrations C_2, C_3, and C_4 are set equal to zero. The concentration $[Na^+]$ in equation (9-25) is also zero in acid-only systems.

which can be rewritten as

$$[H_3O^+] + [Na^+] = \frac{K_w}{[H_3O^+]} + [H_3PO_4]\left(\frac{K_w}{[H_3O^+]} + \frac{2K_1K_2}{[H_3O^+]^2} + \frac{3K_1K_2K_3}{[H_3O^+]^3}\right) \qquad (9\text{-}27)$$

The algebra will be much easier to follow if at this point we make the substitution $x = [H_3O^+]$:

$$x + [Na^+] = \frac{K_w}{x} + [H_3PO_4]\left(\frac{K_1}{x} + \frac{2K_1K_2}{x^2} + \frac{3K_1K_2K_3}{x^3}\right)$$

Multiply through by x^3 and, in several steps, solve for $[H_3PO_4]$:

$$x^4 + [Na^+]x^3 = K_wx^2 + [H_3PO_4](K_1x^2 + 2K_1K_2x + 3K_1K_2K_3)$$

$$x^4 + [Na^+]x^3 - K_wx^2 = [H_3PO_4](K_1x^2 + 2K_1K_2x + 3K_1K_2K_3)$$

$$[H_3PO_4] = \frac{x^4 + [Na^+]x^3 - K_wx^2}{K_1x^2 + 2K_1K_2x + 3K_1K_2K_3} \qquad (9\text{-}28)$$

From (9-22) we eliminate $[H_2PO_4^-]$, $[HPO_4^{2-}]$, and $[PO_4^{3-}]$ using (9-19), (9-20), and (9-21):

$$T = [H_3PO_4] + \frac{K_1[H_3PO_4]}{[H_3O^+]} + \frac{K_1K_2[H_3PO_4]}{[H_3O^+]^2} + \frac{K_1K_2K_3[H_3PO_4]}{[H_3O^+]^3}$$

or

$$T = [H_3PO_4]\left(1 + \frac{K_1}{x} + \frac{K_1K_2}{x^2} + \frac{K_1K_2K_3}{x^2}\right) \qquad (9\text{-}29)$$

We combine (9-28) and (9-29):

$$T = \left(1 + \frac{K_1}{x} + \frac{K_1K_2}{x^2} + \frac{K_1K_2K_3}{x^3}\right)\frac{x^4 + [Na^+]x^3 - K_wx^2}{K_1x^2 + 2K_1K_2x + 3K_1K_2K_3} \qquad (9\text{-}30)$$

Then, subtracting T from both sides of (9-30) and multiplying both sides by the denominator yields

$$0 = \left(1 + \frac{K_1}{x} + \frac{K_1K_2}{x^2} + \frac{K_1K_2K_3}{x^3}\right)(x^4 + [Na^+]x^3 - K_wx^2) - T(K_1x^2 + 2K_1K_2x + 3K_1K_2K_3)$$

The final steps will be somewhat easier if at this stage we multiply both sides by x:

$$0 = \left(1 + \frac{K_1}{x} + \frac{K_1K_2}{x^2} + \frac{K_1K_2K_3}{x^3}\right)(x^5 + [Na^+]x^4 - K_wx^3) - T(K_1x^3 + 2K_1K_2x^2 + 3K_1K_2K_3x)$$

Multiply out and group all terms in powers of x:

$$
\begin{aligned}
&x^5 + [\mathrm{Na^+}]x^4 \quad -K_w x^3 \\
&\quad + K_1x^4 + [\mathrm{Na^+}]K_1x^3 \quad -K_1K_wx^2 \\
&\quad\quad + K_1K_2x^3 + [\mathrm{Na^+}]K_1K_2x^2 \quad -K_1K_2K_wx \\
&\quad\quad\quad + K_1K_2K_3x^2 + [\mathrm{Na^+}]K_1K_2K_3x - K_1K_2K_3K_w \\
&\quad\quad -TK_1x^3 \quad -2TK_1K_2x^2 \quad -3TK_1K_2K_3x \quad = 0
\end{aligned}
\tag{9-31}
$$

It is convenient to write this in standard polynomial form:

$$Ax^5 + Bx^4 + Cx^3 + Dx^2 + Ex + F = 0 \tag{9-32}$$

where

$$
\begin{aligned}
A &= 1 \\
B &= K_1 + [\mathrm{Na^+}] \\
C &= K_1\left([\mathrm{Na^+}] + K_2 - \frac{K_w}{K_1} - T\right) \\
D &= K_1K_2\left([\mathrm{Na^+}] + K_3 - \frac{K_w}{K_2} - 2T\right) \\
E &= K_1K_2K_3\left([\mathrm{Na^+}] - \frac{K_w}{K_3} - 3T\right) \\
F &= -K_1K_2K_3K_w
\end{aligned}
$$

Finally, we eliminate $[\mathrm{Na^+}]$ from the equation. From stoichiometric considerations we know the relationship between $[\mathrm{Na^+}]$ and the initial concentrations (indicated by the subscript "o") of the several different salts.

$$
\begin{aligned}
[\mathrm{Na^+}] &= 3[\mathrm{Na_3PO_4}]_0 + 2[\mathrm{Na_2HPO_4}]_0 + 1[\mathrm{NaH_2PO_4}]_0 \\
&= 3C_4 + 2C_3 + C_2
\end{aligned}
\tag{9-33}
$$

The treatment for diprotic acids such as H_2CO_3 follows similar lines, and leads to a similar result,

$$Ax^4 + Bx^3 + Cx^2 + Dx + E = 0 \tag{9-34}$$

where

$$
\begin{aligned}
A &= 1 \\
B &= K_1 + [\mathrm{Na^+}] \\
C &= K_1\left([\mathrm{Na^+}] + K_2 - \frac{K_w}{K_1} - T\right) \\
D &= K_1K_2\left([\mathrm{Na^+}] - \frac{K_w}{K_2} - 2T\right) \\
E &= -K_1K_2K_w \\
T &= [\mathrm{H_2CO_3}]_0 + [\mathrm{NaHCO_3}]_0 + [\mathrm{Na_2CO_3}]_0 \\
[\mathrm{Na^+}] &= [\mathrm{NaHCO_3}]_0 + 2[\mathrm{Na_2CO_3}]_0
\end{aligned}
$$

The next step is to solve either (9-32) or (9-34) for x, the H_3O^+ concentration, given the initial concentrations of the several compounds in the mixture. Both equations (9-32) and (9-34) are readily programmed in the BASIC language for computer solution; a program which solves both is given in Appendix Section II.4. The program will find the pH in any di- or triprotic acid system, by which is meant an aqueous solution of a weak acid and/or one or more of its salts. Any of the initial concentrations may be set equal to zero, so that pure acid ionization or pure salt hydrolysis problems as well as buffer problems may be solved by the same program.

The calculation method also allows calculation of the concentrations of all species involved, after the pH is calculated. Rearrangement of (9-29) gives the concentration of undissociated phosphoric acid,

$$[H_3PO_4] = \left(\frac{T}{1 + \dfrac{K_1}{[H_3O^+]} + \dfrac{K_1K_2}{[H_3O^+]^2} + \dfrac{K_1K_2K_3}{[H_3O^+]^3}}\right) \tag{9-35}$$

from which the concentrations of the other species are found by successive application of equations (9-19), (9-20), and (9-21):

$$[H_2PO_4^-] = \frac{[H_3PO_4]K_1}{[H_3O^+]} \tag{9-19}$$

$$[HPO_4^{2-}] = \frac{[H_3PO_4]K_1K_2}{[H_3O^+]^2} \tag{9-20a}$$

$$[PO_4^{3-}] = \frac{[H_3PO_4]K_1K_2K_3}{[H_3O^+]^3} \tag{9-21a}$$

Equation (9-35) is also correct for diprotic acid systems, if we let $K_3 = 0$ and $[H_3PO_4] = [H_2CO_3]$.

9.4 CONCENTRATION VERSUS pH GRAPHS

Equations (9-35), (9-19), (9-20a), and (9-21a) can be used to answer such questions as "What species are present?" and "What are the predominant species?" when the pH is known. This information is conveniently presented in the form of graphs such as Figure 9.1, which shows the relative concentration of each of the species H_2A, HA^-, and A^{2-} for a diprotic acid with $K_1 = 4.0 \times 10^{-7}$ ($pK_1 = 6.4$) and $K_2 = 4.0 \times 10^{-11}$ ($pK_2 = 10.4$). (These values are those of carbonic acid.) A striking feature of Figure 9.1 is that there is no value of pH at which all three forms of carbonic acid are present in significant quantities. Throughout the pH range there are at most two forms present. This feature of polyprotic acid systems is a direct consequence of the relatively wide separation in the pK values. The behavior of triprotic acid systems is shown in Figures 9.2 and 9.3, which present relative concentration versus pH information on the species H_3A, H_2A^-, HA^{2-}, and A^{3-}. Figure 9.2 is drawn for an acid with widely separated pK values; $pK_1 = 2.12$, $pK_2 = 7.21$, $pK_3 = 12.66$ (phosphoric acid), while Figure 9.3 is drawn for an acid (citric

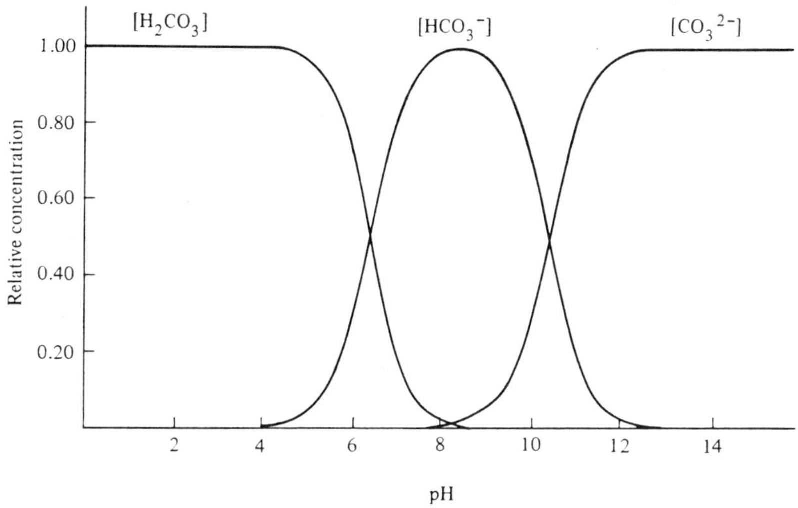

Figure 9.1 pH dependence of the relative concentrations of the species H_2CO_3, HCO_3^-, and CO_3^{2-} in the carbonic acid system. $pK_1 = 6.4$, $pK_2 = 10.4$.

acid) with the more closely spaced pK values characteristic of the larger organic acids; $pK_1 = 3.13$, $pK_2 = 4.76$, $pK_3 = 5.41$.

Figure 9.2 resembles Figure 9.1 in that there is no pH value at which more than two forms are present to any significant extent. In fact, throughout much of the pH range only one form is present at a relative concentration greater than 0.05 (5%). When we examine Figure 9.3, however, we see that there are three forms present at relative concentrations greater than 0.05 whenever the pH is between about 3.5 and 5.5. This phenomenon is due entirely to the close spacing of the pK values and it is important to remember this qualitative difference in the behavior of the mineral acids and the larger organic acids.

Another feature of polyprotic acid systems which is shown in Figures 9.1, 9.2, and 9.3 is that the pH value at which the several curves cross one another is numerically equal to the pK. In Figure 9.2, for instance, the curve for H_3A crosses that of H_2A^- at pH 2.1. At the crossing point the concentrations $[H_3A]$ and $[H_2A^-]$ are equal to one another, and equation (9-8) reduces to $K_1 = [H_3O^+]$, or $pK_1 = pH$. This is quite a general result, true for the subsequent crossing points in the phosphoric acid system at pH 7.2 and 12.7, as well as for the more closely spaced crossing points in the citric acid system, at pH 3.13, 4.76, and 5.41. This rule applies only to the crossing points of the curves for two species which are a conjugate acid/base pair. In Figure 9.3 the crossing point at pH = 5.2, relative concentration = 0.25, does *not* correspond to a pK because the two species, H_2Cit^- and Cit^{3-}, are not a conjugate pair.

The curves of Figures 9.1, 9.2, and 9.3 are convenient sources of specific information about polyprotic acid systems.

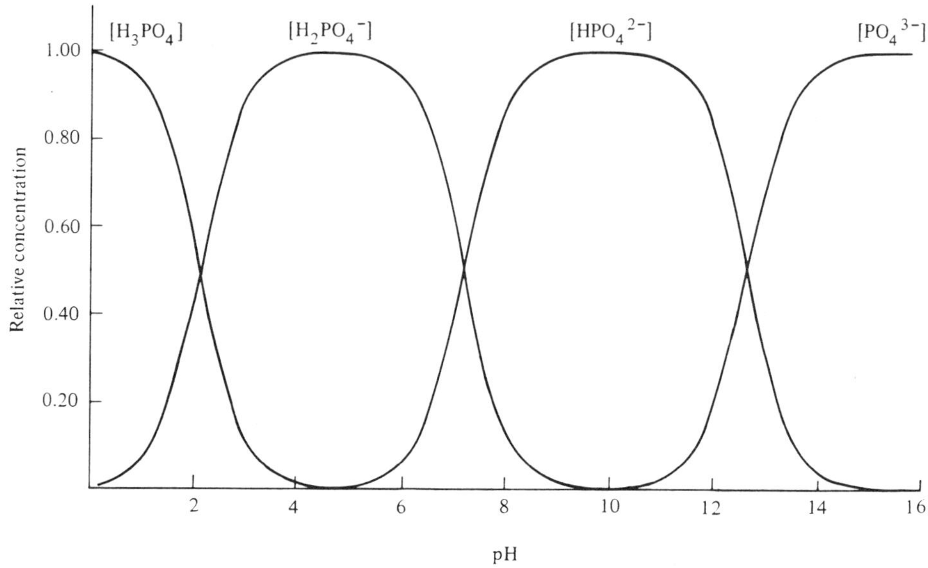

Figure 9.2 pH dependence of the relative concentrations of the species H_3PO_4, $H_2PO_4^-$, HPO_4^{2-}, and PO_4^{3-} in the phosphoric acid system. pK_1 = 2.12, pK_2 = 7.21, pK_3 = 12.66.

Example 9.1

Use Figure 9.2 and describe a solution of phosphoric acid whose pH has been adjusted by the addition of NaOH until the pH is exactly 9.0.

Answer: At pH 9.0, the dihydrogen phosphate ($H_2PO_4^-$) ion is present at 2% of the total, while the relative amount of hydrogen phosphate (HPO_4^{2-}) is about 98%. No other phosphorus-containing species is present in significant amounts.

Furthermore, as was discussed in connection with buffer systems, the same curves apply regardless of whether the solution is prepared by dissolving pure acid plus NaOH, pure sodium salt plus HCl, or any combination of acid and salt(s).

Example 9.2

A solution is prepared by dissolving 3.0 g of KH_2PO_4 and 5.9 g of K_3PO_4 in 1 liter of water, then adding HCl or NaOH as needed to bring the pH to 9.0.[2] Describe the relative concentrations of the phosphorus-containing species.

[2]Although it is not a part of this example, you might think about whether acid or base is needed in this case, and how this question might be answered.

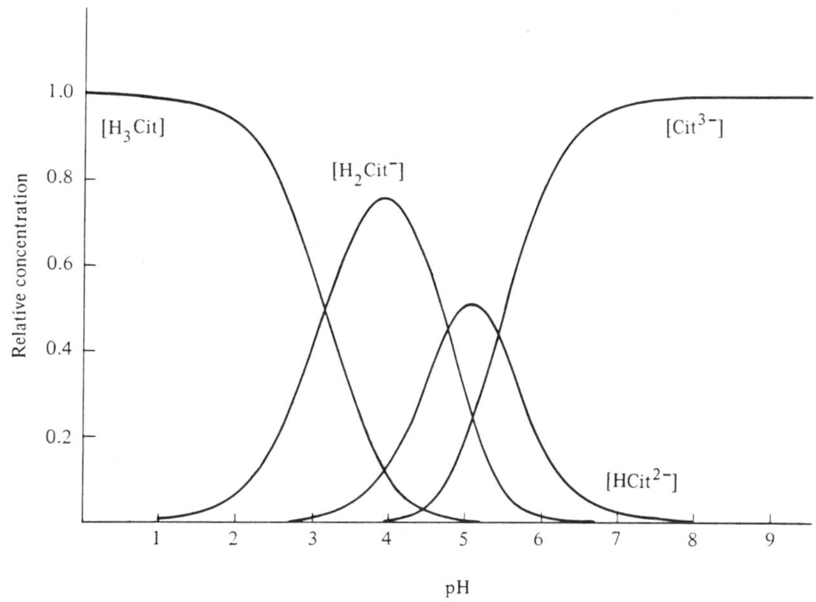

Figure 9.3 pH dependence of the relative concentrations of the species H_3Cit, H_2Cit^-, $HCit^{2-}$, and Cit^{3-} in the citric acid system. $pK_1 = 3.13$, $pK_2 = 4.76$, $pK_3 = 5.41$.

Answer: Since the equilibrium concentrations of the various species depend on pH and not on the method of preparation, the answer is exactly the same as the answer to Example 9.1. We read the following values from Figure 9.2 at pH = 9: $[HPO_4^{2-}] \cong 98\%$, $[H_2PO_4^-] \cong 2\%$; $[H_3PO_4]$ and $[PO_4^{3-}]$ are negligibly small. Note that we do not even need to convert grams solute to molarity.

The general features of polyprotic acid systems that have been discussed in this section suggest an appropriate way of thinking, by which much of the information present in pH profiles like Figures 9.1, 9.2, and 9.3 can be obtained without actually drawing the graph. We base this scheme on three points, of which the third is an approximation valid for mineral acids and less accurate for organic acids.

1. The highly protonated forms, such as H_3A, predominate at low pH, while the unprotonated form A^{3-} predominates at high pH.
2. Two forms have equal concentration whenever pH = pK.
3. Only two forms are present at any given pH.

Example 9.3

Without consulting any graphs, describe the species present in a solution of H_2CO_3 whose pH has been adjusted to (a) 8.4 and (b) 7.0. For H_2CO_3, $pK_1 = 6.4$ and $pK_2 = 10.4$.

Answer: (a) The value pH = 8.4 is exactly halfway between two pK values, and therefore the species that is common to both equilibria (HCO_3^-) will be at its maximum concentration.[3] The solution consists primarily of HCO_3^-, with much lesser amounts of both H_2CO_3 and CO_3^{2-}.

$$H_2CO_3 + H_2O \rightleftharpoons HCO_3^- + H_3O^+ \qquad pK_1 = 6.4$$

$$HCO_3^- + H_2O \rightleftharpoons CO_3^{2-} + H_3O^+ \qquad pK_2 = 10.4$$

(b) The pH is near pK_1, so the only species present in significant quantities will be those of the K_1 equilibrium, H_2CO_3 and HCO_3^-. Since the pH is greater than pK, the less protonated of the two, HCO_3^-, will predominate. If the pH were 1.0 unit higher than the pK, the ratio $[HCO_3^-]/[H_2CO_3]$ would be exactly 10. (Why?) In this example the pH is only 0.6 unit higher than the pK; $[HCO_3^-]$ is in excess, but less than a tenfold excess. (To be precise, if pH is 0.6 unit higher than pK, log $[HCO_3^-]/[H_2CO_3] = 0.6$ and $[HCO_3^-]/[H_2CO_3] = 10^{0.6} = 4$.)

9.5 APPROXIMATE CALCULATIONS

Much of the foregoing section can be condensed into a simplifying principle for treating polyprotic acids. Since, for mineral acids, at least, no more than two species are present at any one time, the system is essentially the same as a monoprotic acid. The simplifying principle is then: *Decide which two forms are present and proceed as for monoprotic acids.*

Example 9.4

Calculate the pH of a 0.01 M solution of K_2HPO_4.

Answer: We need only be concerned with one equilibrium (two forms) and it will be either the ionization equilibrium (9-3) or the hydrolysis equilibrium (9-36).

$$HPO_4^{2-} + H_2O \rightleftharpoons PO_4^{3-} + H_3O^+ \qquad K_3 = 2.2 \times 10^{-13} \tag{9-3}$$

$$HPO_4^{2-} + H_2O \rightleftharpoons H_2PO_4^- + OH^- \qquad K_{2h} = \frac{K_w}{K_2} = \frac{1.0 \times 10^{-14}}{6.23 \times 10^{-8}} \tag{9-36}$$

Since K_{2h} is larger than K_3, it is clear that (9-36) will proceed to a much greater extent

[3] The mathematical demonstration of this intuitive result begins with equation (9-35) modified to a form appropriate to the diprotic acid H_2A: $[H_2A] = T/(1 + K_1/[H_3O^+] + K_1K_2/[H_3O^+]^2)$, which after substituting into $[HA^-] = [H_2A]K_1/[H_3O^+]$ becomes $[HA^-] = T/([H_3O^+]/K_1 + 1 + K_2/[H_3O^+])$. We differentiate, $d[HA^-]/d[H_3O^+] = -T(1/K_1 - K_2/[H_3O^+]^2)([H_3O^+]/K_1 + 1 + K_2/[H_3O^+])^{-2}$ and equate this to zero. The maximum occurs at $(1/K_1 - K_2/[H_3O^+]^2) = 0$ or $[H_3O^+] = \sqrt{K_1K_2}$. Taking logs, we find that the maximum value of $[HA^-]$ occurs at pH = ½(pK_1 + pK_2). [This relationship is exact for diprotic acids; for triprotics the pH at maximum $[H_2A^-]$ is slightly increased, by an amount approximately given by $K_3K_2/(2.3\sqrt{K_1K_2})$. This factor is quite small if the pK values are well separated, and even for citric acid it is less than 0.002 pH unit. It can of course be neglected in the context of Example 9.3.

than (9-3), which may safely be neglected. Since K_{2h} is small, the approximate form may be used to find the hydroxide ion concentration:

$$[OH^-] = \sqrt{K_{2h}C} = \sqrt{(1.6 \times 10^{-7})(0.01)} = 4.0 \times 10^{-5}$$

$$pOH = -\log(4.0 \times 10^{-5}) = 4.4$$

$$pH = 14 - pOH = 9.6$$

If we use the method of the exact solution, equation (9-32), we obtain a pH of 9.559.

Example 9.5

To a sample of 100 mL of 0.0200 M Na_2CO_3 is added 10.0 mL of 0.350 M HCl. What is the pH of the resulting solution? For carbonic $pK_1 = 6.40$ and $pK_2 = 10.40$.

Answer: Carbonate ion is a base (since it is conjugate to the acid HCO_3^-) and it will react with the added HCl via

$$H_3O^+ + CO_3^{2-} \rightarrow HCO_3^- + H_2O$$

If enough acid is added, a second reaction will occur:

$$H_3O^+ + HCO_3^- \rightarrow H_2CO_3 + H_2O$$

Since CO_3^{2-} is a stronger base than HCO_3^- (why?), the second reaction will not take place until all the CO_3^{2-} is consumed. So first we must calculate the extent to which these reactions have occurred.

$$(100 \text{ mL}) \times (0.0200\, M) = 2.00 \text{ mmol } CO_3^{2-}$$

$$(10.0 \text{ mL}) \times (0.350\, M) = 3.50 \text{ mmol HCl}$$

$$\text{excess} = 1.50 \text{ mmol HCl}$$

All of the CO_3^{2-} will react, yielding 2.00 mmol of HCO_3^-. Since 1.50 mmol of HCl is still available, it reacts, producing 1.50 mmol of H_2CO_3 and leaving unreacted 0.50 mmol of HCO_3^-. The amount of unreacted CO_3^{2-} is negligibly small. The pH is found by insertion of the $[HCO_3^-]/[H_2CO_3]$ ratio into the Henderson-Hasselbalch equation (7-30).

$$pH = pK_a + \log\left(\frac{C_s}{C_a}\right)$$

$$= 6.40 + \log\left(\frac{0.50 \text{ mmol}/110 \text{ mL}}{1.50 \text{ mmol}/110 \text{ mL}}\right)$$

$$= 6.40 - 0.477 = 5.92$$

Note particularly that pK_2, which relates to the $[CO_3^{2-}]/[HCO_3^-]$ ratio, is not used in this problem; our simplifying principle tells us that we need consider only one equilibrium.

There are some problems for which this simple approach cannot be used to get accurate numerical results. Suppose that we have a triprotic acid for which $K_1 = 1.0 \times 10^{-2}$, $K_2 = 1.1 \times 10^{-5}$, and $K_3 = 1.1 \times 10^{-9}$, and a solution containing 0.01 M K_2HA is prepared as in Example 9.4. In this case $K_{2h} = K_w/K_2 = 0.91 \times 10^{-9}$, and we can no longer say that one of K_3 and K_{2h} is much larger than the other. In this case we can say only that the reaction

$$HA^{2-} + H_2O \rightarrow A^{3-} + H_3O^+$$

takes place to a *slightly* greater extent than the hydrolysis

$$HA^{2-} + H_2O \rightarrow H_2A^- + OH^-$$

There is a *slight* preponderance of H_3O^+ produced compared to OH^- produced, and the solution pH will be slightly less than it would have been if $K_3 = K_{2h}$. If $K_3 = K_{2h}$, the amounts of H_3O^+ and OH^- produced are equal, and pH = 7.0 (or pH = $pK_w/2$ if the temperature is different from 25°C). Thus all we can say in this example is that the pH is slightly less than 7, and on the basis of the ideas presented so far it is difficult to estimate how much less.[4] The exact calculation gives pH = 6.96.

A different sort of complexity arises if the solution is prepared initially from *two* forms rather than one.

Example 9.6

Estimate the pH of a solution, 1 liter of which contains 0.1 mol of NaH_2PO_4 and 0.2 mol of Na_2HPO_4.

Answer: Since $H_2PO_4^-/HPO_4^{2-}$ are a conjugate acid/base pair, a buffer relationship is established. The Henderson-Hasselbalch equation is used to calculate the pH:

$$pH = pK + \log\left(\frac{C_s}{C_a}\right) \qquad (7\text{-}30)$$

In this case C_a is the concentration of the more protonated form ($H_2PO_4^-$), and C_s is the concentration of the less protonated form (HPO_4^{2-}). The appropriate pK is pK_2 = 7.21.

$$pH = 7.21 + \log\left(\frac{0.2}{0.1}\right) = 7.51$$

If the two forms are *not* members of an acid/base conjugate pair, they differ by more than one proton and cannot enter into a buffer relationship.

Example 9.7

Estimate the pH of a solution prepared by dissolving 0.03 mol of H_3PO_4 and 0.05 mol of K_3PO_4 in 1 liter of water.

[4] K_2HA is an amphiprotic substance, and in Section 9.8 we will present a method for an approximate calculation of the pH of such solutions.

Answer: It is easiest to think about complicated equilibria as if they were actually reached via a series of discrete steps. In the first step we consider the acid-base reaction in which the 0.03 mol of H_3PO_4 reacts with 0.03 mol of the PO_4^{3-}, producing 0.03 mol each of $H_2PO_4^-$ and HPO_4^{2-}:

$$H_3PO_4 + PO_4^{3-} \rightarrow H_2PO_4^- + HPO_4^{2-}$$

The solution's contents are now $H_2PO_4^-$, 0.03 mol; HPO_4^{2-}, 0.03 mol; PO_4^{3-}, 0.02 mol; since no more than two forms can be present at equilibrium, further reaction must occur. The remaining PO_4^{3-} reacts with $H_2PO_4^-$:

$$\underset{(0.02)}{H_2PO_4^-} + \underset{(0.02)}{PO_4^{3-}} \rightarrow \underset{(0.04)}{2HPO_4^{2-}}$$

and the solution's contents are now $H_2PO_4^-$, 0.01 mol; HPO_4^{2-}, 0.07 mol. Once again we have a simple buffer relationship, and

$$\text{pH} = \text{p}K_2 + \log\left(\frac{[HPO_4^{2-}]}{[H_2PO_4^-]}\right) = 7.2 + \log\left(\frac{0.07}{0.01}\right) = 7.2 + 0.8 = 8.0$$

This result is correct even though the reactions take place simultaneously rather than as sequential steps, because equilibrium parameters do not depend on the pathway by which the equilibrium is reached.

At this point we could continue to give examples, and indeed a large number of problems can be constructed. However, our purpose is to illustrate a mode of thinking as an approach to polyprotic acids rather than to present an exhaustive list of different problems that might arise.

9.6 POLYPROTIC BUFFER SOLUTIONS

The most striking feature of buffer solutions prepared from polyprotic acids is that they exhibit more than one buffer range. A monoprotic buffer of the type discussed in Chapter 7 shows a pH range perhaps 2 units wide (centered on the pK of the acid) in which useful buffer action takes place. A polyprotic acid, on the other hand, shows more than one such range. For example, phosphoric acid has three ranges: mixtures of H_3PO_4 and NaH_2PO_4 are buffers in the range pH $\approx$ 1.1 to 3.1 (pK_1 = 2.1), mixtures of NaH_2PO_4 and Na_2HPO_4 are buffers in the range pH $\approx$ 6.2 to 8.2 (pK_2 = 7.2), and mixtures of Na_2HPO_4 and Na_3PO_4 are buffers in the range pH $\approx$ 11.7 to 13.7 (pK_3 = 12.7). Provided that the pK values of the acid are well spaced, and according to the principles developed in Sections 9.4 and 9.5, such buffers are treated as if they were monoprotic.

Example 9.8

Calculate the pH of a buffer prepared by the addition of 10.0 mL of 0.987 M KOH to 500 mL of 0.04 M KH_2PO_4.

Answer: The added base reacts with the acid (KH_2PO_4) by

$$OH^- + H_2PO_4^- \rightarrow H_2O + HPO_4^{2-}$$

$$\text{acid present} = (500\ \text{mL})(0.04\ M) = 20.0\ \text{mmol}$$

$$\text{added base} = (10.0\ \text{mL})(0.987) = 9.87\ \text{mmol}$$

$$\text{acid remaining after reaction} = 10.13\ \text{mmol}$$

$$HPO_4^{2-}\ (\text{"salt"})\ \text{formed} = 9.87\ \text{mmol}$$

$$\text{pH} = \text{p}K + \log\left(\frac{C_s}{C_a}\right)$$

$$= \text{p}K_2 + \log\left(\frac{9.87\ \text{mmol}/510\ \text{mL}}{10.13\ \text{mmol}/510\ \text{mL}}\right)$$

$$= 7.21 + (-0.01) = 7.20$$

One difficulty with these problems is that the terms "acid" and "salt," while appropriate to monoprotic systems, are less so in polyprotic systems. Perhaps, as suggested in Example 9.6, the Henderson-Hasselbalch equation (7-30) ought to be rewritten as (and conceptualized as)

$$\text{pH} = \text{p}K + \log\left(\frac{[\text{less protonated form}]}{[\text{more protonated form}]}\right) \tag{9-37}$$

If the pK values of the polyprotic acid are not well spaced, more than two forms may be present and the Henderson-Hasselbalch equation is not useful. This complicates the calculations and presents us with two choices: either we do an exact, but tedious calculation based on equation (9-32), or we use the approximate approach and accept the inevitable inaccuracy in pH. As a practical laboratory buffer system, however, a close-spaced polyprotic such as citric acid is very useful. The buffer ranges of the successive ionization steps overlap, resulting in a very wide and continuous buffer range. Citric acid can be used as a buffer anywhere between pH 2 and 7. Moreover, since pH is simply and conveniently measured, we can begin with a pure citric acid solution (pH $\approx$ 2) and just add OH^- until the desired pH is reached. Provided that our measurement is accurate, no calculations are necessary.

9.7 TITRATION AND TITRATION CURVES

Titration of a polyprotic acid is carried out in the same way as for monoprotic acids, that is, by the gradual addition of a strong monohydric base until there is a more-or-less abrupt change in pH. The difference is that there is more than one end point. In the titration of H_2CO_3, for example, the first end point corresponds to the stoichiometric equivalence point of the reaction

$$H_2CO_3 + OH^- \rightarrow HCO_3^- + H_2O$$

and the second end point corresponds to the stoichiometric equivalence point of the reaction

$$HCO_3^- + OH^- \rightarrow CO_3^{2-} + H_2O$$

These two neutralization reactions *do not* occur simultaneously. H_2CO_3 ($pK_1 = 6.4$) is a much stronger acid than HCO_3^- ($pK_2 = 10.4$), and thus it competes successfully for the available OH^-. HCO_3^- cannot begin to react until all of the H_2CO_3 has been consumed.

Discussion of titrations is much easier if we can refer to a titration curve, three of which are presented in Figures 9.4, 9.5, and 9.6. The first is for H_2CO_3 and shows the two breaks characteristic of a diprotic acid. These breaks are not particularly sharp, because H_2CO_3 is not a particularly strong acid. (At this point it might be well to look again at Figures 8.4 and 8.6, which display the effect of acid strength on the form of a titration curve.) Note that the second break, at 100 mL of added base and pH $\approx$ 11 to 12, is even less sharp than the first break. It is in general true that the break becomes less sharp as the pK increases, and this is the source of considerable practical difficulty with very weak acids, polyprotic or otherwise. Phosphoric acid, for which the pK values are 2.1, 7.2, and

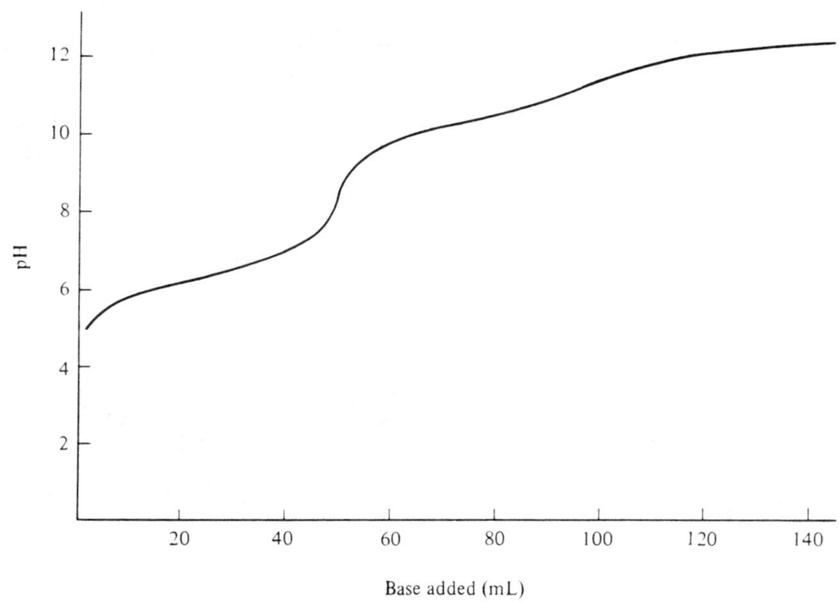

Figure 9.4 Titration of 50 mL of 0.1 *M* H_2CO_3 with 0.1 *M* KOH. Note that, as with monoprotic acids, the pH values at the half-equivalence points (25 and 75 mL) are equal to the pK values (6.4 and 10.4). The break at 100 mL is not well defined because the corresponding acid (HCO_3^-) is so very weak.

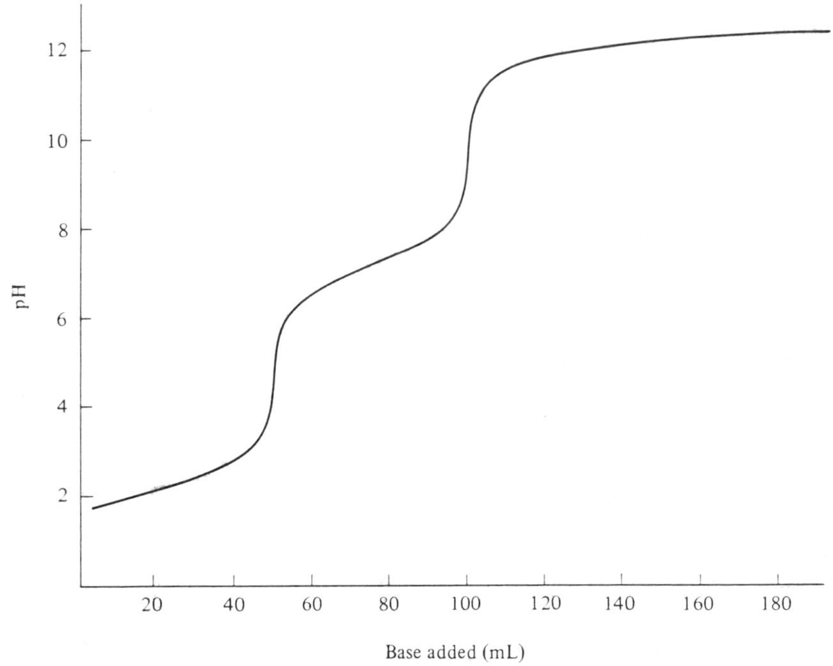

Figure 9.5 Titration of 50 mL 0.1 *M* H_3PO_4 with 0.1 *M* KOH. The first two breaks, at 50 mL and 100 mL, corresponding respectively to $H_3PO_4 + OH^- \rightarrow H_2PO_4^- + H_2O$ and $H_2PO_4^- + OH^- \rightarrow HPO_4^{2-} + H_2O$, are clearly discernible. The third end point, at 150 mL, occurs at such high pH that no break is seen at all. The value of pK_3 cannot be obtained from such a titration.

12.7, shows only two breaks in its titration curve (Figure 9.5.) The third break should occur at 150 mL added base, but it is such a gentle change in the curve that it cannot be discerned.[5]

If a polyprotic acid has well-spaced pK values, its titration curve will show a succession of breaks; of these, some (those at large pK_a values) may be invisible. On the other hand, if a polyprotic acid's pK values are all well below 12, they may not be widely spaced, and the breaks may be invisible for a different reason. Citric acid has pK values of 3.13, 4.76, and 5.41, and its titration curve is shown in Figure 9.6. It would take considerable experience to recognize this as a triprotic acid, since only one break (actually the third) is visible. The explanation, of course, is that the breaks occur over such a narrow pH range that they fade imperceptibly into one another. This curve also helps to explain the wide buffer range of citric acid.

It would seem then that triprotic acids are difficult to titrate. The problems get worse if tetraprotic (and higher-protic) acids are titrated, and we will leave the topic at this point.

[5] As a practical matter, H_3PO_4 could be titrated (to determine its pK values, for example) by first converting it entirely to PO_4^{3-}, and then titrating it as a polyhydric base. Phosphate ion is a moderately strong base with $pK_{b1} = pK_{h3} = 14.0 - pK_3 = 1.3$, and the titration with acid, $PO_4^{3-} + H^+ \rightarrow HPO_4^{2-}$, shows a sharp, easily measurable break.

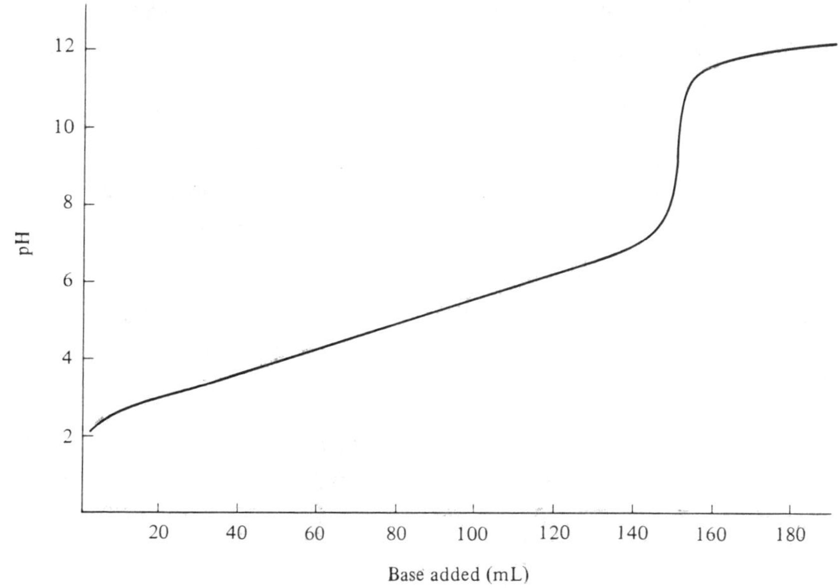

Figure 9.6 Titration of 50 mL of 0.1 *M* citric acid with 0.1 *M* KOH. The only visible break is the third end point. The p*K* values of the first two equivalence points are so close together ($pK_1 = 3.13$, $pK_2 = 4.76$) that no structure is discernible in the titration curve.

9.8 AMPHIPROTIC SUBSTANCES

As we have been using the term, an amphiprotic substance is one that can function either as an acid or a base, that is, as a proton donor or a proton acceptor. There are several common types of amphiprotic substances, and we will discuss them in turn. We have already seen that an intermediate salt of a polyprotic acid (e.g., HCO_3^- or HPO_4^{2-}) is amphiprotic, and we worked a pH problem with such a substance (Example 9.4). In the following example we work a problem similar to Example 9.4, but use terms appropriate to the language of amphiprotic substances.

Example 9.9

An amphiprotic substance HX has $K_a = 6.2 \times 10^{-9}$ and $K_b = 1.7 \times 10^{-9}$. Calculate the pH of a 0.05 *M* solution of HX.

Answer: The appropriate equilibria are

$$\text{base:}\quad HX + H_2O \rightleftharpoons H_2X^+ + OH^- \qquad K_b$$

$$\text{acid:}\quad HX + H_2O \rightleftharpoons H_3O^+ + X^- \qquad K_a$$

Let us make several slight but suggestive changes in the symbols we use to discuss this example. We write HA^- instead of HX, K_2 instead of K_a, and K_{1h} instead of K_b. Then the appropriate equilibria are rewritten as

$$\text{base:}\quad HA^- + H_2O \rightleftharpoons H_2A + OH^- \qquad K_{1h}$$

$$\text{acid:}\quad HA^- + H_2O \rightleftharpoons H_3O^+ + A^{2-} \qquad K_2$$

Although we have changed nothing but the symbols, the problem is now seen to be nothing more than two of the reactions occurring within the diprotic acid system of the acid H_2A; the second ionization constant $K_2 = 6.2 \times 10^{-9}$ and the first ionization constant $K_1 = K_w/K_{1h} = 1.0 \times 10^{-14}/1.7 \times 10^{-9} = 5.9 \times 10^{-6}$. Since there is no way that our choice of algebraic symbols can affect the answer to a mathematical problem or the pH of a solution, it is clear that we can select the notation that is most comfortable for us. We rephrase the problem as: "A diprotic acid H_2A has $K_1 = 5.9 \times 10^{-6}$ and $K_2 = 6.2 \times 10^{-9}$; calculate the pH of a 0.05 M solution of the intermediate salt NaHA." We emphasize that this is not merely a similar problem, it is the *same problem*. The numerical answer is obtained by use of the BASIC program of Appendix II, Section II.4, and is pH = 6.719.

In the following section we develop an approximate approach to this problem.

9.9 AMPHIPROTIC SALTS

A different type of amphiprotism occurs when the substance is an ionic compound, each of whose component ions undergoes acid-base hydrolysis. For example, ammonium nitrite (NH_4NO_2) is the salt of nitrous acid ($K_a = 4 \times 10^{-4}$) and ammonia ($K_b = 1.9 \times 10^{-5}$). Although the ions hydrolyze separately according to

$$NH_4^+ + H_2O \rightleftharpoons NH_3 + H_3O^+ \qquad K_{bh} = \frac{K_w}{K_b} \tag{9-38}$$

and

$$NO_2^- + H_2O \rightleftharpoons HNO_2 + OH^- \qquad K_{ah} = \frac{K_w}{K_a} \tag{9-39}$$

a better way to think of the reaction is as a partial self-neutralization

$$NH_4^+ + NO_2^- \rightleftharpoons NH_3 + HNO_2 \tag{9-40}$$

In this equilibrium K, the equilibrium constant for (9-40), can be calculated from K_w, K_a, and K_b.

The equilibrium expression is set up

$$K = \frac{[NH_3]\,[HNO_2]}{[NH_4^+]\,[NO_2^-]}$$

and the numerator and denominator are multiplied by $[H_3O^+]\,[OH^-]$; of course, this does not change the value of the expression. When the terms are grouped in the appropriate way, it is seen that each term is equal to a simpler expression.

$$K = \frac{[NH_3][HNO_2]}{[NH_4^+][NO_2^-]} \cdot \frac{[H_3O^+][OH^-]}{[H_3O^+][OH^-]}$$

$$= \underbrace{\frac{[NH_3]}{[NH_4^+][OH^-]}}_{\frac{1}{K_b}} \cdot \underbrace{\frac{[HNO_2]}{[H_3O^+][NO_2^-]}}_{\frac{1}{K_a}} \underbrace{([H_3O^+][OH^-])}_{K_w}$$

Once the value of K is known, from K_a, K_b, and K_w, the extent of reaction can be calculated.

Example 9.10

Calculate the extent of self-neutralization in a solution of 0.05 M NH_4NO_2.

Answer: Set up the equilibrium expression in the usual way, using M to represent the molarity of the salt.

$$NH_4^+ + NO_2^- \rightleftharpoons NH_3 + HNO_2$$

	NH_4^+	NO_2^-	NH_3	HNO_2
initially:	M	M	0	0
at equilibrium:	$M-x$	$M-x$	x	x

then

$$K = \frac{(x)(x)}{(M-x)(M-x)}$$

$$K = \frac{x^2}{(M-x)^2}$$

$$\frac{x}{M-x} = \sqrt{K}$$

$$x = M\sqrt{K} - x\sqrt{K}$$

$$x + x\sqrt{K} = M\sqrt{K}$$

$$x(1+\sqrt{K}) = M\sqrt{K}$$

$$x = \frac{M\sqrt{K}}{1+\sqrt{K}}$$

The equilibrium constant is calculated as outlined above,

$$K = \frac{K_w}{K_aK_b} = \frac{10^{-14}}{(4 \times 10^{-4})(1.9 \times 10^{-5})} = 1.3 \times 10^{-6}$$

and, using the given value $M = 0.05$, we find

$$x = 5.73 \times 10^{-5}$$

Expressed as a percent, this reaction proceeds to the extent of

$$\frac{x}{M}(100) = 0.115\%$$

Normally, we are interested in the calculation of pH, and this is not difficult once the extent of reaction of the ionic species has been determined.

Example 9.11

Calculate the pH of the solution of Example 9.10.

Answer: Knowing the extent of reaction $x = 5.73 \times 10^{-5}$, we calculate the equilibrium concentrations:

$$[HNO_2] = x = 5.73 \times 10^{-5}$$

$$[NO_2^-] = 0.05 - x = 0.0499$$

Now, whatever else is occurring in the solution, it is always true that

$$K_a = \frac{[H_3O^+][NO_2^-]}{[HNO_2]} \quad \text{or} \quad [H_3O^+] = K_a \frac{[HNO_2]}{[NO_2^-]}$$

from which

$$[H_3O^+] = \frac{(4 \times 10^{-4})(5.73 \times 10^{-5})}{0.0499} = 4.59 \times 10^{-7}$$

$$pH = 6.34$$

This result could also have been reached by considering the NH_4^+/NH_3 equilibrium:

$$K_b = \frac{[NH_4^+][OH^-]}{[NH_3]} \quad \text{or} \quad [OH^-] = K_b \frac{[NH_3]}{[NH_4^+]}$$

$$[OH^-] = \frac{(1.9 \times 10^{-5})(5.73 \times 10^{-5})}{0.0499} = 2.18 \times 10^{-8}$$

$$pOH = 7.66$$

$$pH = 14 - 7.66 = 6.34$$

The treatment of Examples 9.10 and 9.11 is approximate in that it ignores any contribution to the pH arising from the autoionization of water. As such, it cannot be expected to yield accurate pH values when the concentration of the salt is low. However, before considering the exact method of calculation, let us generalize the above result and indicate a shortcut to the pH calculation.

We begin with the equilibrium expressions for the reactions (9-38) and (9-39):

$$K_{bh} = \frac{[NH_3][H_3O^+]}{[NH_4^+]} \quad \text{and} \quad K_{ah} = \frac{[HNO_2][OH^-]}{[NO_2^-]}$$

Substitute $K_{bh} = K_w/K_b, K_{ah} = K_w/K_a$, and $[OH^-] = K_w/[H_3O^+]$:

$$\frac{K_w}{K_b} = \frac{[NH_3][H_3O^+]}{[NH_4^+]} \quad \text{and} \quad \frac{K_w}{K_a} = \frac{[HNO_2]K_w}{[NO_2^-][H_3O^+]}$$

Rearrange to get two equations for $[H_3O^+]$; then multiply them together:

$$[H_3O^+] = \frac{K_w[NH_4^+]}{K_b[NH_3]} \quad \text{and} \quad [H_3O^+] = K_a\frac{[HNO_2]}{[NO_2^-]}$$

$$[H_3O^+]^2 = K_w\frac{K_a[NH_4^+][HNO_2]}{K_b[NH_3][NO_2^-]}$$

The stoichiometry of reaction (9-40) requires that

$$[NH_4^+] = [NO_2^-] \quad \text{and} \quad [NH_3] = [HNO_2]$$

so that the expression for $[H_3O^+]^2$ can be simplified to

$$[H_3O^+]^2 = K_w\frac{K_a\cancel{[NO_2^-]}\cancel{[NH_3]}}{K_b\cancel{[NH_3]}\cancel{[NO_2^-]}} = K_w\frac{K_a}{K_b}$$

Take logs, multiply both sides by (−1), and convert to "p" notation:

$$2\log[H_3O^+] = \log K_w + \log K_a - \log K_b$$

$$2\text{pH} = \text{p}K_w + (\text{p}K_a - \text{p}K_b)$$

Since $\text{p}K_w = 14$ (at 25°C), we can divide by 2 and get

$$\text{pH} = 7 + \frac{1}{2}(\text{p}K_a - \text{p}K_b) \qquad (9\text{-}41)$$

By substituting $\text{p}K_{ah} = \text{p}K_w - \text{p}K_a$ and $\text{p}K_{bh} = \text{p}K_w - \text{p}K_b$ into (9-41), we obtain the equivalent form,

$$\text{pH} = 7 + \frac{1}{2}(\text{p}K_{bh} - \text{p}K_{ah}) \qquad (9\text{-}42)$$

It is not difficult to show that this treatment also applies to the hydrolysis of intermediate salts of a polyprotic acid. We have already noted that salts such as NaHA are amphiprotic; acid and base reactions of an intermediate salt are shown in reactions (9-38a) and (9-39a).

$$HA^- + H_2O \rightleftharpoons A^{2-} + H_3O^+ \qquad K_2 \tag{9-38a}$$

$$HA^- + H_2O \rightleftharpoons H_2A + OH^- \qquad K_{1h} = \frac{K_w}{K_1} \tag{9-39a}$$

These reactions are the same as (9-38) and (9-39) if we recognize that the conjugate pair HA^-/A^{2-} in (9-38a) corresponds to the conjugate pair NH_4^+/NH_3 in (9-38), and the pair H_2A/HA^- in (9-39a) corresponds to HNO_2/NO_2^- in (9-39). The notation for the ionization constants changes also: K_{bh} in (9-38) becomes K_2 in (9-38a), while K_{ah} in (9-39) becomes K_{1h} in (9-39a). The self-neutralization (9-40) becomes (9-40a).

$$2HA^- \rightleftharpoons H_2A + A^{2-} \tag{9-40a}$$

The equilibrium constant expressions are

$$K_2 = \frac{[H_3O^+]\,[A^{2-}]}{[HA^-]} \qquad \text{and} \qquad K_{1h} = \frac{[H_2A]\,[OH^-]}{[HA^-]}$$

Divide K_2 by K_{1h} and substitute $[OH^-] = K_w/[H_3O^+]$ to get

$$\frac{K_2}{K_{1h}} = \frac{[H_3O^+]\,[A^{2-}]/[HA^-]}{[H_2A]\,K_w/[HA^-]\,[H_3O^+]} = \frac{[H_3O^+]^2\,[A^{2-}]}{K_w\,[H_2A]}$$

The stoichiometry of reaction (9-40a) requires that $[H_2A] = [A^{2-}]$, so that

$$\frac{K_2}{K_{1h}} = \frac{[H_3O^+]^2}{K_w}$$

As before, we rearrange the terms, take logs, and convert to "p" notation to get

$$2\text{pH} = \text{p}K_w + (\text{p}K_2 - \text{p}K_{1h})$$

$$\text{pH} = 7 + \frac{1}{2}(\text{p}K_2 - \text{p}K_{1h}) \tag{9-41a}$$

Equation (9-41a) is entirely equivalent to equation (9-41). We can achieve a further simplification by using $\text{p}K_1$ instead of $\text{p}K_{1h}$. See Problem 9.16.

$$\text{pH} = \frac{1}{2}(\text{p}K_1 + \text{p}K_2) \tag{9-43}$$

The pH of solutions of an intermediate salt of a diprotic acid is simply the average of the pK values of the first and second ionization reactions.

Returning now to the case of an amphoteric salt such as NH_4NO_2, we find that the pleasantly simple result [equation (9-41)] is not hard to interpret. It tells us that the pH is close to neutral, as we might expect. Furthermore, the deviation from neutrality depends on *both* hydrolyses, in the opposite sense. If the hydronium-forming hydrolysis (9-38) is stronger than the hydroxide-forming hydrolysis (9-39), then $K_{bh} > K_{ah}$ and $pK_{ah} > pK_{bh}$: equation (9-42) tells us that the pH decreases, as expected for an acid salt. In a sense the deviation from neutrality is related to the *average* contribution of the acid- and base-forming hydrolyses.

If K_a and K_b for an amphiprotic salt are large, the corresponding K_{ah} and K_{bh} for the hydrolyses are small. In this case the approximate treatment cannot be used, for the autoionization of water will be the dominant source of H_3O^+ and OH^-. The same is true if the salt concentration is very small. In these cases an exact treatment is required. A treatment similar to those in Section 9.3 can be followed, with the result

$$Ax^4 + Bx^3 + Cx^2 + Dx + E = 0 \qquad (9\text{-}44)$$

where

$$A = 1$$

$$B = M + K_a + \frac{K_w}{K_b}$$

$$C = K_w\left(\frac{K_a}{K_b} - 1\right)$$

$$D = -\frac{K_w K_a}{K_b}\left(M + K_b + \frac{K_w}{K_a}\right)$$

$$E = -\frac{K_w{}^2 K_a}{K_b}$$

$x = [H_3O^+]$; M = salt molarity

In deriving this result we have assumed that the salt is a 1:1 salt (of the type MX rather than M_2X or MX_2). The only difference between this and the treatment leading to equation (9-34) is that we need *two* conservation of mass equations rather than the single one (9-22) used previously. In the case of a salt such as NH_4NO_2, these equations are

$$[NH_4{}^+] + [NH_3] = M \qquad \text{and} \qquad [NO_2{}^-] + [HNO_2] = M$$

The BASIC program in Appendix II.4 is written to suit the form of equations (9-32), (9-34), and (9-44); as we have just shown, a separate program for amphiprotic salt systems is not necessary.

9.10 INNER SALTS

There is an important class of organic molecules which are amphoteric as a result of containing both acidic and basic functional groups. A common and important example of this class are the amino acids,

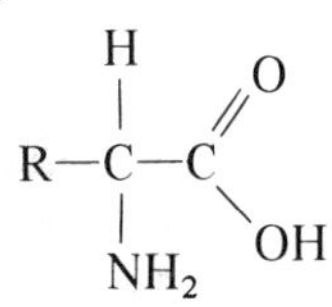

or $H_2NCHRCO_2H$

Although the structural formula of amino acids is most commonly represented as shown above, in fact it has been determined that in the pure, crystalline state there has been an *internal* proton transfer from the acidic functional group to the basic functional group.

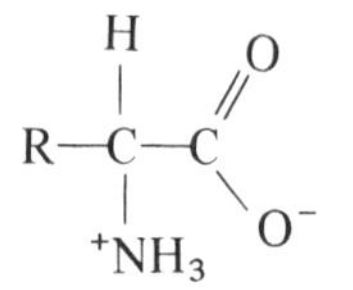

or $H_3\overset{+}{N}CHRCO_2^-$

This dipolar structure is called an *inner salt* or *zwitterion,* and it predominates in solution as well as in the solid state. Typically, only one in about 10^5 dissolved molecules exists in the nonpolar form. In solution the zwitterion can undergo either acidic or basic reaction, and an ionization constant can be determined for each.

$$\text{acid:}\quad H_3\overset{+}{N}CHRCO_2^- + H_2O \rightleftharpoons H_3O^+ + H_2NCHRCO_2^- \qquad K_a \qquad (9\text{-}45)$$

$$\text{base:}\quad H_3\overset{+}{N}CHRCO_2^- + H_2O \rightleftharpoons OH^- + H_3\overset{+}{N}CHRCO_2H \qquad K_b \qquad (9\text{-}46)$$

The pH of a solution of pure amino acid is determined by the relative strengths of these two opposing tendencies. A moment's thought shows that this situation is no different from a solution of an intermediate salt of a polyprotic acid, for instance, $NaHCO_3$. Since this has been discussed and exemplified (Examples 9.4 and 9.9), we will not further discuss pH calculations in zwitterion systems. However, there is one frequently used concept in amino acid and protein chemistry, the isoelectric point, that does merit discussion.

Amphoteric species and diprotic acids share the characteristic of existing in three different degrees of protonation: H_2A, HA^-, and A^{2-} for, say, carbonic acid, and $H_3\overset{+}{N}CHRCO_2H$, $H_3\overset{+}{N}CHRCO_2^-$, and $H_2NCHRCO_2^-$ for an amino acid. As such, the relative amount of each form varies with pH and may be calculated from the equilibrium expressions. The general features of Figure 9.1 apply equally well to diprotic acids and amino acids, provided that the appropriate pK values are well separated. The *isoelectric point* of an amino acid is defined as the pH at which the concentration of the middle form, the zwitterion, is at a maximum. If the pK values are well spaced, this means that at the isoelectric pH there is at most a very small concentration of either of the other two forms. Regardless of the spacing of the pK values it is true that at the maximum concentration of zwitterions the other two forms are present in concentrations equal to one another; the isoelectric point thus can also be described as the pH at which these concentrations are equal. According to equations (9-45) and (9-46),

$$K_a = \frac{[H_3O^+][H_2NCHRCO_2^-]}{[H_3\overset{+}{N}CHRCO_2^-]}$$

and

$$K_b = \frac{[OH^-][H_3\overset{+}{N}CHRCO_2H]}{[H_3\overset{+}{N}CHRCO_2^-]} = \frac{K_w[H_3\overset{+}{N}CHRCO_2H]}{[H_3O^+][H_3\overset{+}{N}CHRCO_2^-]}$$

The quotient K_a/K_b is

$$\frac{K_a}{K_b} = \frac{[H_3O^+][H_2NCHRCO_2^-]}{[H_3\overset{+}{N}CHRCO_2^-]} \cdot \frac{[H_3O^+][H_3\overset{+}{N}CHRCO_2^-]}{K_w[H_3\overset{+}{N}CHRCO_2H]}$$

Since $[H_3\overset{+}{N}CHRCO_2H] = [H_2NCHRCO_2^-]$ at the isoelectric point,

$$\frac{K_a}{K_b} = \frac{[H_3O^+]^2}{K_w}$$

or

$$\frac{K_a K_w}{K_b} = [H_3O^+]^2$$

Take logarithms of both sides:

$$\log K_a + \log K_w - \log K_b = 2 \log [H_3O^+]$$

Multiply by -1 and switch to "p" notation:

$$pK_a + pK_w - pK_b = 2\text{ pH}$$

or

$$\text{pH} = \frac{1}{2}pK_w + \frac{1}{2}(pK_a - pK_b) \qquad (9\text{-}47)$$

This equation is the same as the one used to find the pH of a solution of pure amino acid (9-41); such a solution, if not disturbed by the addition of other acids or bases, will be at its isoelectric point. There are 20 amino acids that occur naturally; these have pK_a values of about 9 to 10, pK_b values of about 11 to 12, and isoelectric points that range from 5 to 7.

PROBLEMS

9.1 Write the ionization constant expression for each of the following acids in water solution. If the species contains more than one ionizable proton, write the ionization constant expression for each successive ionization (K_1, K_2, K_3, etc.) step.

(a) sulfuric acid
(b) perchloric acid
(c) phthalic acid
(d) citric acid
(e) benzoic acid

9.2 Write the ionization constant expression for each of the following bases in water solution.

(a) CH_3NH_2
(b) sodium sulfate
(c) potassium benzoate
(d) $NH_2(CH_2)_6NH_2$
(e) sodium bicarbonate

9.3 The overall ionization constant of H_2S is 1.2×10^{-20}. If K_1 is 1.0×10^{-7}, what is K_2?

9.4 Explain, in your own words, why $K_1 > K_2 > K_3$ for a triprotic acid.

9.5 When H_3AsO_4 is dissolved in water, several arsenic-containing species are formed. Show the structure of each of these substances and write an expression that would give the total arsenic concentration (moles/liter) in the solution.

9.6 Using Figure 9.2, describe the species that are present in a solution of phosphoric acid that has been adjusted to the following pH levels.
(a) pH = 2.0 (c) pH = 9.0 (e) pH = 14.0
(b) pH = 5.0 (d) pH = 10.0

9.7 Using Figure 9.1, describe the carbon-containing species present in a solution that has been prepared by adding 5 g of Na_2CO_3 and 3 g of $NaHCO_3$ to 500 mL of water and then adjusting the pH to 8.0 by adding concentrated HCl.

9.8 Describe the solution prepared in Problem 9.7 after the pH has been adjusted to 11.5 with sodium hydroxide.

9.9 Using Figure 9.3, describe the carbon-containing species present in a water solution of citric acid at each of the following pH values.
(a) pH = 1.0 (d) pH = 4.4 (f) pH = 7.0
(b) pH = 2.0 (e) pH = 5.1 (g) pH = 8.0
(c) pH = 3.5

9.10 What is the maximum molarity of the dihydrogen citrate ion (H_2Cit^- in Figure 9.3) which can be achieved by adjusting the pH of a 0.03 *M* solution of citric acid?

9.11 Use Figure 9.3 to describe the composition of a solution of 0.1 *M* citric acid which has been titrated with 3 *N* NaOH until the pH is 5.0. In what way would the final composition be different if the original solution had been 0.1 *M* sodium citrate, $Na_3(C_6H_5O_7)$, and had been titrated to pH = 5.0 with 3 *N* HCl?

9.12 Without consulting Figure 9.2, determine the pH at which $[HPO_4{}^{2-}]$ is maximized in a phosphoric acid solution. The p*K* values are $pK_1 = 2.12$, $pK_2 = 7.21$, and $pK_3 = 12.66$. Check your answer by consulting Figure 9.2.

9.13 Calculate the pH of the following solutions.
(a) 200 mL of 0.01 *M* NaOH plus 50 mL of 0.02 *M* HCl.
(b) 200 mL of 0.10 *M* Na_2CO_3 plus 20 mL of 0.20 *M* HCl.
(c) 250 mL of 0.10 *M* H_2S plus 50 mL of 0.30 *M* NaOH.
(d) 20 mL of 0.80 *M* KOH plus 300 mL of 0.10 *M* KH_2PO_4.
(e) 200 mL of 0.1 *M* HNO_2 plus 50 mL of 0.04 *M* NH_3.
(f) 0.042 *M* Na_2CO_3.
(g) 0.052 *M* H_2CO_3.
(h) 100 mL of 0.50 *M* H_2CO_3 plus 400 mL of 0.2 *M* KOH.

9.14 What is the pH of a solution prepared by mixing equal volumes of 0.3 *M* $NaHCO_3$ and 0.1 *M* Na_2CO_3?

9.15 Explain, in your own words, why the first break is the sharpest in carbonic acid titrations (Figure 9.4), while in citric acid titrations (Figure 9.6) the final break is the sharpest.

9.16 Show that equations (9-41), (9-41a), (9-42), and (9-43) are equivalent to one another.

Chapter 10

Related Equilibria

10.1 COMPLEXATION

One consequence of the Lewis acid nature of metal ions, especially transition-metal ions, is their ability to form complex ions with electron-rich species known as *ligands.* For example, silver ion forms a complex ion with water as the ligand:

$$AgNO_3 + 2H_2O \rightleftharpoons Ag(H_2O)_2^+ + NO_3^- \qquad (10\text{-}1)$$

Each ligand shares two of its nonbonded electrons with the metal ion, forming a coordinate covalent bond. The number of such electron pairs the metal ion accepts is called the *coordination* number of the metal ion. Silver ion can accept two electron pairs, so its coordination number is 2. It is also said that silver has two *coordination sites*. As is the case with hydrogen ion, there is no such thing as a "bare" transition metal ion in aqueous solution. When a crystalline transition-metal salt dissolves, the metal ion becomes separated from the crystal and simultaneously undergoes a Lewis acid-base addition reaction analogous to (10-1), in which water is the base. Now although the correct formula of the species in an aqueous solution of silver nitrate is $Ag(H_2O)_2^+$, it is usual to represent it as $Ag^+(aq)$ or even as Ag^+. In this discussion we will follow common practice and use M^+ or $M^+(aq)$ to represent an aqueous metal ion unless it is important to stress that a number of water molecules are present as ligands.

Many common neutral and anionic species can function as ligands. Many solvent molecules are acceptable ligands, provided that they, like water, are electron-rich. Hydroxide ion is an electron-rich species and is a ligand in complex ions such as $Zn(OH)_4^{2-}$. Some liquids contain more than one atom having nonbonded electrons and can donate more than one electron pair to a metal ion. Two molecules of ethylenediamine ($H_2NCH_2CH_2NH_2$), for example, can fill all four coordination sites of the Zn^{2+} ion.

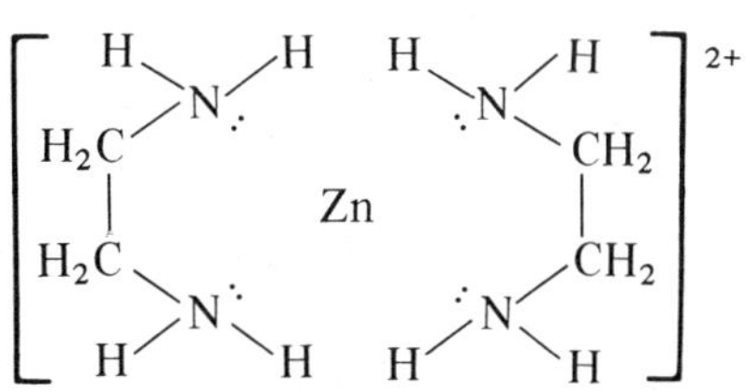

Such ligands are called bidentate, and two additional examples are

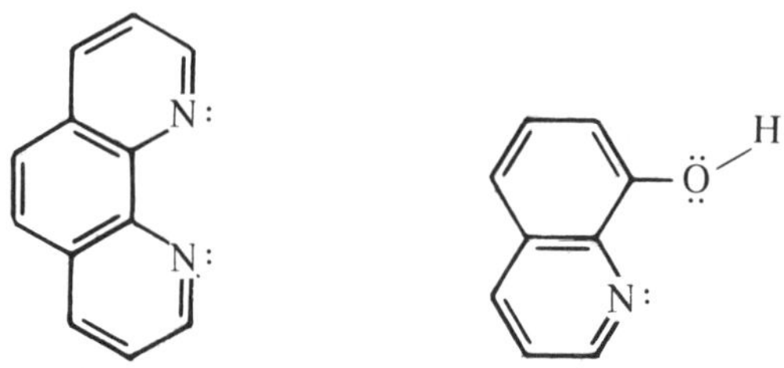

1,10-phenanthroline 8-hydroxyquinoline

Tridentate ligands have three electron-rich atoms, and *tetradentate* ligands have four. Ethylenediaminetetraacetic acid (EDTA) is an example of a hexadentate (six electron-rich atoms) ligand. Coordination complexes containing polydentate ligands are called *chelates* (from the Greek *keta,* crab's claw), and polydentate ligands themselves are sometimes called *chelating agents.*

Complex ions readily undergo nucleophilic displacement reactions in which one ligand (a Lewis base) replaces another.

$$Ag(H_2O)_2^+ + 2NH_3 \rightleftharpoons Ag(NH_3)_2^+ + 2H_2O \qquad (10\text{-}2)$$

$$Cu(NH_3)_4^{2+} + H_2O \rightleftharpoons Cu(NH_3)_3(H_2O)^{2+} + NH_3 \qquad (10\text{-}3)$$

$$Zn(OH)_4^{2-} + H_2O \rightleftharpoons Zn(OH)_3(H_2O)^- + OH^- \qquad (10\text{-}4)$$

Although the bare metal ions are Lewis acids only, and not Brønsted acids, many ligands (such as hydroxide ion) are Brønsted bases as well as Lewis bases. Therefore, in aqueous systems the $H_2O/H^+/OH^-$ equilibrium is involved and complex ion formation with such ligands is dependent on pH. This is shown explicitly in (10-4); in this reaction the position of equilibrium depends on pOH, which in turn depends on pH.

All complex-forming reactions are equilibrium reactions, and as such are characterized by equilibrium constants. Consider the formation of the tetraammine complex of copper(II) ion:

$$Cu^{2+}(aq) + 4NH_3 \rightleftharpoons Cu(NH_3)_4^{2+} \qquad (10\text{-}5)$$

Although the stoichiometry is correctly given by reaction (10-5), the complex ion is actually formed in a series of stepwise nucleophilic displacement reactions.

$$Cu(H_2O)_4^{2+} + NH_3 \rightleftharpoons Cu(NH_3)(H_2O)_3^{2+} + H_2O \quad K_{\text{i}} \qquad (10\text{-}6)$$

$$Cu(NH_3)(H_2O)_3^{2+} + NH_3 \rightleftharpoons Cu(NH_3)_2(H_2O)_2^{2+} + H_2O \quad K_{\text{ii}} \qquad (10\text{-}7)$$

$$Cu(NH_3)_2(H_2O)_2^{2+} + NH_3 \rightleftharpoons Cu(NH_3)_3(H_2O)^{2+} + H_2O \quad K_{\text{iii}} \qquad (10\text{-}8)$$

$$Cu(NH_3)_3(H_2O)^{2+} + NH_3 \rightleftharpoons Cu(NH_3)_4^{2+} + H_2O \quad K_{\text{iv}} \qquad (10\text{-}9)$$

Each of these steps is an equilibrium reaction. As with polyprotic acids, and indeed as with all stepwise equilibria, the overall equilibrium constant is the product of the individual constants.

$$Cu(H_2O)_4^{2+} + 4NH_3 \rightleftharpoons Cu(NH_3)_4^{2+} + 4H_2O \quad K = K_i K_{ii} K_{iii} K_{iv} \tag{10-10}$$

When we write the algebraic expression for this equilibrium constant,

$$K = \frac{[H_2O]^4\,[Cu(NH_3)_4^{2+}]}{[NH_3]^4\,[Cu(H_2O)_4^{2+}]}$$

and make the usual assumption that the concentration of the solvent is constant, we get what is called the *association constant* (or *formation constant* or *stability constant*), K_{assoc}.

$$K_{assoc} = \frac{K}{[H_2O]^4} = \frac{[Cu(NH_3)_4^{2+}]}{[NH_3]^4\,[Cu(H_2O)_4^{2+}]}$$

In using this expression it is neither necessary nor helpful to make explicit recognition of the nature of the aqueous copper ion, $[Cu(H_2O)_4^{2+}]$, so it is usual to write

$$K_{assoc} = \frac{[Cu(NH_3)_4^{2+}]}{[NH_3]^4\,[Cu^{2+}(aq)]} \tag{10-11}$$

Note that this is exactly the expression we would have obtained if we had simply written the equilibrium expression for reaction (10-5). Just as in Chapter 3 with respect to H^+, $H^+(aq)$, and H_3O^+, the notation we use to represent solvated species does not affect equilibrium calculations.

Now, although we have presented the discussion in terms of the stepwise introduction of ligands other than water into the coordination sphere of a complex ion, it is much more common to write these reactions in reverse, as *dissociations*.

$$Cu(NH_3)_4^{2+} \rightleftharpoons Cu(aq)^{2+} + 4NH_3 \tag{10-12}$$

A dissocation equilibrium such as (10-12) is described by a *dissociation constant* (or *instability constant*) K_{diss}.

$$K_{diss} = \frac{[Cu(aq)^{2+}]\,[NH_3]^4}{[Cu(NH_3)_4^{2+}]} \tag{10-13}$$

The usual relationship between the equilibrium constants of reverse reactions holds, that is, $K_{diss} = 1/K_{assoc}$. The individual steps of (10-12) are also represented as dissociation equilibria. For example,

$$Cu(NH_3)_4^{2+} \rightleftharpoons Cu(NH_3)_3^{2+} + NH_3 \quad K_1 \tag{10-14}$$

As usual, $K_{diss} = K_1K_2K_3K_4$. For many reactions only the overall K_{diss} is known, and the individual K values have not been measured. Dissociation constants of several common complex ions are given in Table 10.1.

Table 10.1 Dissociation Constants of Some Complex Ions

Formula	K_{diss}
$Al(OH)_4^-$	1×10^{-33}
$AgCl_2^-$	1×10^{-5}
$Ag(NH_3)_2^+$	4×10^{-8}
$Ag(CN)_2^-$	1×10^{-21}
$Au(CN)_2^-$	3×10^{-29}
$Cu(NH_3)_4^{2+}$	2×10^{-13}
$Cd(NH_3)_4^{2+}$	1×10^{-7}
$Zn(NH_3)_4^{2+}$	1×10^{-9}
$Zn(OH)_4^{2-}$	3×10^{-16}
$Co(NH_3)_6^{3+}$	1×10^{-35}

Example 10.1

Ammonia is added to a 0.5 M solution of $Cu(NO_3)_2$ until $[NH_3] = 0.01\ M$. Calculate what fraction of the copper ion is present as $Cu(aq)^{2+}$. The overall dissociation constant of $Cu(NH_3)_4^{2+}$ is 2×10^{-13}.

Answer: There are several ways of approaching an equilibrium problem. We can assume that initially all Cu^{2+} is in the form of $Cu^{2+}(aq)$, and ask how much $Cu(NH_3)_4^{2+}$ is formed as we add ammonia to a final concentration of 0.01 M. Alternatively, we can assume that initially all Cu^{2+} is in the form of $Cu(NH_3)_4^{2+}$, and ask how much $Cu^{2+}(aq)$ is formed if the solution contains 0.01 M NH_3. Although it might seem that the first of these approaches corresponds more closely to the problem, this is not so. Remember that *an equilibrium system has no memory of the path by which the equilibrium is established,* so that both approaches must lead to the same answer. However, difficulties with significant figures arise with the first approach, whereas the second is straightforward.[1] We assume then that all copper is initially present as $Cu(NH_3)_4^{2+}$, and set up the dissociation equilibrium problem in the usual way.

$$Cu(NH_3)_4^{2+} \rightleftharpoons Cu(aq)^{2+} + 4NH_3$$

	$Cu(NH_3)_4^{2+}$	$Cu(aq)^{2+}$	NH_3
initially:	0.5	0	
at equilibrium:	$0.5 - x$	x	0.01

$$K_{\text{diss}} = \frac{[NH_3]^4\,[Cu(aq)^{2+}]}{[Cu(NH_3)_4^{2+}]}$$

[1] It is interesting and useful to try the first approach, just to see where the trouble starts.

Since K_{diss} is small, the extent of dissociation x is small, and we may approximate $0.5 - x \cong 0.5$; $[NH_3]$ is known to be 0.01 M, and the equation may be solved for x.

$$K_{diss} = 2 \times 10^{-13} \cong \frac{(0.01)^4(x)}{0.5}$$

$$x \cong \frac{(0.5)(2 \times 10^{-13})}{1 \times 10^{-8}} \cong 1 \times 10^{-5}$$

$$\text{fraction of } Cu(aq)^{2+} = \frac{[Cu(aq)^{2+}]}{\text{total}} = \frac{1 \times 10^{-5}}{0.5} = 2 \times 10^{-5}$$

The mathematical approximation used in Example 10.1 is a good one, because (0.5 - 0.00001) = 0.49999, which is very close to 0.5. The approximate method is almost always valid in complex equilibrium problems. Most dissociation constants are small, and in addition are not usually known to more than one-digit accuracy. There is scarcely ever any additional inaccuracy introduced by the mathematical approximation "x is small." We will not develop the exact treatment, which follows the procedure used for polyprotic acid systems in Chapter 9.

10.2 SOLUBILITY

When an ionic substance dissolves in water the process can be represented by a chemical equation such as

$$NaCl(s) \rightarrow Na^+(aq) + Cl^-(aq)$$

If the substance is only sparingly soluble, the reaction will reach a condition, called *saturation*, in which dissolved ions are in equilibrium with undissolved salt. For example,

$$BaF_2 \rightleftharpoons Ba^{2+}(aq) + 2F^-(aq) \qquad K \qquad (10\text{-}15)$$

We note, however, that in the equilibrium expression corresponding to (10-15), the concentration of the solid substance is a constant (equal to its density; the density of BaF_2 is

$$K = \frac{[Ba^{2+}(aq)]\,[F^-(aq)]^2}{[BaF_2(s)]}$$

4.83 g/cm^3 = 27.6 mol/liter). Since the value of any constant concentration, solid or liquid, may be incorporated into the equilibrium constant, we write

$$K[BaF_2(s)] = [Ba^{2+}(aq)]\,[F^-(aq)]^2 \qquad (10\text{-}16)$$

or

$$K_{sp} = [Ba^{2+}(aq)]\,[F^-(aq)]^2$$

and call the new constant K_{sp}, or *solubility product*. It is important to remember that *the solubility product refers to the equilibrium that exists between a solid substance and its dissolved dissociation products (ions); the solubility product has no relevance if the substance is completely dissolved and there is no solid present.* Many examples of solubility equilibrium problems can be written; we offer a selection of problems in the remainder of this chapter.

Example 10.2

Magnesium hydroxide has a solubility of 9.0×10^{-4} g/100 mL at 18°C. Calculate the solubility product K_{sp}.

Answer: The formula weight of $Mg(OH)_2$ is 58.3 g/mol, so that its molar solubility is $9.0 \times 10^{-4}/58.3 = 1.54 \times 10^{-5}$ mol/100 mL or 1.54×10^{-4} *M*. If 1.54×10^{-4} mol dissolves in each liter of solution, then $[Mg^{2+}] = 1.54 \times 10^{-4}$ *M* and $[OH^-] = 3.08 \times 10^{-4}$ *M*, from which the solubility product is found by

$$
\begin{aligned}
K_{sp} &= [Mg^{2+}][OH^-]^2 \\
&= (1.54 \times 10^{-4})(3.08 \times 10^{-4})^2 \\
&= 1.5 \times 10^{-11}
\end{aligned}
$$

This problem may also be worked in reverse, that is, calculation of the solubility from the solubility product.

Example 10.3

Calculate the solubility of magnesium hydroxide, given that $K_{sp} = 1.5 \times 10^{-11}$.

Answer: Set up the equilibrium problem in the usual way.

$$Mg(OH)_2(s) \rightleftharpoons Mg^{2+} + 2OH^-$$

	$Mg(OH)_2(s)$	Mg^{2+}	$2OH^-$
initially:	solid	0	0
at equilibrium:	solid	x	$2x$

$$K_{sp} = [Mg^{2+}][OH^-]^2 = (x)(2x)^2 = 4x^3$$

$$
\begin{aligned}
[Mg^{2+}] = x &= \sqrt[3]{\frac{K_{sp}}{4}} = \sqrt[3]{\frac{1.5 \times 10^{-11}}{4}} \\
&= \sqrt[3]{3.75 \times 10^{-12}} = 1.54 \times 10^{-4}\ M
\end{aligned}
$$

Since the molar solubility of $Mg(OH)_2$ is equal to the molarity of Mg^{2+} in a saturated solution, we find the mass solubility from the molar solubility and the molecular weight.

$$
\begin{aligned}
S &= (1.54 \times 10^{-4}\ \text{mol/liter})(58.3\ \text{g/mol}) \\
&= 9.0 \times 10^{-3}\ \text{g/liter}
\end{aligned}
$$

Example 10.4

Solid silver chloride is in equilibrium with a solution containing both Ag^+ and Cl^- ions. If the chloride ion concentration is 0.025 *M*, what is $[Ag^+]$? K_{sp} for AgCl is 1.1×10^{-10}.

Answer: The problem specifies that the system is in equilibrium, so the relationship

$$K_{sp} = [Ag^+][Cl^-]$$

must hold. We insert the given values of $[Cl^-]$ and K_{sp} to find $[Ag^+]$:

$$[Ag^+] = \frac{K_{sp}}{[Cl^-]}$$

$$= \frac{1.1 \times 10^{-10}}{0.025} = 4.4 \times 10^{-9}\ M$$

In a solution of an ionic substance the concentrations of anions and cations will always have some value, so that we can always attach a numerical value to a product such as $[Mg^{2+}][OH^-]^2$ or $[Ag^+][Cl^-]$. Such a product is called the *concentration product* or the *ion product*. The concentration product is equal to the solubility product *only if the dissolved ions are in equilibrium with undissolved solid*, that is, if the solution is saturated. Imagine slowly dissolving $Mg(OH)_2$ in water; the concentration product, $P = [Mg^{2+}][OH^-]^2$, gradually increases from an initial value of zero as more and more $Mg(OH)_2$ dissolves. Eventually, the limit to solubility is reached: when more $Mg(OH)_2$ is added it does not dissolve, but remains as solid material in equilibrium with the solution. At this point *P* has become equal to K_{sp}. The concentration product may assume any value between 0 and K_{sp}, *but it can never exceed* K_{sp} in an equilibrium system.[2]

Example 10.5

Concentrated NaCl solution is added dropwise to a 0.075 *M* solution of $AgNO_3$. What chloride ion concentration must be reached for the precipitation of AgCl to begin?

Answer: Precipitation begins when solubility equilibrium is attained, that is, when the concentration product equals K_{sp}. At this point no loss of Ag^+ from the solution has yet occurred, so $[Cl^-]$ can be calculated from the initial Ag^+ concentration and K_{sp}.

$$K_{sp} = [Ag^+][Cl^-]$$

$$1.1 \times 10^{-10} = [0.075][Cl^-]$$

$$[Cl^-] = \frac{1.1 \times 10^{-10}}{0.075} = 1.5 \times 10^{-9}\ M$$

[2] Nonequilibrium solutions are possible in which *P* exceeds K_{sp}; these are called *supersaturated solutions*, and they are unstable. Eventually some of the dissolved material precipitates out, and the value of *P* decreases until it becomes equal to K_{sp}. Supersaturated solutions cannot be prepared by simply dissolving extra solute. Instead, an almost-saturated solution is prepared in which $P \approx K_{sp}$; then K_{sp} is decreased, usually by lowering the temperature. After this the concentration product, which does not change unless precipitation occurs, is greater than K_{sp}.

In working this problem we made the tacit assumption that the volume increase is negligible. Suppose that we had 1.00 liter of solution initially; we would need to add 1.5×10^{-9} mole of NaCl to cause precipitation. If the concentrated NaCl solution were 0.3 *M*, this would require no more than $(1.5 \times 10^{-9})/0.3 = 5 \times 10^{-9}$ liter, or about 1/10,000 of a drop. The assumption of negligible volume increase was justified.

Since many hydroxides are only slightly soluble, solubility problems are often intimately related to pH.

Example 10.6

K_{sp} for $Mg(OH)_2$ is 1.5×10^{-11} at 18°C. What is the pH of a saturated solution of $Mg(OH)_2$?

Answer: We have already solved part of this problem in Example 10.3, when we found that at equilibrium

$$[Mg^{2+}] = x = \sqrt[3]{\frac{K_{sp}}{4}} = 1.54 \times 10^{-4}\ M$$

By the stoichiometry of the compound the OH^- concentration is twice the Mg^{2+} concentration:

$$[OH^-] = 2x = 3.08 \times 10^{-4}\ M$$

$$pOH = 3.51$$

At 18°C, $pK_w = 14.24$, so

$$pH = pK_w - pOH = 10.73$$

10.3 SIMULTANEOUS EQUILIBRIA

We are so accustomed to dealing with the relationship between $[H^+]$ and $[OH^-]$ in aqueous solutions, pH + pOH = pK_w, that Example 10.6 might not have seemed like anything more than a simple solubility-product problem. But, in fact, it is considerably more complicated, for it is an example of *simultaneous equilibria.* By this we mean the presence, in the same system, of two or more equilibrium chemical reactions which have a species in common. The two equilibria in Example 10.6 are

$$Mg(OH)_2(s) \rightleftharpoons Mg^{2+} + 2OH^- \qquad K_{sp}$$

$$H_2O + H_2O \rightleftharpoons H_3O^+ + OH^- \qquad K_w$$

The corresponding equilibrium relationships $K_{sp} = [Mg^{2+}][OH^-]^2$ and $K_w = [H_3O^+][OH^-]$ are both true. We used this fact implicitly in solving Example 10.6. A somewhat more involved problem of simultaneous equilibria is given in the next example.

Example 10.7

Calculate the pH of a saturated solution of strontium fluoride, $K_{sp} = 7.9 \times 10^{-10}$. For hydrofluoric acid, $K_a = 6.6 \times 10^{-4}$.

Answer: We begin by assuming that there is no hydrolysis of F^- and calculate $[F^-]$ by the procedure given in Example 10.3.

$$SrF_2(s) \rightleftharpoons Sr^{2+} + 2F^-$$

initially	solid	0	0
at equilibrium:	solid	x	$2x$

$$K_{sp} = [Sr^{2+}][F^-]^2 = x(2x)^2 = 4x^3$$

$$x = \sqrt[3]{\frac{K_{sp}}{4}} = 5.82 \times 10^{-4}$$

$$[F^-] = 2x = 1.16 \times 10^{-3}\ M$$

Next we consider the hydrolysis equilibrium, and calculate $[OH^-]$ in a solution in which $[F^-]$ is initially $1.16 \times 10^{-3}\ M$.

	F^-	$+ H_2O \rightleftharpoons$	HF	$+ OH^-$
initially:	1.16×10^{-3}		0	0
at equilibrium:	$1.16 \times 10^{-3} - y$		y	y

$$K_h = \frac{y^2}{1.16 \times 10^{-3} - y}$$

The hydrolysis constant is found from K_a.

$$K_h = \frac{K_w}{K_a} = \frac{1.0 \times 10^{-14}}{6.6 \times 10^{-4}} = 1.5 \times 10^{-11}$$

We assume that the extent of hydrolysis is small and use the approximate form:

$$K_h \cong \frac{y^2}{1.16 \times 10^{-3}}$$

$$y^2 = 1.16 \times 10^{-3} \frac{K_w}{K_a}$$

$$[OH^-] = y = \sqrt{1.16 \times 10^{-3} \frac{K_w}{K_a}} = 1.33 \times 10^{-7}$$

$$pOH = 6.88$$

$$pH = 7.12$$

Since the extent of hydrolysis of F^- in Example 10.7 is so small (0.01%), the use of the approximate equation for $[OH^-]$ was justified. The SrF_2 solubility equilibrium is essentially undisturbed by the hydrolysis. A similar assumption in Example 10.8, however, is found not to be valid.

Example 10.8

$MgCO_3$ has a solubility product of 4.0×10^{-5} at 25°C. Calculate the pH of a saturated solution.

Answer: As in Example 10.7, we assume that CO_3^{2-} hydrolysis exerts a negligible influence on the solubility equilibrium, and calculate $[CO_3^{2-}]$ from the solubility product.

$$MgCO_3(s) \rightleftharpoons Mg^{2+} + CO_3^{2-}$$

	$MgCO_3(s)$	Mg^{2+}	CO_3^{2-}
initially:	solid	0	0
at equilibrium:	solid	x	x

$$K_{sp} = [Mg^{2+}][CO_3^{2-}] = x^2$$

$$x = [CO_3^{2-}] = \sqrt{K_{sp}} = 6.3 \times 10^{-3}\ M$$

Still assuming that the concentration of CO_3^{2-} established by the $MgCO_3$ solubility equilibrium suffers only a negligible decrease by hydrolysis, we calculate the pH of a $6.3 \times 10^{-3}\ M$ solution of CO_3^{2-}.

$$CO_3^{2-} + H_2O \rightleftharpoons HCO_3^- + OH^-$$

	CO_3^{2-}	HCO_3^-	OH^-
initially:	6.3×10^{-3}	0	0
at equilibrium:	$(6.3 \times 10^{-3} - y)$	y	y

The hydrolysis constant is found from K_a,

$$K_h = \frac{K_w}{K_a} = \frac{10^{-14}}{4.0 \times 10^{-11}} = 2.5 \times 10^{-4}$$

$$= \frac{[HCO_3^-][OH^-]}{[CO_3^{2-}]} = \frac{y^2}{6.3 \times 10^{-3} - y} \cong \frac{y^2}{6.3 \times 10^{-3}}$$

$$[HCO_3^-] = [OH^-] = y \cong \sqrt{(6.3 \times 10^{-3})K_h} = 1.25 \times 10^{-3}$$

$$pOH = 2.9$$

$$pH = 14 - 2.9 = 11.1$$

Although we have finished the pH calculation, as the problem asked for, we should check the validity of our assumption by finding $[CO_3^{2-}]$.

$$[CO_3^{2-}] = 6.3 \times 10^{-3} - 1.25 \times 10^{-3} = 5.0 \times 10^{-3}$$

Hydrolysis apparently causes a decrease of $[CO_3^{2-}]$ from about 0.0063 to about 0.005; such a change can hardly be considered negligible. Thus in this case it is not valid to assume that hydrolysis may be neglected. The problem must be solved in a different way, by successive approximations.

10.4 SUCCESSIVE APPROXIMATIONS

The accuracy of the result in Example 10.8 can be improved as follows. Imagine a hypothetical sequence of events leading to the final state of equilibrium. First, $MgCO_3$ dissolves to the extent predicted by its K_{sp}, but *without* hydrolysis of CO_3^{2-}. Second, hydrolysis takes place, decreasing the amount of CO_3^{2-} and thus decreasing the concentration product to a value less than K_{sp}. Third, more $MgCO_3$ dissolves (without hydrolysis of the additional CO_3^{2-}) until the concentration product $P = K_{sp}$, and fourth, the additional CO_3^{2-} hydrolyzes, decreasing P below K_{sp} once again. Since only a small amount of additional $MgCO_3$ dissolves in step 3, the change in the overall $[CO_3^{2-}]$ in step 4 is considerably less than it was in step 2. Finally, we imagine this process to be repeated over and over again, with the change in $[CO_3^{2-}]$ becoming smaller each time, until at last $[CO_3^{2-}]$ reaches its final equilibrium value. Although, of course, the system does not behave this way, we stress again that equilibrium calculations give the same results no matter what path is followed to the equilibrium state. As it happens, the path described above is the description, in words, of the mathematical procedure known as the *method of successive approximations.*

Let us use the method of successive approximations to arrive at accurate values for the pH, pOH, and the concentrations of Mg^{2+}, CO_3^{2-}, and HCO_3^- in Example 10.8. We wish to parallel, in mathematical terms, the recursive approach to equilibrium described above. It is always true that

$$K_{sp} = [Mg^{2+}][CO_3^{2-}] \tag{10-17}$$

or

$$[CO_3^{2-}] = \frac{K_{sp}}{[Mg^{2+}]}$$

Similarly, it is true that $[Mg^{2+}] = [CO_3^{2-}] + [HCO_3^-]$, since the only source of HCO_3^- is CO_3^{2-} hydrolysis, and the only source of CO_3^{2-} is $MgCO_3$. Thus we can write

$$[CO_3^{2-}] = \frac{K_{sp}}{[CO_3^{2-}] + [HCO_3^-]} \tag{10-18}$$

Our first approximation is to ignore the hydrolysis (as we did in Example 10.8), by setting $[HCO_3^-] = 0$ in (10-18). We then solve (10-18); the resulting value of $[CO_3^{2-}]$, which is the same as that obtained in Example 10.8, is called "$[CO_3^{2-}]_1$":

$$[CO_3^{2-}]_1 = \frac{K_{sp}}{[CO_3^{2-}]_1 + 0} = \frac{K_{sp}}{[CO_3^{2-}]_1}$$

$$[CO_3^{2-}]_1^2 = K_{sp}$$

$$[CO_3^{2-}]_1 = \sqrt{K_{sp}} = \sqrt{4.0 \times 10^{-5}}\ M = 6.32 \times 10^{-3}\ M$$

Improvement in accuracy is gained by including the nonzero value of $[HCO_3^-]$ in (10-18). We recognize that $K_h = [OH^-][HCO_3^-]/[CO_3^{2-}]$ and that, by stoichiometry, $[OH^-] = [HCO_3^-]$. Then

$$K_h = \frac{[HCO_3^-]^2}{[CO_3^{2-}]}$$

or

$$[HCO_3^-] = \sqrt{[CO_3^{2-}]K_h}$$

Substituting this relationship into equation (10-18), we get a new expression for $[CO_3^{2-}]$,

$$[CO_3^{2-}] = \frac{K_{sp}}{[CO_3^{2-}] + \sqrt{[CO_3^{2-}]K_h}} \qquad (10\text{-}19)$$

This expression is easier to use if it is rewritten:

$$[CO_3^{2-}]([CO_3^{2-}] + \sqrt{[CO_3^{2-}]K_h}) = K_{sp}$$

$$[CO_3^{2-}]^2 + [CO_3^{2-}]\sqrt{[CO_3^{2-}]K_h} = K_{sp}$$

$$[CO_3^{2-}]^2 = K_{sp} - [CO_3^{2-}]\sqrt{[CO_3^{2-}]K_h}$$

$$[CO_3^{2-}] = \sqrt{K_{sp} - [CO_3^{2-}]\sqrt{[CO_3^{2-}]K_h}} \qquad (10\text{-}20)$$

where $K_{sp} = 4.0 \times 10^{-5}$ and $K_h = 2.5 \times 10^{-4}$.

Although this expression looks cumbersome, its use is surprisingly easy. What we must do is insert, on the right-hand side, the approximate value $[CO_3^{2-}]_1 = 6.32 \times 10^{-3}\ M$; the result will be a new and better approximation, called $[CO_3^{2-}]_2$, which takes into account the fact that $[CO_3^{2-}]$ is slightly less than $[Mg^{2+}]$ because of hydrolysis.

$$[CO_3^{2-}]_2 = \sqrt{K_{sp} - [CO_3^{2-}]_1\sqrt{[CO_3^{2-}]_1 K_h}} = 5.66 \times 10^{-3}$$

We continue this process as many times as we like, each time using approximation number n on the right-hand side to generate approximation number $(n + 1)$ on the left. The results are

$$[CO_3^{2-}]_1 = 0.00632456 \quad (= \sqrt{K_{sp}})$$

$$[CO_3^{2-}]_2 = 0.00566103$$

$$[CO_3^{2-}]_3 = 0.00576761$$

$$[CO_3^{2-}]_4 = 0.00575103$$

$$[CO_3^{2-}]_5 = 0.00575362$$

$$[CO_3^{2-}]_6 = 0.00575321$$

$$[CO_3^{2-}]_7 = 0.00575328$$

$$[CO_3^{2-}]_{8,9\ \ldots\ \infty} = 0.00575327$$

Here we have expressed the result to six significant figures in order to illustrate the way in which this procedure operates. The successive approximations get closer and closer to the correct answer, and it is up to the problem solver to stop when the answer is sufficiently accurate. In this case we need go no further than $[CO_2^{2-}]_3 = 0.0058$, since K_h and K_{sp} are known to only two significant figures.

Once the value of $[CO_3^{2-}]$ is known, the concentrations of the other species are found. Again we use an inappropriately high number of significant figures, in order to better demonstrate the power of the method.

$$[HCO_3^-] = \sqrt{K_h [CO_3^{2-}]} = \sqrt{(2.50000 \times 10^{-4})(0.00575327)} = 1.19930 \times 10^{-3}$$

$$[Mg^{2+}] = [CO_3^{2-}] + [HCO_3^-]$$

$$= 5.75327 \times 10^{-3} + 1.19930 \times 10^{-3} = 6.95257 \times 10^{-3}$$

$$[OH^-] = [HCO_3^-] = 1.19930 \times 10^{-3}$$

$$pOH = 2.92$$

$$pH = 11.08$$

If our value of $[CO_3^{2-}]$ is correct, the product $[Mg^{2+}][CO_3^{2-}]$ will equal the solubility product. In fact it does, to at least six significant figures: $(0.00575327)(0.00695257) = 4.00000 \times 10^{-5}$.

The general problem illustrated in Examples 10.7 and 10.8 occurs whenever the anion of a sparingly soluble salt is also the anion of a weak acid. The same problem occurs, although less commonly, if the cation of a sparingly soluble salt is also the cation of a weak base.

10.5 PRECIPITATION OF HYDROXIDES BY pH CONTROL

In Example 10.6 the concentration of a saturated solution of $Mg(OH)_2$ was calculated and found to be 1.5×10^{-4} *M*, with pH = 10.7. If to such a solution we add hydroxide from another source (NaOH, for example), the $[OH^-]$ will increase and $[Mg^{2+}]$ must decrease in order to maintain the solubility equilibrium. The only way this can happen is by the formation of more solid $Mg(OH)_2$; that is, the solubility equilibrium is forced to the left and precipitation occurs (compare Example 10.5). Precipitation can also be induced in solutions that were not originally saturated, by increasing the hydroxide concentration.

Example 10.9

The pH of a solution of 5.0×10^{-7} *M* $Mg(OH)_2$ is increased by the dropwise addition of 1 *M* NaOH. Assuming a negligible change in volume, at what pH will precipitation begin?

Answer: Precipitation begins when the concentration product equals the solubility product. Since $[Mg^{2+}]$ is 5.0×10^{-7} *M*, the required $[OH^-]$ is given by

$$K_{sp} = [Mg^{2+}][OH^-]^2$$

$$[OH^-] = \sqrt{\frac{K_{sp}}{[Mg^{2+}]}} = \sqrt{\frac{1.5 \times 10^{-11}}{5.0 \times 10^{-7}}}$$

$$= 5.5 \times 10^{-3}$$

$$pOH = 2.26$$

$$pH = 11.74$$

In this problem (and others of this sort) it is not necessary that the solution be prepared originally from the material that precipitates, $Mg(OH)_2$. Magnesium ion from any source will behave in the same way.

Example 10.10

One liter of a solution, originally 3.0×10^{-4} *M* $MgCl_2$, is concentrated by boiling until its volume is reduced to 25 mL. At this point 25 mL of 1.0 *M* NaOH is added. Calculate (a) the equilibrium concentration of Mg^{2+} and (b) the fraction of the total magnesium which is present in solid form.

Answer: (a) As a first approximation we can say that the $[OH^-]$ concentration of the final 50-mL solution is 0.5 *M*, since the NaOH has been diluted from 25 mL to 50 mL. The maximum concentration of Mg^{2+} that can remain in solution is found from the solubility product:

$$[Mg^{2+}] = \frac{K_{sp}}{[OH^-]^2} = \frac{1.5 \times 10^{-11}}{0.5^2} = 6.0 \times 10^{-11}$$

(b) The solution contains $(6.0 \times 10^{-11}\ M)(0.050\ \text{liter}) = 3.0 \times 10^{-12}$ mol of dissolved magnesium. Since the original amount was $(3.0 \times 10^{-4}\ M)(1\ \text{liter}) = 3.0 \times 10^{-4}$ mol, the fraction $3.0 \times 10^{-12}/3.0 \times 10^{-4} = 10^{-8}$ remains in solution; 99.999999% of the magnesium is precipitated as solid $Mg(OH)_2$.

The hydroxide concentration calculated in (a) is approximate, since the quantity of

OH^- (about 2% of the total OH^-) that reacts to form the precipitate was not considered. An exact answer may be obtained by successive approximations. See Problem 10.24.

10.6 PRECIPITATION OF SULFIDES BY pH CONTROL

The weak acid H_2S is soluble in water to the extent of about 0.10 *M*. In solution it ionizes to give H_3O^+, HS^-, and S^{2-}, according to its ionization constants $K_1 = 1.0 \times 10^{-7}$ and $K_2 = 1.2 \times 10^{-13}$. The equilibrium concentrations are related by the expression

$$K = K_1K_2 = \frac{[H_3O^+]^2[S^{2-}]}{[H_2S]} \tag{10-21}$$

Since $[H_2S]$ is 0.10 *M* in a saturated solution, we can rearrange (10-21) and write

$$[S^{2-}] = \frac{[H_2S]K_1K_2}{[H_3O^+]^2}$$

$$= \frac{(0.10)K_1K_2}{[H_3O^+]^2}$$

$$= \frac{1.2 \times 10^{-21}}{[H_3O^+]^2} \tag{10-22}$$

In solution, the concentration of S^{2-} can be manipulated by pH control.

Example 10.11

Calculate $[S^{2-}]$ in a saturated solution of H_2S after the pH has been adjusted to 6.5.

Answer:

$$[H_3O^+] = 10^{-pH} = 10^{-6.5} = 3.16 \times 10^{-7}$$

$$[S^{2-}] = \frac{1.2 \times 10^{-21}}{[H_3O^+]^2} = 1.2 \times 10^{-8}\ M$$

The fact that $[S^{2-}]$ can be easily adjusted through pH control is of great importance to qualitative analysis of metals, because almost all metal sulfides are sparingly soluble. Suppose that we have a solution which is 0.01 *M* in both $CdCl_2$ and $NiCl_2$; both of these, like almost all chlorides, are quite soluble. We wish to treat the solution in such a way as to remove (as a precipitate) one metal ion completely, while leaving the other completely in solution. Precipitation of the sulfide will achieve this because, although both CdS and NiS are sparingly soluble, CdS is much less soluble than NiS. K_{sp} for NiS is 1.8×10^{-21}, and K_{sp} for CdS is 7.8×10^{-27}. Because of this difference it is possible to saturate the solution with H_2S and adjust $[S^{2-}]$ so that CdS precipitates and NiS does not.[3] We wish to adjust $[S^{2-}]$ so that the 0.01 M Ni^{2+} ions remain in solution, that is, so that the concentration product $[Ni^{2+}][S^{2-}]$ does not exceed K_{sp}.

$$K_{sp} = [Ni^{2+}][S^{2-}]$$

$$[S^{2-}] = \frac{K_{sp}}{[Ni^{2+}]} = \frac{1.8 \times 10^{-21}}{0.01} = 1.8 \times 10^{-19}\ M$$

Provided that $[S^{2-}]$ remains less than 1.8×10^{-19} M, no NiS will precipitate. This upper limit to sulfide concentration is achieved through pH control. From equation (10-22),

$$[H_3O^+]^2 = \frac{1.2 \times 10^{-21}}{[S^{2-}]} = \frac{1.2 \times 10^{-21}}{1.8 \times 10^{-19}} = 6.7 \times 10^{-3}$$

$$[H_3O^+] = \sqrt{6.7 \times 10^{-3}} = 8.2 \times 10^{-2}$$

$$\text{pH} = 1.1$$

Thus, at a pH of 1.1, $[S^{2-}]$ cannot exceed 1.8×10^{-19}; as much as 0.01 M Ni^{2+} can remain in solution without the precipitation of NiS.

The cadmium ions, however, will be almost completely precipitated. When $[S^{2-}] = 1.8 \times 10^{-19}$, the maximum possible concentration of Cd^{2+} in solution is found by

$$K_{sp} = [Cd^{2+}][S^{2-}]$$

$$[Cd^{2+}] = \frac{K_{sp}}{[S^{2-}]} = \frac{7.8 \times 10^{-27}}{1.8 \times 10^{-19}} = 4.3 \times 10^{-8}\ M$$

Since the solution originally contained 0.01 M, only $(4.3 \times 10^{-8}/0.01) \times 100 = 0.0004\%$ remains in solution. Clearly, a separation of Ni^{2+} and Cd^{2+} has been achieved, and equally clearly, the greater the difference between the two K_{sp} values, the greater will be the degree of separation possible by this method. A pH of 1.1 is best for this separation. If the pH is higher, $[S^{2-}]$ will be greater and NiS will precipitate also. If it is lower, $[S^{2-}]$ will be lower and more Cd^{2+} will remain in solution.

10.7 OTHER pH-DEPENDENT PRECIPITATIONS

The use of pH control to adjust the solubility of ionic substances is by no means limited to hydroxides and sulfides. The solubility of any sparingly soluble salt *of a weak acid* can be controlled in this way, because the anion concentration depends on pH through the weak acid equilibrium. Decrease of pH will cause a decrease in anion concentration, which in turn will increase the solubility (*not* the solubility product) of the

[3] Saturation of a solution with H_2S can be done by bubbling gaseous H_2S through the solution. However, this is rarely done, since H_2S is quite toxic. Instead, H_2S is produced *in situ* by adding thioacetamide and heating gently. Thioacetamide is an unstable compound that yields H_2S as a product of thermal decomposition.

salt. There are many salts in this category, because most substances have limited solubility and most acids are weak. Almost all sulfides, most carbonates, many phosphates, bromates, fluorides, arsenates, and so on, as well as the salts of most organic acids, have low solubility products. Their solubilities can therefore be manipulated by pH control.

PROBLEMS

10.1 Prove that for any complex-forming reaction, $K_{assoc} = 1/K_{diss}$.

10.2 Write the stepwise equations for the formation of tetrahydroxozincate(II) ion, $Zn(OH)_4^{2-}$, by the addition of NaOH to an aqueous solution of $Zn(H_2O)_4^{2+}$.

10.3 Calculate the concentration of Ag^+(aq) in equilibrium with 0.2 M NH_3 and 0.15 M $Ag(NH_3)_2^+$.

10.4 Cyanide ion can react with Ag^+, forming a complex, and also with H^+, forming HCN. Suppose that CN^- is added to a solution in which Ag^+ and H^+ have the same concentration. With which ion will CN^- react to a greater extent?

10.5 Some processes for refining gold involve the intermediate step of formation of the cyanide complex $Au(CN)_2^-$. Suppose that to a solution of Au^+(aq), originally 3×10^{-5} M, is added NaCN so that the $[CN^-]$ becomes 5×10^{-3} M. What fraction of the gold remains present as $Au(aq)^+$?

10.6 Calculate the solubility products for each of the following ionic substances, given the solubility in grams per 100 mL.
(a) silver bromate, $AgBrO_3$; 0.18 g/100 mL
(b) silver carbonate, Ag_2CO_3; 3.2×10^{-3} g/100 mL
(c) magnesium carbonate, $MgCO_3$; 0.060 g/100 mL
(d) aluminum hydroxide, $Al(OH)_3$; 1.9×10^{-8} g/100 mL

10.7 Calculate the solubilities in grams per liter of the following substances, given the solubility products.
(a) lead chromate, $PbCrO_4$; $K_{sp} = 2 \times 10^{-14}$
(b) barium fluoride, BaF_2; $K_{sp} = 1.7 \times 10^{-6}$
(c) silver dichromate, $Ag_2Cr_2O_7$; $K_{sp} = 2.0 \times 10^{-7}$
(d) chromium hydroxide, $Cr(OH)_3$; $K_{sp} = 1 \times 10^{-30}$

10.8 A saturated solution of a metal hydroxide, $M(OH)_2$, has a pH of 10.12. Calculate K_{sp} for $M(OH)_2$.

10.9 Calculate the pH of a saturated solution of $Cr(OH)_3$, given $K_{sp} = 1.0 \times 10^{-30}$.

10.10 Calculate the pH of a saturated solution of $Pb(OH)_2$, given $K_{sp} = 2.5 \times 10^{-16}$.

10.11 Calculate the pH of a saturated solution of AgOH, given $K_{sp} = 1.6 \times 10^{-8}$.

10.12 Calculate the solubility, in grams per liter, of $Cd(OH)_2$ in a solution whose pH is maintained at 9.0. Repeat the calculation for pH = 8.0.

10.13 Calculate the pH of a saturated solution of silver acetate, $K_{sp} = 2.5 \times 10^{-3}$.

10.14 Calculate the pH of a saturated solution of barium carbonate, $K_{sp} = 5.0 \times 10^{-9}$. Compare the results obtained by the approximate treatment with the result of successive approximations.

10.15 Ag_2S has $K_{sp} = 5.5 \times 10^{-51}$, and CuS has $K_{sp} = 8.0 \times 10^{-37}$. Which sulfide has the greater solubility (a) in moles per liter, and (b) in grams per liter?

10.16 Suppose that a solution of $SnCl_2$ (0.05 *M*) and $ZnCl_2$ (0.05 *M*) is made strongly acid with HCl and then saturated with H_2S. Now if the pH is increased by addition of NaOH, at what pH will the first precipitate appear? What compound precipitates first? At what pH will the second precipitate appear? What concentration of the first cation remains in solution when the second one first begins to precipitate? $K_{sp}(SnS) = 1.2 \times 10^{-25}$, $K_{sp}(ZnS) = 1.1 \times 10^{-21}$.

10.17 Aqueous NaCl was added to a solution containing a trace of $Ag^+(aq)$ and a small amount of precipitate formed. When more NaCl was added, the precipitate dissolved. Explain these observations.

10.18 Aluminum sulfide dissolves in water with the release of hydrogen sulfide and the formation of a white precipitate. Use balanced chemical reactions to explain these observations.

10.19 Calculate the solubility of CdS ($K_{sp} = 1.0 \times 10^{-28}$) in 0.01 *M* HCl saturated with H_2S.

10.20 What is the minimum concentration of H^+ necessary to prevent precipitation of ZnS from a solution of 0.05 *M* $ZnCl_2$ when the solution is saturated with H_2S? K_{sp} for ZnS = 1.1×10^{-21}.

10.21 Suppose that a solution which is 0.1 *M* in hydrochloric acid and also 0.1 *M* in $MnCl_2$ is saturated with H_2S. Does a precipitate of MnS form? $K_{sp}(MnS) = 5.1 \times 10^{-15}$.

10.22 The solubility product of $Mg(OH)_2$ is 1.5×10^{-11}. Calculate the solubility of $Mg(OH)_2$ in a solution which is 0.3 *M* in NH_4Cl and 0.1 *M* in NH_3.

10.23 If 0.005 mol of $AgNO_3$ is dissolved in 1 liter of 0.3 *M* NH_3 [thus forming $Ag(NH_3)_2{}^+$] and the concentration of Cl^- is brought to 0.05 *M* by the addition of NaCl, will AgCl precipitate? $K_{sp} = 1.1 \times 10^{-10}$.

10.24 Use the method of successive approximations to calculate the final hydroxide-ion concentration in Example 10.10.

Chapter 11

Experimental Determination of pK

11.1 WEAK ACIDS

In this chapter we discuss some of the considerations involved in the experimental determination of pK_a values. For the student who wishes to perform such a measurement, we can recommend no better source than A. Albert and E. P. Serjeant's monograph, *The Determination of Ionization Constants* (2nd ed., Chapman & Hall, London, 1971).

The measurement of pK_a is based on the Henderson-Hasselbalch equation (11-1), either in its original form as derived in Section 7.4 or in one of its modified versions.

$$\text{p}K_a = \text{pH} + \log\left(\frac{[\text{protonated form}]}{[\text{unprotonated form}]}\right) \tag{11-1}$$

In general, dilute solutions are used for pK_a determination. If the acid concentration does not exceed 0.01 M, pK_a values accurate to ±0.05 unit can be obtained from equation (11-1). For greater accuracy, however, the equation must be written in terms of solute *activities* rather than concentrations.[1]

$$\text{p}K_a = \text{pH} + \log\left(\frac{a_{\text{protonated form}}}{a_{\text{unprotonated form}}}\right) \tag{11-2}$$

These equations show that the pK_a of a monoprotic acid can be determined by measurement of the pH and of the concentration or activity ratio of the two forms of the substance under study.

It is apparent from equation (11-1) and the titration curves discussed in Chapter 9 that at the half-equivalence point of a titration of a weak acid, where [HA] = [A$^-$], pK_a = pH. This suggests a simple method of pK_a determination; add half the equivalent amount of base to the acid and measure the pH. However, a single measurement of any quantity is unreliable because errors, from any source, are difficult to detect. A common method for improving the precision of a pK_a determination is as follows. Equation (11-1) shows that the sum of the pH and the logarithm of the ratio [HA]/[A$^-$] is equal to pK_a.

[1] A terse account of solute activity is presented in Appendix IV. Students unfamiliar with the concepts of activity and activity coefficients should study it before proceeding with this chapter.

The pK_a can therefore be calculated from any point on the titration curve, not just the point of half neutralization. Measuring the pH at nine points during a titration, calculating the pK_a for each, and averaging the results yields a pK_a value that is three times as reliable as the result of a single measurement. In addition, common errors such as misreading the pH or the volume of titrant are at once apparent from discrepancies in the pK_a values calculated from the different measurements. In the following sections we discuss several different methods of pH and concentration measurements.

11.2 POTENTIOMETRIC TITRATION

If the acid has a pK_a value in the range 2 to 11 and is sufficiently soluble so that a 0.01 *M* solution can be prepared, a carefully performed potentiometric titration can yield a pK_a value accurate to about ±0.05. In the potentiometric method it is the electrochemical potential of the solution which is measured. This potential changes as the activity of H_3O^+ changes, and the measured potential may readily be converted to pH. The instrument used in these determinations is in fact nothing more than an accurate pH meter. Titrations should be carried out in solutions no more concentrated than 0.01 *M*, and the normality of the titrant should be at least 10 times that of the acid. The temperature must be held constant, and CO_2 should be excluded.[2]

To perform the pK_a measurement, a solution of the acid of accurately known concentration is prepared, and successive additions of 1/10 of the equivalent amount of base are made; the pH is measured after each addition. The result of these measurements is a set of nine pH values, each paired with a ratio of

$$\frac{[\text{protonated form, HA}]}{[\text{nonprotonated form, A}^-]}$$

equal to 9/1, 8/2, 7/3, 6/4, 5/5, 4/6, 3/7, 2/8, and 1/9. Each of these pairs is converted to a value of pK_a by means of equation (11-1). Ideally, these nine values of pK_a should be identical, but in practice there is always some scatter in the results. The values are averaged and reported as pK_a = x.xx ±0.yy, where the estimated error ±0.yy is equal to the largest deviation of any of the nine values from the average value. If |0.yy| is larger than 0.05, the entire determination is suspect and should be repeated. Among the possible reasons for poor results are faulty laboratory technique or equipment, impure sample or incorrectly known molecular weight, or an acid that is chemically unstable under the conditions of the titration.

The procedure outlined above may be used without modification, provided that all pH measurements are in the range 4 to 10. Within this range the neutralization reaction of the titration goes to completion,

$$HA + OH^- \rightarrow H_2O + A^- \tag{11-3}$$

[2] Since it is a base, the normality of the titrant will gradually change due to neutralization by carbon dioxide absorbed from the atmosphere:

$$CO_2 + OH^- \rightarrow HCO_3^-$$

Textbooks of analytical chemistry give instructions for minimizing the decrease in accuracy from this cause.

and at any point during the titration (up to the equivalence point) the amount of A^- produced is exactly equal to the amount of OH^- that has been added.

If the pH lies outside the range 4 to 10, there are two reasons why the concentration of A^- might not be given by the amount of base added. If the pH is less than 4, it is because the acid is moderately strong ($pK_a \leqslant 3$) and therefore ionizes to a considerable extent even before OH^- is added. As long as the pH is less than 4, the amount of A^- is significantly greater than the amount of added base. On the other hand, if the pH is greater than 10, it must be that the acid is very weak. If that is so, reaction (11-3) does not go to completion, and the amount of A^- is less than the amount of added base. If the pH is outside the range 4 to 10 the concentrations of A^- and HA are not calculated from the amount of titrant added, but rather from the requirement of overall electrical neutrality of the solution. If the titrant is KOH, for instance, it is always true that

$$[H_3O^+] + [K^+] = [A^-] + [OH^-]$$

or

$$[A^-] = [K^+] + [H_3O^+] - [OH^-] \qquad (11\text{-}4)$$

The concentration $[K^+]$ is equal to the amount of base added, $[H_3O^+]$ is measured, and $[OH^-]$ is calculated from $[H_3O^+]$ and K_w. The concentration $[A^-]$ is calculated from these values. The amount of HA is found by subtracting the calculated amount of A^- from the amount of HA originally present.

Example 11.1

After exactly 2/10 of the equivalent amount of 0.100 *N* KOH has been added to 50.0 mL of a 0.0100 *N* solution of a weak acid, the pH is found to be 10.50. Calculate pK_a on the assumption that the titration equilibrium lies far to the right.

Answer: The concentrations of sample and titrant are low enough so that equation (11-1) applies; that is, concentration rather than activity may be used. At the 2/10 point the concentration of HA has fallen to 80% of its initial value, and an amount of A^- equal to the decrease in HA has been formed. Thus

$$\frac{[HA]}{[A^-]} = \frac{80}{20}$$

and

$$pK_a = pH + \log\left(\frac{[HA]}{[A^-]}\right)$$

$$= 10.50 + \log(4.00)$$

$$= 10.50 + 0.60 = 11.10$$

Example 11.2

Repeat the calculation in Example 11.1, this time taking into consideration the large concentration of $[OH^-]$ that exists at pH = 10.50.

Answer: Since equation (11-4) must be used to calculate $[A^-]$, we must first calculate $[K^+]$, $[H_3O^+]$, and $[OH^-]$. The initial amount of acid was

$$(0.0100\ N)(50.0\ \text{mL}) = 0.500\ \text{meq}$$

so that 2/10 neutralization requires (2/10)(0.500) = 0.100 meq base. The titrant concentration is 0.100 *N*, and the volume that contains 0.100 meq OH^- is therefore

$$\frac{0.100\ \text{meq}}{0.100\ \text{meq/mL}} = 1.00\ \text{mL KOH}$$

The solution volume is 50.0 + 1.00 = 51.0 mL. The concentration of K^+ is

$$[K^+] = \frac{0.100\ \text{meq}}{51.0\ \text{mL}} = 0.00196\ N$$

From the pH we calculate

$$[H_3O^+] = 10^{-pH} = 10^{-10.50} = 3.16 \times 10^{-11}\ N$$

and

$$[OH^-] = \frac{K_w}{[H_3O^+]} = \frac{1.00 \times 10^{-14}}{3.16 \times 10^{-11}} = 0.000316\ N$$

The concentration of A^- is

$$[A^-] = [K^+] + [H_3O^+] - [OH^-] \tag{11-4}$$

$$= 0.00196 + 3 \times 10^{-11} - 0.000316$$

$$= 0.00164$$

Note that $[H_3O^+]$ is negligible in equation (11-4). This is always so at high pH, and $[OH^-]$ is always negligible at low pH. There is no pH at which both $[H_3O^+]$ and $[OH^-]$ are significant.

The *total* concentration ([HA] + $[A^-]$) is found from the initial amount of HA and the new volume.

$$\text{total}\ (HA + A^-) = \text{initial HA} = (0.0100\ \text{meq/mL})(50.0\ \text{mL}) = 0.500\ \text{meq}$$

$$\text{total concentration} = [HA] + [A^-] = \frac{0.500\ \text{meq}}{51.0\ \text{mL}} = 0.00980\ N$$

The concentration [HA] is found by subtracting $[A^-]$ from the total.

$$[HA] = 0.00980 - [A^-]$$

$$= 0.00980 - 0.00164$$

$$= 0.00816\ N$$

Finally, the pK_a is calculated from equation (11-1).

$$\mathrm{p}K_a = \mathrm{pH} + \log\left(\frac{[\mathrm{HA}]}{[\mathrm{A}^-]}\right)$$

$$= 10.50 + \log\left(\frac{0.00816}{0.00164}\right)$$

$$= 10.50 + \log(4.98)$$

$$= 10.50 + 0.697 = 11.20$$

Comparison of the $\mathrm{p}K_a$ values obtained in Examples 11.1 and 11.2 demonstrates the importance of correcting for $[\mathrm{OH}^-]$ at high pH. If the example had been at low pH, we would have found the correction for $[\mathrm{H_3O^+}]$ to be significant.

The calculations in Example 11.2 can be summarized as follows. Let C be the total concentration of acid, $C = [\mathrm{HA}] + [\mathrm{A}^-]$; note that C is not constant but decreases during the titration owing to the increase in volume. Then

$$[\mathrm{HA}] = C - [\mathrm{A}^-]$$

$$= C - [\mathrm{K}^+] - [\mathrm{H_3O^+}] + [\mathrm{OH}^-] \tag{11-5}$$

Equation (11-1) becomes

$$\mathrm{p}K_a = \mathrm{pH} + \log\left(\frac{C - [\mathrm{K}^+] - [\mathrm{H_3O^+}] + [\mathrm{OH}^-]}{[\mathrm{K}^+] + [\mathrm{H_3O^+}] - [\mathrm{OH}^-]}\right) \tag{11-6}$$

For a 0.01 *N* sample, use of equation (11-6) is necessary only when the pH lies outside the range 4 to 10. Outside this range either $[\mathrm{H_3O^+}]$ or $[\mathrm{OH}^-]$ is at least 0.0001 *N*, large enough to affect the accuracy of the determination.

11.3 SPECTROPHOTOMETRIC METHOD

The potentiometric method is fast and accurate but cannot be used if the unknown substance is insufficiently soluble or if the $\mathrm{p}K_a$ lies outside the range 2 to 11. In such cases, provided that either HA or A^- (or both) absorbs visible or ultraviolet light, the necessary concentrations can be determined with a spectrophotometer. No corrections need be made for especially large values of $[\mathrm{H_3O^+}]$ or $[\mathrm{OH}^-]$, since the method measures the concentrations $[\mathrm{HA}]$ and $[\mathrm{A}^-]$ directly.

In principle the method is simple enough; a set of seven or eight buffer solutions, ranging in pH from about one unit less than the $\mathrm{p}K_a$ of the unknown to about one unit above, is prepared. The concentrations of the buffer components should not exceed 0.01 *N*. Equal amounts of a solution of the unknown acid are added to each buffer solution. The concentration of the unknown acid *must* be the same in each buffer solution and is usually in the range 10^{-5} to 10^{-4} *M*. The light transmittance or absorbance of each of the resulting solutions is measured, and the pH is measured as well. The concentration ratio $[\mathrm{HA}]/[\mathrm{A}^-]$ is calculated from the absorbance data. In practice there are some difficulties that make this method more time consuming than the potentiometric method. The most obvious of these is the time required to prepare seven or eight buffer solutions.

A second problem is that one must know at the outset an approximate value of pK_a in order to prepare buffers in the pH range pK_a ±1. Usually, what is done is to estimate the pK_a by analogy to similar, known compounds. The estimate is based on the effect of the functional groups in the molecule on the pK_a. The estimate can be confirmed with a single, hasty measurement of the pH of a half-neutralized sample. A third difficulty involves the selection of the wavelength at which the spectrophotometric determination is to be made. Ideally, one should choose a wavelength at which one of the species HA or A^- absorbs strongly and the other absorbs only weakly. A comparison is made of the absorbance of each species, separately, at many different wavelengths. This comparison requires adjustment of the pH so that only one form is present. At a pH 3 units below the pK_a, 99.9% of the acid is present as HA, and the effect of A^- on absorbance is usually negligible. Similarly, when the pH is 3 units larger than the pK_a, the sample contains 99.9% A^- and the effect of HA on the absorbance is negligible.

The determination of pK_a proceeds with a careful measurement of the pH and the absorbance of the set of solutions. The transmittance of a dissolved substance is given by Beer's law.

$$I = I_0 \times 10^{-\epsilon cl} \tag{11-7}$$

where

I = intensity of light transmitted by the sample
I_0 = intensity of light incident on the sample
ϵ = absorption coefficient of the dissolved substance, liter-mol^{-1}-cm^{-1}
c = molarity of the dissolved substance
l = thickness of sample cell, cm

Equation (11-7) is usually written in logarithmic form, as

$$\log\left(\frac{I}{I_0}\right) = -(\epsilon cl)$$

or

$$\log\left(\frac{I_0}{I}\right) = \epsilon cl \tag{11-8}$$

The term on the left of equation (11-8), $\log(I_0/I)$, is called the absorbance and is measured directly by a spectrophotometer. The Henderson-Hasselbalch equation,

$$\mathrm{p}K_a = \mathrm{pH} + \log\left(\frac{[\mathrm{HA}]}{[\mathrm{A^-}]}\right) \tag{11-9}$$

can be rewritten in terms of absorbances as

$$\mathrm{p}K_a = \mathrm{pH} + \log\left(\frac{A_{\mathrm{ion}} - A}{A - A_{\mathrm{mol}}}\right) \tag{11-10}$$

The derivation of equation (11-10) is given in Appendix V. In equation (11-10) the symbol A stands for the absorbance of a solution containing both HA and A^- at a total concentration C = [HA] + [A^-]; A_{ion} is the absorbance of a solution in which [A^-] = C

and [HA] = 0, while A_{mol} is the absorbance of a solution in which [HA] = C and $[A^-] = 0$.

Equation (11-10) is applied to the results of the measurements performed on each solution. When the measurements are averaged, the result is a pK_a determination accurate to perhaps ±0.02.[3]

11.4 OTHER METHODS

Ionization-constant determinations are not limited to potentiometric and spectrophotometric methods; in principle any measurable property of a solution that varies with the extent or rate of a neutralization reaction can be used. An early method (which has been largely supplanted by potentiometric techniques) is that of conductimetric titrations. In conductimetry the electrical conductance of a solution is measured, and concentrations of various solute species are inferred from the overall conductance of the solution. At the beginning of a conductimetric titration of a weak acid the solution contains very few ions, since the substance HA is only slightly ionized. The conductance is low because it is proportional to the concentration of ions in the solution. During the course of the titration the conductance rises smoothly as the acid is converted to its salt (e.g., $K^+ A^-$). When the titration reaches the equivalence point there is an increase in slope. In the undertitration region the ions being added to the solution are K^+ and A^-, whereas in the overtitration region the additional ions are K^+ and OH^-. Since OH^- has the highest intrinsic conductance of all anions, the conductance rises more steeply in the overtitration region.

A conductimetric pK_a determination proceeds like a potentiometric one; a careful titration is carried out, and after each successive 1/10 addition of titrant, both pH and conductance are measured. The pK_a is then calculated from the results of the measurements by a form of the Henderson-Hasselbalch equation that is almost the same as equation (11-10). (The equation differs only in that conductance is used in place of absorbance.) Again, the results are averaged to improve the precision of the determination.

Neutralization of an acid is an exothermic reaction; in the simplest case, the neutralization of a strong acid with a strong base,

$$H_3O^+ + OH^- \rightarrow 2H_2O$$

the reaction releases about 55 kJ per mole of H_3O^+. Neutralizations of weak acids

$$HA + OH^- \rightarrow H_2O + A^-$$

release various, usually lesser, amounts of heat. In either case the amount of heat released during a titration is a measure of the extent of reaction, and the concentrations of HA and A^- can be calculated from the temperature rise. The pK_a is obtained from the values of [HA], $[A^-]$, and pH, using equation (11-1). This method is used most frequently by those whose primary interest is in thermochemistry, in which measurement of the heat

[3] The chief limitation to the accuracy is the interference of the components of the buffer solution with the spectrophotometric measurement. This interference can be minimized by the use of a double-beam spectrophotometer.

released in a protonation reaction is a major objective. The heat of protonation of a substance is determined by a calorimetric titration in which temperature is measured as a function of added base. If the pH is measured as well, the same data can be used to establish the value of pK_a.

11.5 STRONG ACIDS

The "leveling effect" of the solvent (water) was discussed in Chapter 2. A strong acid is by definition one for which the equilibrium

$$HA + H_2O \rightleftharpoons H_3O^+ + A^- \tag{11-11}$$

lies far to the right. All strong acids are leveled to the same strength in aqueous solution, because, regardless of inherent differences in proton-donating tendency, all are essentially 100% ionized. Thus the relative strengths of strong acids cannot be determined in aqueous solution. Measurements can be made in two ways—in the absence of *any* solvent (i.e., gas phase) or in a nonaqueous solvent such as glacial acetic acid or concentrated sulfuric acid.

11.6 NONAQUEOUS SOLVENTS

In a solvent S an acid HA ionizes according to

$$HA + S \rightleftharpoons SH^+ + A^- \tag{11-12}$$

Whether this reaction goes to completion depends on the acid strength of SH^+ as well as that of HA. If the solvent acid SH^+ is particularly strong, then the equilibrium lies to the left and HA is incompletely ionized. Thus HA could be "strong" in aqueous solution [i.e., completely ionized via (11-11)] and "weak" in the solvent S [i.e., only slightly ionized via (11-12)]. A series of strong acids, of different inherent strengths but all completely ionized in water, might be ionized to different extents in a different solvent S. If so, measurement of the extent of ionization yields the relative strength, expressed as the ionization constant K^S. We could define the K^S (and pK^S) as

$$K^S = \frac{[SH^+][A^-]}{[HA]} \tag{11-13}$$

using exactly the same reasoning that led us to define K_a as

$$K_a = \frac{[H_3O^+][A^-]}{[HA]}$$

in aqueous solution.

In practice a more complicated form of equation (11-13) must be used. Solute concentrations and activities are usually not approximately equal to one another in nonaqueous solutions of very strong acids. Therefore, activities and activity coefficients must be used rather than concentrations (see Appendix IV).

Equation (11-13) is rewritten in terms of activities

$$K^{S} = \frac{(a_{SH^+})(a_{A^-})}{(a_{HA})} \tag{11-14}$$

Use of the general form $a_X = \gamma_X[X]$, where γ_X is the activity coefficient of the substance X, allows (11-14) to be written as

$$K^{S} = \left(\frac{[A^-]}{[HA]}\right)(a_{SH^+})\left(\frac{\gamma_{A^-}}{\gamma_{HA}}\right) \tag{11-15}$$

We define h_- and H_- by

$$h_- = (a_{SH^+})\left(\frac{\gamma_{A^-}}{\gamma_{HA}}\right) \tag{11-16}$$

$$H_- = -\log(h_-)$$

$$= -\log(a_{SH^+}) + \log\left(\frac{\gamma_{HA}}{\gamma_{A^-}}\right) \tag{11-17}$$

Recall that in aqueous solutions pH is defined by $pH = -\log(a_{H_3O^+})$ (which in dilute solutions is the same as $pH = -\log[H_3O^+]$) and that $a_{H_3O^+}$ is the *activity of the solvated proton*. Similarly, in a nonaqueous solvent S, a_{SH^+} is the activity of the solvated proton. After substituting $pH = -\log(a_{SH^+})$, equation (11-17) becomes

$$H_- = pH + \log\left(\frac{\gamma_{HA}}{\gamma_{A^-}}\right) \tag{11-18}$$

It is an experimental fact that, for a given solvent, the ratio γ_{HA}/γ_{A^-} is approximately the same for a variety of acids HA, HA′, HA″, and so on (provided that these acids are not too dissimilar in structure). Thus, if we refer always to the same solvent, we can treat γ_{HA}/γ_{A^-} as a constant and write (11-18) as

$$H_- = pH + \text{const.} \tag{11-19}$$

Equation (11-19) applies to any solvent in which the equilibrium (11-12) lies to the left even for a strong acid HA. This means that SH^+ itself is a strong acid. In such strongly acidic media, H_-, which is called the *acidity function*, is a measure of acid strength in exactly the same way that pH measures acid strength in aqueous media. Taking the negative logarithm of equation (11-15) gives

$$-\log K^{S} = -\log(a_{SH^+}) + \log\left(\frac{\gamma_{HA}}{\gamma_{A^-}}\right) + \log\left(\frac{[HA]}{[A^-]}\right)$$

Incorporating the definition (11-17) results in

$$pK^{S} = H_- + \log\left(\frac{[HA]}{[A^-]}\right) \tag{11-20}$$

which is the form of the Henderson-Hasselbalch equation appropriate to strongly acidic, nonaqueous media.

In this text we will not discuss experimental details, but it is clear from equation (11-20) how the pK of a strong acid can be measured. The ratio $[HA]/[A^-]$ is measured, perhaps spectrophotometrically, in a set of solutions of known H_-.

Our discussion so far has been for a very strong acid HA, which requires a strongly acidic medium in order to prevent the ionization (11-12) from being complete. We could equally well have discussed the ionization of a base B which is so weak that the ionization

$$B + H_2O \rightleftharpoons BH^+ + OH^- \tag{11-21}$$

does not take place at all. In an aqueous solution of such a base, the pH is controlled by the autoionization of water, and is equal to 7. This is analogous to the leveling effect, in that there is no way to measure the relative strengths of such weak bases in aqueous solution. What is needed is a solvent so strongly acidic that it can force a proton onto an "unwilling" base. By the Brønsted–Lowry theory a salt (e.g., the chloride BH^+Cl^-) of a very weak base is a very strong acid. It follows that in aqueous solution the equilibrium

$$BH^+ + H_2O \rightleftharpoons B + H_3O^+ \tag{11-22}$$

lies far to the right. Complete dissociation of the strong acid BH^+ can be prevented only in a strongly acidic solvent, in which the equilibrium

$$BH^+ + S \rightleftharpoons B + SH^+ \tag{11-23}$$

does not lie completely to the right. Reactions (11-12) and (11-23) are identical if we make the substitution BH^+ for HA and B for A^-. The notation is different only because BH^+ is a cationic acid and HA is a neutral acid. The development of equations (11-13) through (11-20) can now be repeated with these substitutions. However, *different notations for the acidity functions are used for the two cases.* If the acid is a neutral molecule HA, the acidity function is denoted by H_- and defined by equations (11-16) through (11-18). The pK^S of a neutral acid is calculated using equation (11-20). If the acid is a cationic species BH^+, the acidity function is denoted by H_0 and defined by equations (11-24) through (11-26).

$$h_0 = (a_{SH^+})\left(\frac{\gamma_B}{\gamma_{BH^+}}\right) \tag{11-24}$$

$$H_0 = -\log(a_{SH^+}) + \log\left(\frac{\gamma_{BH^+}}{\gamma_B}\right) \tag{11-25}$$

$$H_0 = pH + \log\left(\frac{\gamma_{BH^+}}{\gamma_B}\right) \tag{11-26}$$

For cationic acids K^S is defined in the usual way as

$$K^S = \frac{(a_{SH^+})(a_B)}{a_{BH^+}}$$

$$= \frac{[B]}{[BH^+]} a_{SH^+} \frac{\gamma_B}{\gamma_{BH^+}} \tag{11-27}$$

Take the negative logarithm of both sides of (11-27), and use the definition of (11-25)

$$-\log K^S = -\log a_{SH^+} - \log \frac{\gamma_B}{\gamma_{BH^+}} - \log \frac{[B]}{[BH^+]}$$

$$= -\log a_{SH^+} + \log \frac{\gamma_{BH^+}}{\gamma_B} + \log \frac{[BH^+]}{[B]}$$

$$pK^S = H_0 + \log \frac{[BH^+]}{[B]} \qquad (11\text{-}28)$$

11.7 VERY WEAK ACIDS: CARBON ACIDS AND CARBANIONS

When a hydrocarbon such as methane ionizes according to the acid reaction

$$CH_4 + H_2O \rightarrow CH_3^- + H_3O^+ \qquad (11\text{-}29)$$

it is called a *carbon acid* because the proton is lost by heterolytic cleavage of a C–H bond. Methane, like all hydrocarbons, is an exceedingly weak acid. Reaction (11-29) proceeds to an extent that is far too small to be measurable. Nonetheless, if methane (or another hydrocarbon) is substituted with enough electron-withdrawing groups, its conjugate base can be sufficiently stabilized to permit a measurable extent of ionization. Thus dicyanomethane [$CH_2(CN)_2$, usually called malononitrile] ionizes via

$$CH_2(CN)_2 + H_2O \rightleftharpoons CH(CN)_2^- + H_3O^+$$

and has a pK_a of about 12. Tricyanomethane [$CH(CN)_3$] is a strong acid with $pK_a < 1$.

In order to measure the pK_a of a very weak acid ($pK_a > 12$) a solvent other than water must be used. The solvent itself must be a very strong base, able to pull a proton away from a very unwilling donor

$$CH_4 + S \rightleftharpoons CH_3^- + SH^+ \qquad (11\text{-}30)$$

The mathematical foundation for the determination of pK_a in nonaqueous solvents has already been developed in Section 11.6. Since carbon acids are neutral species, H_- is the appropriate notation for the acidity function. The pK^S is determined from measurements based on equation (11-20), which in the case of the acid methane is

$$pK^S = H_- + \log\left(\frac{[CH_4]}{[CH_3^-]}\right) \qquad (11\text{-}31)$$

Determinations of pK^S must be carried out under conditions such that both forms are present in significant concentrations, for only then can their concentrations be measured accurately enough to apply equation (11-31). Thus the experimental problem is one of finding solvents that have values of H_- roughly the same (say within 2 units) as the pK^S to be measured. (For carbon acids pK^S can be considerably greater than 20.) Such solvents are known, and usually consist of mixtures rather than pure substances. For example, mixtures of KOH/H_2O/DMSO (DMSO = dimethyl sulfoxide) can be prepared in which H_- is as large as 26. Determination of the pK values of very weak acids (and of very strong acids as well) is experimentally difficult and open to somewhat controversial opinions. Research in this field continues, and we may expect progress in both experiment and theory.

11.8 GAS-PHASE ACIDITIES

The gas-phase *basicity* of a substance B is related to its proton affinity (P.A.), defined as the heat of reaction when a substance accepts a proton

$$B(g) + H^+(g) \rightarrow BH^+(g) \qquad \Delta E = \text{P.A.} \tag{11-32}$$

The greater the proton affinity, the stronger the base. The gas-phase acidity of a substance is related to the energy required to remove a proton from a substance. For example, the heat of heterolytic dissociation of gaseous HCl is +1393 kJ/mol.

$$HCl(g) \rightarrow H^+(g) + Cl^-(g) \qquad \Delta E = +1393 \text{ kJ/mol} \tag{11-33}$$

The heat of the reverse reaction (11-34) is the proton affinity of the conjugate base, Cl^- ion.

$$Cl^-(g) + H^+(g) \rightarrow HCl(g) \qquad \Delta E = -1393 \text{ kJ/mol} = \text{P.A.} \tag{11-34}$$

Reactions like (11-32) are called *ion-molecule reactions*, and provided that they take place rapidly, can be studied in a mass spectrometer. Energies (ΔE) of many such reactions have been measured in mass spectrometric investigations.

Comparison of the proton affinities of a variety of substances leads to a ranking of "intrinsic" basicities, uncomplicated by the distorting effects of solvent interaction.

The recent development of ion cyclotron resonance (ICR) has expanded the scope of measurements of the heats of gaseous ion reactions, by allowing measurements of relatively slow reactions. In ICR, gaseous ions are steered into a tight circular orbit by a combination of electrical fields. Orbiting ions can react with neutral reagent molecules present (but not orbiting) in the apparatus. In ICR the energy of the orbiting ions can be controlled, allowing measurement of the heat of reactions.

It should be noted that most chemistry is practiced in solution rather than in the gas phase. Consequently gas-phase acidities are, to most chemists, less useful than acidities that are measured in aqueous or other solutions.

11.9 CONCLUSION

In this chapter we have presented an overview of methods by which pK values may be determined. The technique can be as simple as the acid-base titration which is encountered very early in the study of chemistry. On the other hand, the more difficult determinations of acidities outside the range measurable in aqueous solution, in the gas phase, and so on, are the subject of ongoing research involving sophisticated apparatus and techniques. Although we have come to the final chapter of this text, the final chapter of research in the chemistry of acids and bases is a long way from being written.

Appendix I

Autoionization Constant of Water

t (°C)	K_w	pK_w
0	1.14×10^{-15}	14.944
10	2.92×10^{-15}	14.535
15	4.51×10^{-15}	14.346
20	6.81×10^{-15}	14.167
25	1.01×10^{-14}	13.997
30	1.47×10^{-14}	13.833
35	2.09×10^{-14}	13.680
40	2.92×10^{-14}	13.335
50	5.47×10^{-14}	13.262
60	9.62×10^{-14}	13.017

Appendix II

Computer Programs in BASIC

The several computer programs used to make the calculations in this text are presented here for student use. They are written in the source language BASIC, and they are written in the most general way possible. Each type of computer, whether large or of tabletop size, will respond to certain specific commands or statements which may be rejected as meaningless by other computers. The programs that follow use as few of these statements as possible. One exception to this generality is the use of both upper- and lower-case type in the REM[ark] and PRINT statements; this is a convenience available on many home computers and can be ignored when using the programs on other equipment. A second exception is the use of an 80-column format; although the programs will run on a 40-column display, the 80-column width is easier to read and understand.

PROGRAM II.1 ERRORS OF APPROXIMATION

If the degree of dissociation is small, the hydrogen ion concentrations in a solution of a weak monoprotic acid is given approximately by $[H_3O^+] = \sqrt{K_aM}$, where M is the molarity of the acid. If the degree of dissociation is not small, $[H_3O^+]$ is found from the relationship $[H_3O^+] = [-K_a + \sqrt{K_a(K_a + 4M)}]/2$. The BASIC program given below compares the results of these two calculations and compares the difference between the two with the error estimated by equation (VI-7).

PROGRAM II.2 THE EXACT SOLUTION

The cubic equation $ax^3 + bx^2 + cx + d = 0$ (where a, b, c, and d are constants) has three roots, that is, three values of x for which the equation is true. These roots may be real or complex, positive or negative. If the autoionization of water is taken into account, the equation by which the pH of a weak monoprotic acid is calculated is of this form; the coefficients are $a = 1, b = K_a, c = -(K_aC + K_w)$, and $d = -K_wK_a$. Descartes's rule of signs[1] tells us that the equation has one only root that is both real and positive,

[1] Descartes's rule is discussed in most college algebra textbooks. See, for example, G. Fuller, *College Algebra* (4th ed., Van Nostrand, Reinhold, New York, 1977).

whereas the other two are either both complex or both negative real. Analysis of the first and second derivatives shows these to be negative rather than complex. Since the concentration of a chemical species cannot be a negative number, any calculational scheme must reject or ignore a negative answer and search only for the one positive root.

Of the several different methods for solving polynomial equations on a computer, the simplest and most trouble-free is the Newton-Raphson method. Like all iterative methods, it begins with a guessed value and repeatedly improves the guess until some designated degree of accuracy is achieved. In the BASIC program given below, chemical knowledge is used in two ways. First, the initial guess at the H_3O^+ concentration is chosen to be the value obtained by ignoring the contribution of autoionization, that is, by solving $[H_3O^+] = [-K_a + \sqrt{K_a(K_a + 4\ M)}]/2$. Second, since in an acid solution $[H_3O^+]$ can never be less than it is in pure water, the initial guess is constrained to be at least 10^{-7}.

PROGRAM II.3 BUFFER SYSTEMS

This program will calculate the pH and concentration of all species present in a mixture of a weak monoprotic acid and one of its salts. Calculations can also be made on a buffer system composed of a weak base and one of its salts. The autoionization of water is specifically included. Hydrolysis of a counterion, such as Na^+ in a $HNO_2/NaNO_2$ buffer or Cl^- in a NH_3/NH_4Cl buffer, is assumed to be negligible. The program is based on the exact treatment given in Section 7.7.

PROGRAM II.4 POLYPROTIC ACID CALCULATIONS

This program will calculate the pH and the concentrations of all species in the following solutions:

1. A diprotic acid (or a dihydric base) and/or any combination of its salts.
2. A triprotic acid (or a trihydric base) and/or any combination of its salts.
3. An amphoteric substance.

The program is based on the development in Chapter 9; specifically, equations (9-32), (9-34), and (9-44) are solved.

Program II.1 Errors of Approximation

```
100 REM ****** Approximation and Error in pH Calculations *******************
110 REM *                                                                   *
120 REM *   Calculation of H-ion (or OH-ion) concentration in solutions of  *
130 REM *  weak acids, given the ionization constant K and the molarity M.  *
140 REM *  Two methods are used, the approximate form                       *
150 REM *                                                                   *
160 REM *                 H = (K*M)**(1/2)                                  *
170 REM *                                                                   *
180 REM *  and the quadratic form                                           *
190 REM *                                                                   *
200 REM *                 H = [-K + (K*K+4*K*M)**(1/2)]/2                   *
210 REM *                                                                   *
220 REM *   The error incurred by the approximation is estimated by the     *
230 REM *  equation                                                         *
240 REM *                                                                   *
250 REM *          % error = 100*[(K/M)**(1/2)]/2                           *
260 REM *                                                                   *
270 REM *  and the actual error is calculated according to                  *
280 REM *                                                                   *
290 REM *          % error = 100*[H(approx)-H(quadratic)]/[H(quadratic)]    *
300 REM *                                                                   *
310 REM *   These calculated errors are compared to the "rule of thumb"     *
320 REM *  that the approximate calculation is safe if K/M < 0.0001          *
330 REM *                                                                   *
340 REM *     The following symbols are defined and used in the program:    *
350 REM *                                                                   *
360 REM *               HQ =   molarity of H-ion as calculated              *
370 REM *                    by the quadratic equation                      *
380 REM *               HP =   molarity of H-ion as calculated              *
390 REM *                    by the approximate expression                  *
400 REM *                K =   acid ionization constant                     *
410 REM *                M =   molarity of acid solution                    *
420 REM *               PH =   true pH of solution                          *
430 REM *               PP =   approximate pH of solution                   *
440 REM *               EE =   estimated error                              *
450 REM *               EA =   actual error                                 *
460 REM *               RT =   "rule of thumb" parameter, K/M               *
470 REM *                                                                   *
480 REM *   Calculations for weak bases may also be done by this program,   *
490 REM *  provided that the input is Kb rather than Ka, and that the re-   *
500 REM *  sults are interpreted as pOH rather than pH,  and  OH-ion        *
510 REM *  rather than H-ion concentration.  Similarly, the program will    *
520 REM *  calculate the pH of a solution of a salt of a weak base, if K is *
530 REM *  interpreted as Kh.  Finally, if K is interpreted as Kh for the   *
540 REM *  salt of as weak acid, result is the pOH of such a solution.      *
550 REM *********************************************************************
560 REM
570 REM ***************** Data Input Section *******************************
580 HOME
590 INPUT "ENTER THE VALUE OF THE IONIZATION CONSTANT ",K
600 INPUT "ENTER THE MOLARITY OF THE WEAK ACID ",M
610 REM
620 REM ********************* Calculations ********************************
630 HP=SQR(K*M)
640 PP=(-1)*LOG(HP)/2.30259
650 HQ=(-K+SQR(K*K+4*K*M))/2
660 PH=(-1)*LOG(HQ)/2.30259
670 EA=50*SQR(K/M)
680 EE=100*(HP-HQ)/HQ
690 RT=K/M
700 REM
```

```
710 REM ****************** Output Operations ******************************
720 REM
730 PRINT:PRINT"      APPROXIMATE AND QUADRATIC ACIDITY CALCULATIONS"
740 PRINT:PRINT   "IONIZATION CONSTANT =";K
750 PRINT     "ACID CONCENTRATION  =";M;" MOLAR"
760 PRINT:PRINT   "APPROXIMATE H-ION =";HP;" MOLAR";"   PH =";PP
770 PRINT     "QUADRATIC H-ION   =";HQ;" MOLAR";"   PH =";PH
780 PRINT     "ESTIMATED ERROR   =";EE;" PERCENT"
790 PRINT     "ACTUAL ERROR      =";EA;" PERCENT"
800 PRINT
810 PRINT  "K/M =";RT;", SHOULD BE < 1.0E-4 FOR APPROXIMATE METHOD"
820 IF ((PH<6) AND (PP<6)) THEN GOTO 870
830 PRINT "*** NOTE: BOTH THE APPROXIMATE AND THE QUADRATIC METHODS  YIELD"
840 PRINT "INACCURATE RESULTS WHEN THE PH IS GREATER THAN ABOUT 6, BECAUSE"
850 PRINT "THEY IGNORE THE AUTOIONIZATION  OF WATER.  USE ONE OF THE EXACT"
860 PRINT "PROGRAMS, APPENDIX II.2, II.3, OR II.4.      ******************"
870 PRINT
880 PRINT:PRINT"DO YOU WISH TO DO ANOTHER CALCULATION? (Y/N) "
890 GET A$
900 IF((A$="Y")OR(A$="y")) THEN GOTO 570
910 PRINT"END OF CALCULATIONS"
920 END
```

Sample Output from Program II.1

```
       APPROXIMATE AND QUADRATIC ACIDITY CALCULATIONS

ENTER THE VALUE OF THE IONIZATION CONSTANT  .0035
ENTER THE MOLARITY OF THE WEAK ACID  .08

APPROXIMATE H-ION = .0167332  MOLAR   PH = 1.77642
QUADRATIC H-ION   = .0150745  MOLAR   PH = 1.82175
ESTIMATED ERROR   = 11.0037  PERCENT
ACTUAL ERROR      = 10.4583  PERCENT

K/M = .04375 , SHOULD BE < 1.0E-4 FOR APPROXIMATE METHOD


DO YOU WISH TO DO ANOTHER CALCULATION? (Y/N) Y
ENTER THE VALUE OF THE IONIZATION CONSTANT  .0000035
ENTER THE MOLARITY OF THE WEAK ACID  .08

APPROXIMATE H-ION = 5.2915E-04  MOLAR   PH = 3.27641
QUADRATIC H-ION   = 5.27403E-04  MOLAR   PH = 3.27785
ESTIMATED ERROR   = .331287  PERCENT
ACTUAL ERROR      = .330719  PERCENT

K/M = 4.375E-05 , SHOULD BE < 1.0E-4 FOR APPROXIMATE METHOD


DO YOU WISH TO DO ANOTHER CALCULATION? (Y/N) N
END OF CALCULATIONS
```

```
100 REM *************** Dilute Acid, Cubic **********************************
110 REM *                                                                   *
120 REM *   Exact calculation of pH for very weak or very dilute mono-      *
130 REM *  protic acids (or bases, or salts of weak acids or bases).        *
140 REM *  The exact equation, which takes explicit account of water        *
150 REM *  autoionization, is cubic in H-ion concentration:                 *
160 REM *                                                                   *
170 REM *       H**3 + K*H**2 - (M*K+W)*H - W*K = 0                         *
180 REM *                                                                   *
190 REM *   where      H = Hydrogen ion concentration                       *
200 REM *              K = Acid ionization constant, Ka                     *
210 REM *              M = Molarity of acid                                 *
220 REM *              W = Autoionization constant of water, 1.0E-14        *
230 REM *                                                                   *
240 REM *   This equation has a single positive root, which is found by     *
250 REM *  Newton-Raphson iteration to an iterate change of less than Z      *
260 REM *  (set usually at 1E-6).  The method requires an initial guess      *
270 REM *  of the H-ion concentration; the guess is then improved until      *
280 REM *  no further improvement is necessary.   The initial guess is       *
290 REM *  set equal to the result of the quadratic solution, which ig-      *
300 REM *  nores autoionization.  (Since any acid, however weak or dilute,  *
310 REM *  cannot give a solution of pH>7, the initial guess is constrained *
320 REM *  to be at least 1.0E-7.  If the calculation accidentally con-     *
330 REM *  verges on one of the two negative roots of the cubic equation,   *
340 REM *  it begins again with a larger initial guess.  The program also   *
350 REM *  calculates the approximate solution from K=(H**2)/M.  Following  *
360 REM *  the calculation some computational information is displayed,     *
370 REM *  including the "function residual".  This is the final calculated *
380 REM *  value of the left-hand side of the cubic equation (statement     *
390 REM *  170, above):  for an exact value of H the function residual      *
400 REM *  equals zero.                                                     *
410 REM *                                                                   *
420 REM *   In addition to those above, the following symbols are           *
430 REM *  defined and used in the program:                                 *
440 REM *                                                                   *
450 REM *           A,B,C,D =   Coefficients of the cubic equation          *
460 REM *                HQ =   H-ion molarity as calculated                *
470 REM *                     by the quadratic form                         *
480 REM *                HP =   value of H from approximate expression      *
490 REM *                HI =   value of H, after improvement by iteration  *
500 REM *                 F =   function value (l.h.s. of cubic)            *
510 REM *                DV =   derivative of F                             *
520 REM *                 Z =   convergence limit: iteration ceases when    *
530 REM *                     next-cycle improvement is less than Z         *
540 REM *                 L =   iteration limit: stop after L cycles        *
550 REM *                 N =   number of tries required to reach           *
560 REM *                     positive root (almost always = 1)             *
570 REM *                PH =   -log(H)                                     *
580 REM *                EP =   % error in approximate calculation of H     *
590 REM *                EQ =   % error in quadratic calculation of H       *
600 REM *                                                                   *
610 REM *   To convert the calculation to a weak base or a salt, the        *
620 REM *  parameters should be interpreted as Kb or Kh rather than Ka,     *
630 REM *  and OH-ion concentration and pOH rather than H-ion concentration *
640 REM *  and pH.                                                          *
650 REM *                                                                   *
660 REM *********************************************************************
670 REM
680 REM *********** Input and Initialization Operations ********************
690 REM
700 HOME
710 INPUT "ENTER VALUE OF IONIZATION CONSTANT";K
720 INPUT "ENTER MOLARITY OF ACID SOLUTION";M
```

```
730 W=1E-14
740 Z=.000001
750 L=20
760 N=1
770 REM SET COEFFICIENTS OF CUBIC EQUATION
780 A=1
790 B=K
800 C=-(M*K+W)
810 D=-W*K
820 REM FIND THE APPROXIMATE AND QUADRATIC H-ION CONCENTRATIONS
830 HP=SQR(K*M)
840 HQ=(-K+SQR(K*K+4*K*M))/2
850 REM SET INITIAL GUESS OF H
860 G=HQ
870 IF G<.0000001 THEN G=.0000001
880 REM
890 REM ****************** Newton-Raphson Iteration ************************
900 REM
910 REM SET INITIAL VALUE OF H TO GUESS
920 H=G
930 FOR I=1 TO L
940 REM CALCULATE FUNCTION AND DERIVATIVE
950 F=((A*H+B)*H+C)*H+D
960 DV=(3*A*H+2*B)*H+C
970 IF DV <> 0 THEN GOTO 1000
980 H=1.01*H
990 GOTO 950
1000 HI=H-F/DV
1010 REM EXIT LOOP IF CONVERGENCE IS ATTAINED
1020 IF ABS(1-HI/H)<Z THEN 1090
1030 H=HI
1040 NEXT I
1050 PRINT:PRINT"ROOT NOT FOUND IN";L;" ITERATIONS.  CHECK INPUT DATA"
1060 PRINT"AND/OR PERMIT MORE ITERATIONS (INCREASE L)"
1070 GOTO 1370
1080 REM CHECK THAT ROOT IS POSITIVE
1090 IF HI>0 THEN GOTO 1180
1100 REM NEGATIVE ROOT FOUND.  INCREASE INITIAL GUESS AND TRY AGAIN.
1110 G=2*G
1120 N=N+1
1130 IF N<5 THEN GOTO 920
1140 PRINT:PRINT N;" NEGATIVE ROOTS FOUND, NO POSITIVE ROOT YET"
1150 PRINT"CHECK INPUT DATA AND TRY AGAIN"
1160 GOTO 1370
1170 REM *
1180 REM *************** Output Operations **********************************
1190 REM
1200 PRINT:PRINT"                  HYDROGEN ION MOLARITY"
1210 PRINT
1220 PH=(-1)*LOG(H)/2.303
1230 PRINT"EXACT VALUE (CUBIC)        =";H;" PH =";PH
1240 EQ=100*(HQ-H)/H
1250 PRINT"QUADRATIC VALUE            =";HQ;" %ERROR =";EQ
1260 EA=100*(HP-H)/H
1270 PRINT"APPROXIMATE VALUE          =";HP;" %ERROR =";EA
1280 PRINT:PRINT
1290 PRINT"IONIZATION CONSTANT         =";K
1300 PRINT"MOLARITY OF ACID            =";M
1310 PRINT"FUNCTION RESIDUAL           =";F
1320 PRINT"NUMBER OF ITERATIONS        =";I
1330 PRINT:PRINT"DO YOU WISH TO DO ANOTHER CALCULATION (Y/N)";
1340 GET B$
1350 IF ((B$="Y") OR (B$="y")) THEN GOTO 700
1360 PRINT:PRINT"END OF CALCULATIONS"
1370 END
```

```
  EXACT CALCULATION, VERY WEAK AND/OR DILUTE ACIDS

ENTER VALUE OF IONIZATION CONSTANT  .0000004
ENTER MOLARITY OF ACID SOLUTION  6E-08

                HYDROGEN ION MOLARITY

EXACT VALUE (CUBIC)        = 1.25414E-07  PH = 6.90041
QUADRATIC VALUE            = 5.29822E-08  %ERROR =-57.7542
APPROXIMATE VALUE          = 1.54919E-07  %ERROR = 23.5263


IONIZATION CONSTANT         = .0000004
MOLARITY OF ACID            = 6E-08
FUNCTION RESIDUAL           =-8.07794E-28
NUMBER OF ITERATIONS        = 5

DO YOU WISH TO DO ANOTHER CALCULATION (Y/N) Y



  EXACT CALCULATION, VERY WEAK AND/OR DILUTE ACIDS

ENTER VALUE OF IONIZATION CONSTANT  4E-12
ENTER MOLARITY OF ACID SOLUTION  .006

                HYDROGEN ION MOLARITY

EXACT VALUE (CUBIC)        = 1.84389E-07  PH = 6.73305
QUADRATIC VALUE            = 1.54917E-07  %ERROR =-15.9836
APPROXIMATE VALUE          = 1.54919E-07  %ERROR =-15.9825


IONIZATION CONSTANT         = 4E-12
MOLARITY OF ACID            = .006
FUNCTION RESIDUAL           =-1.68958E-29
NUMBER OF ITERATIONS        = 5

DO YOU WISH TO DO ANOTHER CALCULATION (Y/N) N

END OF CALCULATIONS
```

```
100 REM *************** Monoprotic Buffer Systems ***************************
110 REM *                                                                    *
120 REM *    Exact calculation of pH for mixtures of monoprotic acids       *
130 REM *  (or monohydric bases) and their salts.  The method includes,     *
140 REM *  explicitly, the autoionization of water.  The exact equation     *
150 REM *  is a third-degree polynomial (cubic) in H-ion concentration:     *
160 REM *                                                                    *
170 REM *       A*H**3 + B*H**2 + C*H + D = 0                                *
180 REM *                                                                    *
190 REM *  where         A = 1.0                                             *
200 REM *                B = K + M1                                          *
210 REM *                C = -(M0*K + W)                                     *
220 REM *                D = -K*W                                            *
230 REM *                                                                    *
240 REM *  and           H = Hydrogen ion concentration                      *
250 REM *                K = Acid ionization constant, Ka                    *
260 REM *               M0 = Molarity of acid                                *
270 REM *               M1 = Molarity of salt                                *
280 REM *                W = Autoionization  constant of water, 1.0E-14      *
290 REM *                                                                    *
300 REM *    In addition to those defined above, the following symbols      *
310 REM *  are defined and used in the program:                              *
320 REM *                                                                    *
330 REM *       HQ =   H-ion molarity as calculated                          *
340 REM *             by the quadratic form for pure acid                    *
350 REM *       HH =   value of H from quadratic form, pure salt             *
360 REM *       HP =   value of H from approximate expression                *
370 REM *       HB =   value of H from Henderson-Hasselbalch equation        *
380 REM *       HI =   value of H, after improvement by iteration            *
390 REM *       C1 =   equilibrium concentration of nonionized acid          *
400 REM *       C0 =   equilibrium concentration of anion                    *
410 REM *        F =   function value (l.h.s. of cubic)                      *
420 REM *       DV =   derivative of F                                       *
430 REM *        G =   initial guess at H-ion concentration                  *
440 REM *        Z =   convergence limit: iteration ceases when              *
450 REM *             next-cycle improvement is less than Z                  *
460 REM *        L =   iteration limit: stop after L cycles                  *
470 REM *        N =   number of tries required to reach                     *
480 REM *             positive root (almost always = 1)                      *
490 REM *       PH =   -log(H)                                               *
500 REM *                                                                    *
510 REM *    This equation has a single positive root, which is found by    *
520 REM *  Newton-Raphson iteration to an iterate change of less than Z     *
530 REM *  (set usually at 1E-6).  The method requires an initial guess     *
540 REM *  of the H-ion concentration; the guess is then improved until     *
550 REM *  no further improvement is necessary.   The initial guess G is    *
560 REM *  generated in three different ways, depending on the type of      *
570 REM *  solution:                                                         *
580 REM *       (1)  Pure acid;  G is obtained from the quadratic equation,  *
590 REM *  which ignores autoionization.  (Since any acid, however weak or   *
600 REM *  dilute, cannot give a solution of pH>7, the initial guess is     *
610 REM *  constrained to be at least 1.0E-7.                               *
620 REM *       (2)  Pure salt;  G is obtained from the quadratic equation   *
630 REM *  using Kh=Kw/Ka, and constrained to be less than 1.0E-7            *
640 REM *       (3)  Mixture of acid and salt; G is obtained from the       *
650 REM *  Henderson-Hasselbalch equation, pH = pK + log(M1/M0).            *
660 REM *                                                                    *
670 REM *    If the calculation accidentally converges on one of the two    *
680 REM *  negative roots of the cubic equation, it begins again with a     *
690 REM *  larger initial guess.                                            *
700 REM *                                                                    *
```

```
710 REM *  Notes:                                                          *
720 REM *        (1)      The symbols are chosen to conform to an acid     *
730 REM *        buffer.  To apply to a basic system such as ammonia/      *
740 REM *        ammonium ion, substitute [OH] for [H], Kb for Ka, and     *
750 REM *        pOH for pH.                                               *
760 REM *        (2)      The program may be used for pure acid solutions  *
770 REM *        or pure salt solutions, by entering M1 or M0 = 0.         *
780 REM *        (3)      The program assumes that a 1:1 salt such as      *
790 REM *        sodium acetate is used;  however correct answers will be  *
800 REM *        obtained from di- and trivalent cations as long as M1     *
810 REM *        refers to the concentration of the anion.  For example, if *
820 REM *        the salt is 0.3 M Magnesium Acetate [Mg(Ac)2], M1 should  *
830 REM *        be entered as 0.6 M rather than 0.3 M.                    *
840 REM *        (4)      The autoionization of water may be ignored, if   *
850 REM *        desired, by setting W=0.                                  *
860 REM *                                                                  *
870 REM ********************************************************************
880 REM
890 REM ************ Data Input and Initialization ************************
900 HOME
910 PRINT"                 MONOPROTIC BUFFER SYSTEMS ":PRINT
920 INPUT "ENTER THE IONIZATION CONSTANT OF THE ACID ",K
930 INPUT "ENTER THE MOLARITY OF THE WEAK ACID ",M0
940 INPUT "ENTER THE MOLARITY OF THE SALT ",M1
950 REM Autoionization constant for room temperature
960 W=1E-14
970 REM define coefficients
980 A=1
990 B=K+M1
1000 C=(-1)*(M0*K+W)
1010 D=(-1)*K*W
1020 REM set convergence limit and maximum iterations
1030 Z=.000001
1040 L=20
1050 REM set initial guess of [H], depending on solution type
1060 IF M1=0 THEN GOTO 1120
1070 IF M0=0 THEN GOTO 1170
1080 REM both acid and salt present
1090 HB=K*M0/M1
1100 G=HB
1110 GOTO 1220
1120 REM pure acid solution
1130 HQ=(-K+SQR(K*K+4*K*M0))/2
1140 G=HQ
1150 IF G<.000001 THEN G=.000001
1160 GOTO 1220
1170 REM pure salt solution
1180 KH=W/K
1190 OH=(-KH+SQR(KH*KH+4*KH*M1))/2
1200 HH=W/OH
1210 G=HH
1220 REM ********************* Calculations *********************************
1230 REM
1240 REM perform Newton-Raphson iteration to convergence Z
1250 H=G
1260 FOR I=1 TO L
1270 F=((A*H+B)*H+C)*H+D
1280 DV=(3*A*H+2*B)*H+C
1290 HI=H-F/DV
```

```
1300 IF ABS(1-HI/H)<Z THEN GOTO 1370
1310 H=HI
1320 NEXT I
1330 PRINT"*** ROOT NOT FOUND IN";I;"ITERATIONS.  CHECK"
1340 PRINT"INPUT DATA AND/OR PERMIT MORE ITERATIONS (INCREASE L)"
1350 PRINT"TYPE 'RUN' TO TRY AGAIN"
1360 GOTO 1810
1370 IF HI>0 THEN GOTO 1410
1380 PRINT"NEGATIVE ROOT FOUND, TRYING AGAIN WITH LARGER INITIAL GUESS"
1390 G=5*G
1400 GOTO 1240
1410 REM calculate concentrations of species other than H
1420 REM solution must be neutral, that is H + M1(cation) = OH + CO
1430 CO=H+M1-W/H
1440 REM find nonionized acid from equilibrium expression
1450 C1=H*CO/K
1460 REM
1470 REM *************** Output Operations ********************************
1480 REM
1490 REM acid, salt, or buffer?
1500 IF M1=0 THEN GOTO 1630
1510 IF MO=0 THEN GOTO 1710
1520 REM solution is a buffer
1530 PRINT:PRINT"                 BUFFER SOLUTION, K=";K
1540 PRINT:PRINT" ","INITIAL", "APPROXIMATE", "EXACT"
1550 PRINT"[H]","ZERO",HB,H
1560 PRINT"ACID",MO,MO,C1
1570 PRINT"ANION",M1,M1,CO
1580 PRINT"EXACT PH =";(-1)*LOG(H)/2.30259;";  ";
1590 PRINT"APPROXIMATE PH =";(-1)*LOG(HB)/2.30259
1600 PRINT"ERROR IN [H] CALCULATED BY THE HENDERSON-HASSELBALCH EQUATION  = ";
1610 PRINT 100*(1-HB/H);"%"
1620 GOTO 1810
1630 REM  solution is a pure acid
1640 PRINT:PRINT"H-ION MOLARITY IN A";MO;"M SOLUTION OF WEAK ACID, K =";K
1650 PRINT:PRINT" "," ","   [H]", "   PH", "% ERROR IN [H]"
1660 PRINT"EXACT CALCULATION",H,(-1)*LOG(H)/2.30259
1670 PRINT"QUADRATIC EQUATION",HQ,(-1)*LOG(HQ)/2.30259,100*(HQ-H)/H
1680 HP=SQR(K*MO)
1690 PRINT"APPROXIMATE EQUATION",HP,(-1)*LOG(HP)/2.30259,100*(HP-H)/H
1700 GOTO 1810
1710 REM Solution contains salt only, no acid
1720 PRINT:PRINT"  H-ION MOLARITY IN A";M1;"M SOLUTION CONTAINING ONLY THE"
1730 PRINT"SALT OF A WEAK ACID WHOSE IONIZATION CONSTANT Ka IS";K
1740 PRINT:PRINT " "," ","    [H]","    PH", "% ERROR IN [H]"
1750 PRINT"EXACT CALCULATION",H,(-1)*LOG(H)/2.30259
1760 PRINT"QUADRATIC EQUATION",HH,(-1)*LOG(HH)/2.30259,100*(HH-H)/H
1770 OH=SQR(M1*W/K)
1780 HP=W/OH
1790 PRINT"APPROXIMATE EQUATION",HP,(-1)*LOG(HP)/2.30259, 100*(HP-H)/H
1800 GOTO 1810
1810 PRINT:PRINT"DO YOU WISH TO MAKE ANOTHER CALCULATION? (Y/N) "
1820 GET B$
1830 IF((B$="Y") OR (B$="y")) THEN GOTO 920
1840 PRINT:PRINT"END OF CALCULATIONS"
1850 END
```

```
                 MONOPROTIC BUFFER SYSTEMS

ENTER THE IONIZATION CONSTANT OF THE ACID  .00002
ENTER THE MOLARITY OF THE WEAK ACID  .06
ENTER THE MOLARITY OF THE SALT  .03

            BUFFER SOLUTION, K= .00002

             INITIAL         APPROXIMATE   EXACT
[H]          ZERO            .00004          3.99203E-05
ACID          .06            .06            .0599601
ANION         .03            .03            .0300399
EXACT PH = 4.3988 ;  APPROXIMATE PH = 4.39793
ERROR IN [H] AS CALCULATED BY
     THE HENDERSON-HASSELBALCH EQUATION = -.199735 %

DO YOU WISH TO MAKE ANOTHER CALCULATION? (Y/N) Y


ENTER THE IONIZATION CONSTANT OF THE ACID  .00002
ENTER THE MOLARITY OF THE WEAK ACID  .01
ENTER THE MOLARITY OF THE SALT  0

H-ION MOLARITY IN A .01 M SOLUTION OF WEAK ACID, K = .00002

                               [H]          PH         % ERROR IN [H]
EXACT CALCULATION           4.37326E-04   3.35919
QUADRATIC EQUATION          4.37326E-04   3.35919        0
APPROXIMATE EQUATION        4.47214E-04   3.34948        2.26107

DO YOU WISH TO MAKE ANOTHER CALCULATION? (Y/N) Y


ENTER THE IONIZATION CONSTANT OF THE ACID  .00002
ENTER THE MOLARITY OF THE WEAK ACID  0
ENTER THE MOLARITY OF THE SALT  .01

  H-ION MOLARITY IN A .01 M SOLUTION CONTAINING ONLY THE
SALT OF A WEAK ACID WHOSE IONIZATION CONSTANT Ka IS .00002

                               [H]          PH         % ERROR IN [H]
EXACT CALCULATION           4.46817E-09   8.34985
QUADRATIC EQUATION          4.47264E-09   8.34942        .0999759
APPROXIMATE EQUATION        4.47214E-09   8.34947        .0887846

DO YOU WISH TO MAKE ANOTHER CALCULATION? (Y/N) N


END OF CALCULATIONS
```

```
100 REM ******** Polyprotic Acid Systems and Amphoteric Salts ***************
110 REM *                                                                   *
120 REM *   Exact calculation of pH for systems consisting of mixtures of   *
130 REM *  a triprotic acid H3A and its salts NaH2A, Na2HA, and Na3A, or     *
140 REM *  of a diprotic acid H2B and its salts NaHB and Na2B.   The        *
150 REM *  method includes, explicitly, the autoionization of water.        *
160 REM *   For a triprotic acid, the exact equation is a 5th degree        *
170 REM *  polynomial in H-ion concentration:                               *
180 REM *                                                                   *
190 REM *       A*H**5 + B*H**4 + C*H**3 + D*H**2 + E*H + F = 0             *
200 REM *                                                                   *
210 REM *  with coefficients                                                *
220 REM *                                                                   *
230 REM *       A = 1.0                                                     *
240 REM *       B = K1 - N                                                  *
250 REM *       C = K1*N - K1*T + K1*K2 - W                                 *
260 REM *       D = K1*(K2*N - 2*K2*T + K2*K3 - W)                          *
270 REM *       E = K1*K2*(K3*N -3*K3*T - W)                                *
280 REM *       F = -K1*K2*K3*W                                             *
290 REM *                                                                   *
300 REM *  where                                                            *
310 REM *                                                                   *
320 REM *        H = H-ion molarity at equilibrium                          *
330 REM *       K1 = ionization constant for H3A-->H+H2A or H2B-->H+HB      *
340 REM *       K2 = ionization constant for H2A-->H+HA  or HB-->H+B        *
350 REM *       K3 = ionization constant for HA  --> H + A                  *
360 REM *        W = autoionization constant of water H2O --> H + OH        *
370 REM *       M0 = initial molarity of acid H3A       or H2B              *
380 REM *       M1 = initial molarity of salt NaH2A     or NaHB             *
390 REM *       M2 = initial molarity of salt Na2HA     or Na2B             *
400 REM *       M3 = initial molarity of salt Na3A                          *
410 REM *       C0 = equilibrium molarity of acid H3A     or H2B            *
420 REM *       C1 = equilibrium molarity of anion H2A-   or HB-            *
430 REM *       C2 = equilibrium molarity of anion HA--   or B--            *
440 REM *       C3 = equilibrium molarity of anion A---                     *
450 REM *        Q = 1/(mole fraction of A--- or B--)                       *
460 REM *        T = total molarity, T=M0+M1+M2+M3                          *
470 REM *        N = total cationic charge (other than H-ion)               *
480 REM *                                                                   *
490 REM *  Notes:                                                           *
500 REM *      (1)     The equation is unchanged if di- or trivalent        *
510 REM *         cations are used rather than Na salts, provided that      *
520 REM *         M1, M2, and M3 refer to the molarity of the anion         *
530 REM *         rather than the salt.  For example if a salt is 0.004 M   *
540 REM *         magnesium phosphate, Mg3(PO4)2, M1 = 0.008.               *
550 REM *      (2)     If K3 is 0, the acid is diprotic, and the equation   *
560 REM *         is reduced to 4th degree.                                 *
570 REM *                                                                   *
580 REM *    The program may be used for buffer systems involving poly-     *
590 REM *  hydric bases such as ethylene diamine, by appropriate inter-     *
600 REM *  pretation of the input and output (pH is read as pOH, [H] as     *
610 REM *  [OH], K1 as K1b rather than K1a, etc).  It may be used to        *
620 REM *  calculate the pH of solutions of an amphoteric substance such    *
630 REM *  an amino acid or a 1:1 salt like ammonium acetate. This is       *
640 REM *  done as a diprotic acid, letting the amphoteric substance be     *
650 REM *  represented by NaHB, Kb by W/K1, and Ka by K2.  K3, M0, M2       *
660 REM *  are all zero for amphoteric substances.                          *
```

```
670 REM *                                                                  *
680 REM *   The equations hava a single positive root, which is found by   *
690 REM *  Newton-Raphson iteration to an iterate change of less than Z    *
700 REM *  (set usually at 1E-6).  The method requires an initial guess    *
710 REM *  of the H-ion concentration; the guess is then improved until    *
720 REM *  no further improvement is necessary.   The initial guess G is   *
730 REM *  set equal to the sum of the component concentrations.  G is     *
740 REM *  almost always very much larger than it needs to be, and many    *
750 REM *  iterations are required for convergence.  However, it always    *
760 REM *  works, and only rarely converges to a negative root.  Program   *
770 REM *  operation can be greatly speeded up by choosing a smaller value *
780 REM *  of G, but at the cost of added complexity and the risk of con-  *
790 REM *  vergence to physically meaningless answers.                     *
800 REM *                                                                  *
810 REM *   In addition to those defined above, the following symbols      *
820 REM *  are defined and used in the program:                            *
830 REM *                                                                  *
840 REM *       HI =   value of H, after improvement by iteration          *
850 REM *       FH =   function value (l.h.s. of equation )                *
860 REM *       DV =   derivative of FH                                    *
870 REM *        G =   initial guess at H-ion concentration                *
880 REM *        Z =   convergence limit: iteration ceases when            *
890 REM *                next-cycle improvement is less than Z             *
900 REM *        L =   iteration limit: stop after L cycles                *
910 REM *       PH =   -log(H)                                             *
920 REM *                                                                  *
930 REM ********************************************************************
940 REM
950 REM ************* Data Input and Initialization ***********************
960 HOME
970 PRINT"      GENERAL PROGRAM FOR POLYPROTIC SYSTEMS ":PRINT
980 W=1E-14
990 K1=0: K2=0: K3=0: M0=0: M1=0: M2=0: M3=0: C0=0: C1=0: C2=0: C3=0
1000 PRINT:INPUT "ENTER TYPE:  (D)IPROTIC OR (T)RIPROTIC ",B$
1010 REM convert to upper case letters if necessary
1020 IF B$="d" THEN B$="D"
1030 IF B$="t" THEN B$="T"
1040 IF B$="D" THEN GOTO 1080
1050 IF B$="T" THEN GOTO 1160
1060 PRINT"YOU MUST CHOOSE D OR T"
1070 GOTO 1000
1080 REM ****** diprotic acid, salt, or buffer; or amphoteric compound
1090 PRINT:PRINT"            DIPROTIC ACID SYSTEM"
1100 INPUT"ENTER FIRST IONIZATION CONSTANT K1 ". K1
1110 INPUT"ENTER SECOND IONIZATION CONSTANT K2 ",K2
1120 INPUT"ENTER INITIAL MOLARITY OF ACID H2B ", M0
1130 INPUT"ENTER INITIAL MOLARITY OF SALT NaHB ", M1
1140 INPUT"ENTER INITIAL MOLARITY OF SALT Na2B ", M2
1150 GOTO 1270
1160 REM ******** triprotic acid, salt or buffer
1170 PRINT:PRINT"            TRIPROTIC ACID SYSTEM"
1180 INPUT"ENTER FIRST IONIZATION CONSTANT K1 ",K1
1190 INPUT"ENTER SECOND IONIZATION CONSTANT K2 ",K2
1200 INPUT"THIRD IONIZATION CONSTANT K3 ",K3
1210 INPUT"ENTER INITIAL MOLARITY OF ACID H3A ", M0
1220 INPUT"ENTER INITIAL MOLARITY OF SALT NaH2A ",M1
1230 INPUT"ENTER INITIAL MOLARITY OF SALT Na2HA ",M2
1240 INPUT"ENTER INITIAL MOLARITY OF SALT Na3A ", M3
1250 GOTO 1270
1260 REM
```

```
1270 REM******************** Calculations ********************************
1280 REM
1290 REM set convergence criterion and limit to number of iterations
1300 L=99
1310 Z=.000001
1320 REM define coefficients of polynomial expression
1330 N = M1 + 2*M2 + 3*M3
1340 T = M0 + M1 + M2 + M3
1350 A = 1
1360 B = K1 + N
1370 C = K1*(N - T + K2) - W
1380 D = K1*(K2*(N - 2*T + K3) - W)
1390 E = K1*K2*(K3*(N - 3*T) - W)
1400 F = -K1*K2*K3*W
1410 G=T
1420 REM set initial guess of [H]
1430 H=G
1440 PRINT"** ITERATION #";
1450 REM ************************* Newton-Raphson Iteration ***************
1460 FOR I=1 TO L
1470 HTAB 15:PRINT I;
1480 FH = ((((A*H+B)*H+C)*H+D)*H+E)*H+F
1490 DV = (((5*A*H+4*B)*H+3*C)*H+2*D)*H+E
1500 IF DV <> 0 THEN GOTO 1530
1510 H=1.01*H
1520 GOTO 1480
1530 H1=H-FH/DV
1540 IF ABS(H1-H)/H <Z THEN GOTO 1600
1550 H=H1
1560 NEXT I
1570 PRINT"NO CONVERGENCE AFTER";I;"ITERATIONS"
1580 PRINT"CHECK INPUT DATA AND/OR PERMIT MORE ITERATIONS"
1590 GOTO 2000
1600 IF H>1E-15 THEN GOTO 1680
1610 PRINT"*** IMPOSSIBLY SMALL OR NEGATIVE ROOT FOUND:"
1620 PRINT"TRYING AGAIN WITH LARGER INITIAL GUESS *****"
1630 G=3*G
1640 IF G<10 THEN 1670
1650 PRINT:PRINT"CALCULATION HALTED, CHECK INPUT DATA"
1660 GOTO 2000
1670 GOTO 1430
```

```
1680 REM ******************* calculate other concentrations
1690 IF B$="T" THEN GOTO 1760
1700 REM ******* diprotic system
1710 Q = 1 + (H/K2)*(1 + (H/K1))
1720 C2=T/Q
1730 C1=C2*H/K2
1740 C0=C1*H/K1
1750 GOTO 1830
1760 REM ***** triprotic system
1770 Q= 1 + (H/K3)*(1 + (H/K2)*(1 + (H/K1)))
1780 C3=T/Q
1790 C2=C3*H/K3
1800 C1=C2*H/K2
1810 C0=C1*H/K1
1820 REM
1830 REM ********************** Output Section ******************************
1840 REM
1850 PRINT:HTAB 28:PRINT"CONCENTRATIONS"
1860 IF B$="D" THEN GOTO 1910
1870 PRINT"            ","[A---]","[HA--]","[H2A-]","[H3A]"
1880 PRINT"INITIALLY   ",M3,M2,M1,M0
1890 PRINT"EQUILIBRIUM",C3,C2,C1,C0
1900 GOTO 1940
1910 PRINT"            ","[B--]","[HB-]","[H2B]
1920 PRINT"INITIALLY   ", M2,M1,M0
1930 PRINT"EQUILIBRIUM", C2,C1,C0
1940 PRINT:PRINT"                PH ="; (-1)*LOG(H)/2.30259;"   [H] =";H
1950 GOTO 1970
1960 GOTO 2000
1970 PRINT: INPUT"DO YOU WISH TO DO ANOTHER CALCULATION? (Y/N) ",A$
1980 IF((A$="Y") OR (A$="y")) THEN GOTO 980
1990 PRINT:PRINT"END OF CALCULATIONS"
2000 END
```

Sample Output 1 from Program II.4

```
      GENERAL PROGRAM FOR POLYPROTIC SYSTEMS


ENTER TYPE:  (D)IPROTIC OR (T)RIPROTIC D

          DIPROTIC ACID SYSTEM
ENTER FIRST IONIZATION CONSTANT K1  .0004
ENTER SECOND IONIZATION CONSTANT K2  4E-11
ENTER INITIAL MOLARITY OF ACID H2B  .03
ENTER INITIAL MOLARITY OF SALT NaHB  .06
ENTER INITIAL MOLARITY OF SALT Na2B  0

CONCENTRATIONS
               [B--]          [HB-]          [H2B]
INITIALLY      0               .06            .03
EQUILIBRIUM    1.21596E-08    .060198        .029802

              PH = 3.70327     [H] = 1.98026E-04

DO YOU WISH TO DO ANOTHER CALCULATION? (Y/N) Y



ENTER TYPE:  (D)IPROTIC OR (T)RIPROTIC D

          DIPROTIC ACID SYSTEM
ENTER FIRST IONIZATION CONSTANT K1  .0004
ENTER SECOND IONIZATION CONSTANT K2  4E-11
ENTER INITIAL MOLARITY OF ACID H2B  0
ENTER INITIAL MOLARITY OF SALT NaHB  .06
ENTER INITIAL MOLARITY OF SALT Na2B  0

CONCENTRATIONS
               [B--]          [HB-]          [H2B]
INITIALLY      0               .06           0
EQUILIBRIUM    1.89853E-05    .0599621       1.89381E-05

              PH = 6.89847     [H] = 1.26334E-07

DO YOU WISH TO DO ANOTHER CALCULATION? (Y/N) N



END OF CALCULATIONS
```

Sample Output 2 from Program II.4

```
      GENERAL PROGRAM FOR POLYPROTIC SYSTEMS


ENTER TYPE:  (D)IPROTIC OR (T)RIPROTIC D

          DIPROTIC ACID SYSTEM
ENTER FIRST IONIZATION CONSTANT K1  .0004
ENTER SECOND IONIZATION CONSTANT K2  .0000008
ENTER INITIAL MOLARITY OF ACID H2B  0
ENTER INITIAL MOLARITY OF SALT NaHB  0
ENTER INITIAL MOLARITY OF SALT Na2B  .05

CONCENTRATIONS
              [B--]          [HB-]          [H2B]
INITIALLY      .05            0              0
EQUILIBRIUM    .049975        2.49935E-05    2.49996E-11

              PH = 9.39782    [H] = 4.00096E-10

DO YOU WISH TO DO ANOTHER CALCULATION? (Y/N) Y



ENTER TYPE:  (D)IPROTIC OR (T)RIPROTIC T

         TRIPROTIC ACID SYSTEM
ENTER FIRST IONIZATION CONSTANT K1  .00752
ENTER SECOND IONIZATION CONSTANT K2  6.23E-08
THIRD IONIZATION CONSTANT K3  2.2E-13
ENTER INITIAL MOLARITY OF ACID H3A  .05
ENTER INITIAL MOLARITY OF SALT NaH2A  .01
ENTER INITIAL MOLARITY OF SALT Na2HA  .3
ENTER INITIAL MOLARITY OF SALT Na3A  .003

CONCENTRATIONS
              [A---]         [HA--]         [H2A-]         [H3A]
INITIALLY      .003           .3             .01            .05
EQUILIBRIUM    2.16276E-06    .255996        .107002        3.70527E-07

              PH = 7.58434    [H] = 2.60403E-08

DO YOU WISH TO DO ANOTHER CALCULATION? (Y/N) N


END OF CALCULATIONS
```

Sample Output 3 from Program II.4

```
        GENERAL PROGRAM FOR POLYPROTIC SYSTEMS


ENTER TYPE:  (D)IPROTIC OR (T)RIPROTIC T

         TRIPROTIC ACID SYSTEM
ENTER FIRST IONIZATION CONSTANT K1  .00752
ENTER SECOND IONIZATION CONSTANT K2  6.23E-08
THIRD IONIZATION CONSTANT K3  2.2E-13
ENTER INITIAL MOLARITY OF ACID H3A  .05
ENTER INITIAL MOLARITY OF SALT NaH2A  0
ENTER INITIAL MOLARITY OF SALT Na2HA  0
ENTER INITIAL MOLARITY OF SALT Na3A  0

CONCENTRATIONS
               [A---]         [HA--]         [H2A-]        [H3A]
INITIALLY      0              0              0              .05
EQUILIBRIUM    8.57049E-19    6.22995E-08    .0159918       .0340081

               PH = 1.79609     [H] = .015992

DO YOU WISH TO DO ANOTHER CALCULATION? (Y/N) Y


ENTER TYPE:  (D)IPROTIC OR (T)RIPROTIC T

         TRIPROTIC ACID SYSTEM
ENTER FIRST IONIZATION CONSTANT K1  .00074
ENTER SECOND IONIZATION CONSTANT K2  .000017
THIRD IONIZATION CONSTANT K3  .0000039
ENTER INITIAL MOLARITY OF ACID H3A  .15
ENTER INITIAL MOLARITY OF SALT NaH2A  0
ENTER INITIAL MOLARITY OF SALT Na2HA  0
ENTER INITIAL MOLARITY OF SALT Na3A  .15

CONCENTRATIONS
               [A---]         [HA--]         [H2A-]        [H3A]
INITIALLY      .15            0              0              .15
EQUILIBRIUM    .0194314       .116182        .159364        5.02181E-03

               PH = 4.63229     [H] = 2.33185E-05

DO YOU WISH TO DO ANOTHER CALCULATION? (Y/N) N


END OF CALCULATIONS
```

Appendix III

Ionization Constants (25° C) of Selected Acids and Bases

Most of the values listed are taken from G. D. Fasman, ed., *CRC Handbook of Biochemistry and Molecular Biology*, Sec. D: *Physical and Chemical Data*, Vol. 1, 3rd ed., CRC Press, Boca Raton, Fla., 1976.

Acids

Acid	pK_a	K_a
Acetic acid	4.76	1.8×10^{-5}
Acetylacetone	8.95	1.1×10^{-9}
Acetylene	25.0	1×10^{-25}
o-Aminobenzenesulfonic acid	2.48	3.3×10^{-3}
m-Aminobenzenesulfonic acid	3.73	1.85×10^{-4}
p-Aminobenzenesulfonic acid	3.24	5.8×10^{-4}
Benzenesulfonic acid	0.70	2.0×10^{-1}
Benzoic acid	4.20	6.3×10^{-5}
Bromoacetic acid	2.86	1.4×10^{-3}
o-Bromobenzoic acid	2.85	1.4×10^{-3}
m-Bromobenzoic acid	3.81	1.6×10^{-4}
p-Bromobenzoic acid	3.97	1.1×10^{-4}
o-Bromophenol	8.42	3.8×10^{-9}
m-Bromophenol	8.87	1.4×10^{-9}

Acid	pK_a	K_a
p-Bromophenol	9.26	5.5×10^{-10}
Butanoic acid	4.82	1.51×10^{-5}
Carbonic acid		
K_1	6.40	4.0×10^{-7}
K_2	10.40	4.0×10^{-11}
Chloric acid	−1	1.0×10^{1}
Chloroacetic acid	2.86	1.4×10^{-3}
o-Chlorobenzoic acid	2.94	1.1×10^{-3}
m-Chlorobenzoic acid	3.83	1.5×10^{-4}
p-Chlorobenzoic acid	4.0	1.0×10^{-4}
2-Chlorobutanoic acid	2.84	1.4×10^{-3}
3-Chlorobutanoic acid	4.06	8.7×10^{-5}
4-Chlorobutanoic acid	4.52	3.0×10^{-5}
o-Chlorophenol	8.48	3.3×10^{-9}
m-Chlorophenol	9.02	9.6×10^{-10}
p-Chlorophenol	9.38	4.2×10^{-10}
2-Chloropropanoic acid	2.80	1.58×10^{-3}
3-Chloropropanoic acid	4.08	8.3×10^{-5}
Chlorous acid	2	1×10^{-2}
cis-Cinnamic acid	3.89	1.3×10^{-4}
trans-Cinnamic acid	4.44	3.6×10^{-5}
Citric acid		
K_1	3.13	7.4×10^{-4}
K_2	4.76	1.7×10^{-5}
K_3	5.41	3.9×10^{-6}
o-Cyanobenzoic acid	3.14	7.2×10^{-4}

Acid	pK_a	K_a
m-Cyanobenzoic acid	3.64	2.3×10^{-4}
p-Cyanobenzoic acid	3.55	2.8×10^{-4}
p-Cyanophenol	7.95	1.1×10^{-8}
Diacetylacetone	7.42	3.8×10^{-8}
Dichloroacetic acid	1.29	5.1×10^{-2}
o-Dihydroxybenzene		
K_1	9.45	3.5×10^{-10}
K_2	12.8	2×10^{-13}
2,2-Dimethylpropanoic acid	5.03	9.3×10^{-6}
2,4-Dinitrophenol	4.09	8.1×10^{-5}
Ethanol	16.0	1×10^{-16}
Ethene	~44.0	$\sim 10^{-44}$
o-Ethylbenzoic acid	3.79	1.6×10^{-4}
m-Ethylbenzoic acid	4.27	5.4×10^{-5}
p-Ethylbenzoic acid	4.35	4.5×10^{-5}
Fluoroacetic acid (20°C)	2.57	2.7×10^{-3}
o-Fluorobenzoic acid	3.27	5.4×10^{-4}
m-Fluorobenzoic acid	3.86	1.4×10^{-4}
p-Fluorobenzoic acid	4.14	7.2×10^{-5}
o-Fluorophenol	8.81	1.5×10^{-9}
m-Fluorophenol	9.28	5.2×10^{-10}
p-Fluorophenol	9.81	1.5×10^{-10}
Formic acid	3.75	1.8×10^{-4}
Hydrogen bromide	−9	10^{9}

Acid	pK_a	K_a
Hydrogen chloride	−7	10^{7}
Hydrogen fluoride	3.18	6.6×10^{-4}
Hydrogen iodide	−10	10^{10}
Hydrogen selenide		
K_1	4	10^{-4}
Hydrogen sulfide		
K_1	7.0	1×10^{-7}
K_2	12.92	1.2×10^{-13}
Hydrogen telluride		
K_1	3	10^{-3}
o-Hydroxybenzoic acid	2.98	1.0×10^{-3}
m-Hydroxybenzoic acid	4.08	8.3×10^{-5}
p-Hydroxybenzoic acid	4.57	2.7×10^{-5}
Iodoacetic acid	3.12	7.6×10^{-4}
o-Iodobenzoic acid	2.86	1.4×10^{-3}
m-Iodobenzoic acid	3.85	1.4×10^{-4}
p-Iodobenzoic acid	4.00	1.0×10^{-4}
o-Iodophenol	8.46	3.5×10^{-9}
m-Iodophenol	8.88	1.3×10^{-9}
p-Iodophenol	9.20	6.3×10^{-10}
Malonic acid		
K_1	2.83	1.5×10^{-3}
K_2	5.69	2.0×10^{-6}
Methane	~49	$\sim 10^{-49}$
Methanol	15.54	2.9×10^{-16}
o-Methoxybenzoic acid	4.09	8.1×10^{-5}

Acid	pK_a	K_a
m-Methoxybenzoic acid	4.09	8.1×10^{-5}
p-Methoxybenzoic acid	4.47	3.4×10^{-5}
o-Methoxyphenol	9.98	1.0×10^{10}
m-Methoxyphenol	9.65	2.2×10^{-10}
p-Methoxyphenol	10.21	6.2×10^{-11}
o-Methylbenzoic acid	3.91	1.2×10^{-4}
m-Methylbenzoic acid	4.27	5.4×10^{-5}
p-Methylbenzoic acid	4.38	4.2×10^{-5}
o-Methylphenol	10.29	5.1×10^{-11}
m-Methylphenol	10.08	8.3×10^{11}
p-Methylphenol	10.26	5.5×10^{-11}
2-Methylpropanoic acid	4.80	1.6×10^{-5}
Methyl sulfonic acid	−1.2	16
p-(Methylthio)phenol (*p*-$CH_3SC_6H_4OH$)	9.53	3.0×10^{-10}
1-Naphthalenesulfonic acid	0.57	2.7×10^{-1}
1-Naphthoic acid	3.70	2.0×10^{-4}
2-Naphthoic acid	4.16	6.9×10^{-5}
1-Naphthol (20°C)	9.85	1.4×10^{-10}
2-Naphthol (20°C)	9.63	2.3×10^{-10}
Nitric acid	−1.64	44
Nitroacetic acid (18°C)	1.68	2.1×10^{-2}
o-Nitrobenzoic acid	2.21	6.2×10^{-3}
m-Nitrobenzoic acid	3.49	3.2×10^{-4}

Acid	pK_a	K_a
p-Nitrobenzoic acid	3.42	3.8×10^{-4}
o-Nitrophenol	7.22	6.0×10^{-8}
m-Nitrophenol	8.39	4.1×10^{-9}
p-Nitrophenol	7.15	7.1×10^{-8}
Nitrous acid	3.4	4×10^{-4}
Oxalic acid		
K_1	1.27	5.40×10^{-2}
K_2	4.27	5.40×10^{-5}
Perchloric acid	−8	10^8
Phenol	10.0	1.0×10^{-10}
Phenylacetic acid	4.31	4.9×10^{-5}
o-Phenylbenzoic acid	3.46	3.47×10^{-4}
m-Phenylbenzoic acid	4.26	5.5×10^{-5}
p-Phenylbenzoic acid	4.21	6.2×10^{-5}
Phosphoric acid		
K_1	2.12	7.6×10^{-3}
K_2	7.21	6.2×10^{-8}
K_3	12.66	2.2×10^{-13}
o-Phthalic acid (phthalic acid)		
K_1	2.95	1.1×10^{-3}
K_2	5.41	3.9×10^{-6}
m-Phthalic acid (isophthalic acid)		
K_1	3.62	2.4×10^{-4}
K_2	4.60	2.5×10^{-5}
p-Phthalic acid (terephthalic acid)		
K_1	3.54	2.9×10^{-4}
K_2	4.46	3.5×10^{-5}

Acid	pK_a	K_a
Picric acid (2,4,6-trinitrophenol)	0.25	5.6×10^{-1}
Propanoic acid	4.88	1.3×10^{-5}
Salicylic acid		
(25°C) K_1	3.00	1.0×10^{-3}
(20°C) K_2	13.82	1.5×10^{-14}
Succinic acid		
K_1	4.21	6.2×10^{-5}
K_2	5.64	2.3×10^{-6}
Sulfuric acid		
K_1	−3	10^3
K_2	2.0	1×10^{-2}
Thiophenol	6.52	3.0×10^{-7}
Triacetylmethane	5.81	1.5×10^{-6}
Trichloroacetic acid	0.65	2.2×10^{-1}
Trifluoroacetic acid	0.23	5.9×10^{-1}
m-Trifluoromethylbenzoic acid	3.77	1.70×10^{-4}
p-Trifluoromethylbenzoic acid	3.66	2.19×10^{-4}
2,4,6-Trinitrophenol (picric acid)	0.25	5.6×10^{-1}

Bases

Base	pK_b	K_b
Acetamide	~14	~10^{-14}
Ammonia	4.73	1.9×10^{-5}
Aniline	9.42	3.8×10^{-10}
Aziridine	5.96	1.1×10^{-6}
Benzamide	~16	~10^{-16}
Benzamidine	2.4	4×10^{-3}
p-Bromoaniline	10.15	7.1×10^{-11}
o-Chloroaniline	11.36	4.4×10^{-12}
m-Chloroaniline	10.54	2.9×10^{-11}
p-Chloroaniline	10.0	1×10^{-10}
1,4-Diaminobutane		
K_1	3.2	6×10^{-4}
K_2	4.7	2×10^{-5}
1,2-Diaminoethane (ethylenediamine)		
K_1	3.9	1×10^{-4}
K_2	7.0	1×10^{-7}
1,8-Diaminooctane		
K_1	3.0	1×10^{-3}
K_2	3.9	1×10^{-4}
Dimethylamine	3.23	5.9×10^{-4}
2,6-Dimethylaniline	10.11	7.8×10^{-11}
N,N-Dimethylaniline	8.93	1.2×10^{-9}
Diphenylamine	13.12	7.6×10^{-14}
Ethanolamine	4.50	3.2×10^{-5}
Ethylamine	3.37	4.3×10^{-4}

Base	pK_b	K_b
p-Fluoroaniline	9.37	4.3×10^{-10}
p-Fluorobenzamide	16.2	6.3×10^{-17}
Formamide	16.60	2.5×10^{-17}
Hydrazine		
K_1	5.9	1×10^{-6}
Hydroxylamine	8.03	9.3×10^{-9}
p-Iodoaniline	10.22	6.0×10^{-11}
N-Isopropylaniline	8.50	3.2×10^{-9}
Methylamine	3.38	4.2×10^{-4}
N-Methylaniline	9.15	7.1×10^{-10}
p-Methoxyaniline	9.71	2.0×10^{-10}
o-Methylaniline (*o*-toluidine)	9.61	2.5×10^{-10}
p-Methylaniline (*p*-toluidine)	8.92	1.2×10^{-9}
2-Methylpyridine	8.03	9.3×10^{-9}
3-Methylpyridine	8.32	4.8×10^{-9}
4-Methylpyridine	7.98	1.0×10^{-8}
o-Nitroaniline	14.29	5.1×10^{-15}
m-Nitroaniline	11.55	2.8×10^{-12}
p-Nitroaniline	13.00	1.0×10^{-13}
Octylamine	3.35	4.5×10^{-4}
Piperidine	2.78	1.7×10^{-3}
Pyridine	8.86	1.4×10^{-9}
Pyrrole	14.27	5.4×10^{-15}
Trimethylamine	4.20	6.3×10^{-5}

Appendix IV

Activity Coefficients

In an ideal solution, properties such as vapor pressure, boiling point, and equilibrium concentrations depend in a regular and predictable way on solute concentration. In most real solutions, however, deviations from ideal behavior are noticed. The concept of "effective concentration," or *solute activity*, has been developed to deal with these nonidealities. In particular, the acid ionization expression is numerically exact for an ideal

$$K_a = \frac{[H_3O^+]\,[A^-]}{[HA]} \tag{IV-1}$$

solution, but may be in error to a greater or lesser extent in a real solution. If we substitute *activities* for concentrations and write

$$K_a^T = (a_{H_3O^+})\frac{a_{A^-}}{a_{HA}} \tag{IV-2}$$

the equation is then exact. The superscript T stands for "thermodynamic" and means that the "thermodynamic ionization constant" K_a^T is written in terms of solute activity rather than molarity.

Activity is a measurable quantity, although the measurements are not easily carried out. The ratio of activity to concentration, γ, is called the *activity coefficient* and is defined by

$$a_X = \gamma_X C_X = \gamma_X\,[X] \tag{IV-3}$$

where X is the solute and C_X is its concentration in moles per liter. The value of γ for a given solute depends on the nature of the solvent, the nature and concentration of the solute, and the nature and concentration of any additional solutes that may be present. Since both a and C are measurable, γ is also an experimentally determined quantity. The value of γ gives immediate insight into the extent of the solution's deviation from ideal behavior. For an ideal solution $a = C$ and $\gamma = 1$. For real solutions the greater the deviation of γ from 1, the greater the deviation from ideal behavior.

Solution nonideality is due both to interactions between solute and solvent molecules, and to interactions among the solute species themselves. The electrical charge of ionic solutes induces interactions which are stronger than those observed with molecular (neutral) solutes. Consequently, ionic solutions deviate more from ideality than molecular solutions of the same concentration. In both types of solution the deviations from ideal behavior increase as the concentration increases.

Activity coefficients of ions in solution may be calculated by means of the Debye-Hückel limiting law,[1] equation (IV-4), which gives values in close agreement with experiment, provided that the concentration is not too high. For an ionic solute X, the activity coefficient γ_X is

$$\log_{10} \gamma_X = -|z_X| A \sqrt{I} \qquad \text{(IV-4)}$$

where z_X is the ionic charge (+1, −2, etc.). The constant A is a composite number dependent on the dielectric constant of the solvent and the temperature; for aqueous solutions at room temperature its value is 0.51. The factor I is called *ionic strength* and depends on the concentration and charge of *all* ions in the solution. Ionic strength is defined as

$$I = \frac{1}{2} \sum C_i z_i^2 \qquad \text{(IV-5)}$$

In equation (IV-5) C_i and z_i are the molarity and charge of the ith ionic solute, and the sum is taken over all types of ion in solution. For a 1:1 electrolyte such as NaCl, ionic strength is equal to molarity.

The Debye-Hückel law is accurately obeyed at low concentrations. It follows from equation (IV-4) that as concentration and ionic strength approach zero, log γ approaches zero and γ approaches 1. Thus even an ionic solution becomes ideal at sufficiently low concentration. Equation (IV-4) becomes increasingly inaccurate as the ionic concentration rises, and becomes unreliable at about 0.01 *M*.

Solutes that are electrically neutral, such as nonionized acids, deviate from ideality much less than ionic solutes. At concentrations for which the Debye-Hückel limiting law is valid, the activity coefficient of an uncharged solute is very close to 1. This fact leads to the modified version of the ionization-constant expression (IV-6).

$$\begin{aligned} K_a^T &= \frac{(\gamma_{H_3O^+})[H_3O^+](\gamma_{A^-})[A^-]}{(\gamma_{HA})[HA]} \\ &= K_a \frac{(\gamma_{H_3O^+})(\gamma_{A^-})}{(\gamma_{HA})} \\ &= K_a(\gamma_{H_3O^+})(\gamma_{A^-}) \end{aligned} \qquad \text{(IV-6)}$$

Equation (IV-6) shows that as the concentration decreases and all γ's approach 1, K_a^T and K_a become equal.

[1] Most undergraduate textbooks of physical chemistry include a derivation of this law.

$$\lim_{C \to 0} K_a^T = K_a \tag{IV-7}$$

Since a pH meter measures the activity of H_3O^+ rather than its concentration,

$$\mathrm{pH} = -\log a_{\mathrm{H_3O^+}} \tag{IV-8}$$

a relationship between pH and K_a^T can be obtained by taking negative logarithms of both sides of (IV-2)

$$-\log K_a^T = -\log (a_{\mathrm{H_3O^+}}) - \log \left(\frac{a_{\mathrm{A^-}}}{a_{\mathrm{HA}}}\right) \tag{IV-9}$$

and substituting equations (IV-8) and (IV-3) into (IV-9)

$$\mathrm{p}K_a^T = \mathrm{pH} - \log\left(\frac{a_{\mathrm{A^-}}}{a_{\mathrm{HA}}}\right) = \mathrm{pH} + \log\left(\frac{a_{\mathrm{HA}}}{a_{\mathrm{A^-}}}\right)$$

$$= \mathrm{pH} + \log\left(\frac{[\mathrm{HA}]}{[\mathrm{A^-}]}\right) + \log\left(\frac{\gamma_{\mathrm{HA}}}{\gamma_{\mathrm{A^-}}}\right) \tag{IV-10}$$

Normally, if one wishes to determine a pK to an accuracy of better than 0.05, equation (IV-10) must be used in place of the Henderson-Hasselbalch equation. Since $\gamma_{\mathrm{HA}} = 1$, and $\gamma_{\mathrm{A^-}}$ can be calculated (in solutions of concentration $\leqslant 0.01\ M$) from equation (IV-4), equation (IV-10) becomes

$$\mathrm{p}K_a^T = \mathrm{pH} + \log\left(\frac{[\mathrm{HA}]}{[\mathrm{A^-}]}\right) - (-0.51)\,|z_{\mathrm{A^-}}|\sqrt{I})$$

and since for a 1:1 electrolyte $|z_{\mathrm{A^-}}| = 1$ and ionic strength = molarity,

$$\mathrm{p}K_a^T = \mathrm{pH} + \log\left(\frac{[\mathrm{HA}]}{[\mathrm{A^-}]}\right) + 0.51\sqrt{[\mathrm{A^-}]} \tag{IV-11}$$

For a solution in which $[\mathrm{A^-}] = 0.01\ M$, for example, $0.51\sqrt{[\mathrm{A^-}]} = 0.05$; $\mathrm{p}K_a^T$ is larger than $\mathrm{p}K_a$ by this amount. Thus a calculation (for a neutral acid) that ignores activity gives a pK which is lower than the thermodynamic value. If the acid is a cationic species $\mathrm{BH^+}$, equation (IV-11) becomes

$$\mathrm{p}K_a^T = \mathrm{pH} + \log\left(\frac{[\mathrm{BH^+}]}{[\mathrm{B}]}\right) - 0.51\sqrt{[\mathrm{BH^+}]} \tag{IV-12}$$

The correction term $-0.51\sqrt{[\mathrm{BH^+}]}$ is negative, and $\mathrm{p}K_a$ values calculated from concentration rather than activity are larger than the thermodynamic values.

Appendix V

Absorbance

Equation (11-10), relating absorbance to pK_a, is derived from the Henderson-Hasselbalch equation as follows. Absorbance A is related to solute concentration C (molarity), absorption coefficient[1] ϵ, and optical path length l by

$$A = \epsilon Cl \tag{V-1}$$

In a solution containing both the protonated form HA and the unprotonated form A^- of a weak acid, the absorbance is the sum of the individual contributions:

$$A = \epsilon_{A^-}[A^-]\,l + \epsilon_{HA}[HA]\,l + \epsilon_{other}[other]\,l \tag{V-2}$$

The term $\epsilon_{other}[other]\,l$ is intended to include contributions to the absorbance from all other sources that may be present; normally experiments are set up so that this term is negligibly small. Since the proportions of $[A^-]$ and $[HA]$ change as the pH of the solution changes, the absorbance A varies also.[2] However, the total amount of acid is constant. Let

$$[HA] + [A^-] = C \tag{V-3}$$

At very high pH all of the acid is in ionic form, so that $[A^-] = C$ and $[HA] \cong 0$. The absorbance of the solution at high pH is denoted A_{ion}, and (V-2) becomes

$$A_{ion} = \epsilon_{A^-}Cl + \epsilon_{other}[other]\,l \tag{V-4}$$

Similarly, at low pH all of the acid is in molecular form, so that $[HA] = C$ and $[A^-] \cong 0$. Under these circumstances the absorbance of the solution becomes

$$A_{mol} = \epsilon_{HA}Cl + \epsilon_{other}[other]\,l \tag{V-5}$$

[1] The magnitude of the absorption coefficient ϵ of any substance depends on the wavelength of the light used in the measurement. In this discussion we assume that a single, fixed wavelength is used, so that ϵ is a constant.

[2] If $\epsilon_{HA} = \epsilon_{A^-}$, of course, the absorbance will *not* change as the relative amounts of HA and A^- change. No determination of pK_a can be made under these circumstances. It is best to choose a wavelength at which one of ϵ_{A^-} and ϵ_{HA} is very large, while the other is very small.

It follows from (V-2), (V-3), and (V-4) that

$$\begin{aligned} A_{\text{ion}} - A &= \epsilon_{\text{A}^-} Cl - \epsilon_{\text{A}^-} [\text{A}^-]\, l - \epsilon_{\text{HA}} [\text{HA}]\, l \\ &= \epsilon_{\text{A}^-} l (C - [\text{A}^-]) - \epsilon_{\text{HA}} [\text{HA}]\, l \\ &= \epsilon_{\text{A}^-} [\text{HA}]\, l - \epsilon_{\text{HA}} [\text{HA}]\, l \\ &= [\text{HA}]\, l (\epsilon_{\text{A}^-} - \epsilon_{\text{HA}}) \end{aligned} \tag{V-6}$$

and from (V-2), (V-3), and (V-5) that

$$A - A_{\text{mol}} = [\text{A}^-]\, l (\epsilon_{\text{A}^-} - \epsilon_{\text{HA}}) \tag{V-7}$$

We combine (V-6) and (V-7) to get

$$\frac{A_{\text{ion}} - A}{A - A_{\text{mol}}} = \frac{[\text{HA}]\, l (\cancel{\epsilon_{\text{A}^-} - \epsilon_{\text{HA}}})}{[\text{A}^-]\, l (\cancel{\epsilon_{\text{A}^-} - \epsilon_{\text{HA}}})} \tag{V-8}$$

Substitution of (V-8) into the Henderson-Hasselbalch equation

$$\text{p}K_a = \text{pH} + \log\left(\frac{[\text{HA}]}{[\text{A}^-]}\right) \tag{V-9}$$

results in

$$\text{p}K_a = \text{pH} + \log\left(\frac{A_{\text{ion}} - A}{A - A_{\text{mol}}}\right) \tag{V-10}$$

Note that although the two absorption coefficients ϵ_{HA} and ϵ_{A^-} must be different, their numerical values need not be known.

Appendix VI

Approximations, Errors, and the Exact Solution

VI.1 ERRORS OF APPROXIMATION

The usual way of assessing the reliability of a quantity is by its *percent error*. If we guess the number of students in a class to be 30 and a count shows that there are only 28 students, our guess was in error. The percent error is $100[(30 - 28)/28] = 7.1\%$. In acidity calculations, the percent error is given by

$$E = 100\left(\frac{[H_3O^+]_{\text{approx}} - [H_3O^+]_{\text{exact}}}{[H_3O^+]_{\text{exact}}}\right) \qquad \text{(VI-1)}$$

Note that, by definition:

$$[H_3O^+]_{\text{approx}} \approx [H_3O^+]_{\text{exact}}$$

Make this substitution in the denominator of equation (VI-1), so that percent error becomes

$$E \approx 100\left(\frac{[H_3O^+]_{\text{approx}} - [H_3O^+]_{\text{exact}}}{[H_3O^+]_{\text{approx}}}\right) \qquad \text{(VI-2)}$$

Into this expression we substitute the previously derived formulas,[1]

$$[H_3O^+]_{\text{exact}} = \frac{-K + \sqrt{K^2 + 4KC}}{2} \qquad \text{(VI-3)}$$

$$[H_3O^+]_{\text{approx}} = \sqrt{KC} \qquad \text{(VI-4)}$$

to obtain

$$E \approx 100\left(\frac{\sqrt{KC} + \frac{1}{2}K - \frac{1}{2}\sqrt{K^2 + 4KC}}{\sqrt{KC}}\right) \qquad \text{(VI-5)}$$

[1] Equations (VI-3) and (VI-4) are derived in Chapter 3 and numbered (3-31) and (3-33).

After some manipulation equation (VI-6) is obtained.

$$E \approx 100\left(\frac{\sqrt{KC}}{\sqrt{KC}} + \frac{\tfrac{1}{2}\sqrt{K^2}}{\sqrt{KC}} - \frac{\tfrac{1}{2}\sqrt{K^2 + 4KC}}{\sqrt{KC}}\right)$$

$$\approx 100\left(1 + \frac{1}{2}\sqrt{\frac{K^2}{KC}} - \frac{1}{2}\sqrt{\frac{K^2}{KC} + \frac{4KC}{KC}}\right)$$

$$\approx 100\left(1 + \frac{1}{2}\sqrt{\frac{K}{C}} - \frac{1}{2}\sqrt{\frac{K}{C} + 4}\right) \qquad \text{(VI-6)}$$

Many acid and base ionization constants are *less* than 10^{-4}, and many chemically useful concentrations are *greater* than 10^{-3}. The quotient K/C therefore rarely exceeds 0.1, and we rarely introduce significant error by saying that $(K/C + 4)$ is approximately equal to 4. Hence

$$E \approx 100\left(1 + \frac{1}{2}\sqrt{\frac{K}{C}} - \frac{1}{2}\sqrt{4}\right)$$

and our final expression for the *estimated error* is

$$E \approx 100\left(\frac{1}{2}\sqrt{\frac{K}{C}}\right) \qquad \text{(VI-7)}$$

The foregoing analysis was given in terms of acid ionization, but the results are identical for base ionization. For bases, however, the percent error refers to calculation of $[OH^-]$ rather than $[H_3O^+]$.

Equation (VI-7), which is useful over most of the range of concentrations and ionization constants of interest to chemists, offers a quick estimate of the error involved in calculating $[H_3O^+]$ with the approximate method of equation (VI-4).

Example VI.1

Using equation (VI-7), estimate the error in the approximate calculation of $[OH^-]$ in a 0.15 M solution of methylamine, $K_b = 4.17 \times 10^{-4}$.

Answer:

$$E \approx 100\left(\frac{1}{2}\sqrt{\frac{K}{C}}\right) = 100\left(\frac{1}{2}\sqrt{\frac{4.17 \times 10^{-4}}{0.15}}\right)$$

$$\approx 2.6\%$$

This calculation was treated in Example 3.11: the *actual* error was calculated to be 2.7%, so the error *estimated* here is very close. If the estimated error is unacceptably large, equation (VI-3), suitably modified for base rather than acid solutions, should be used in calculating $[OH^-]$.

There are three points to remember about the approximate solution and its error. (1) Over a substantial range of chemically interesting concentrations and equilibrium constants, no serious error results from the use of the approximate equation (VI-4) (see Section VI.3). (2) The estimated error is very close to the actual error. They do not differ significantly unless the error is greater than 100%, in which case the approximate equation (VI-4) would not be used anyway. (3) The error is a positive quantity: that is, the $[H_3O^+]$ calculated by the approximate form (VI-4) cannot be less than the $[H_3O^+]$ calculated from the quadratic equation (VI-3). This fact may be used to improve the value of the approximate pH or $[H_3O^+]$, as in Example VI.2.

Example VI.2

Estimate the error in calculating $[OH^-]$ in a 2.00 M solution of methylamine, $K_b = 4.17 \times 10^{-4}$, and use the estimated error to improve the approximation.

Answer: Using equation (VI-7) the error is estimated as

$$E \approx 100\left(\frac{1}{2}\sqrt{\frac{K}{C}}\right) = 100\ \frac{1}{2}\sqrt{\frac{4.17 \times 10^{-4}}{2.00}} = 0.72\%$$

The approximate concentration of hydroxide ion is

$$[OH^-] \approx \sqrt{KC} = 0.0289\ M$$

which we know to be high by 0.72%; 0.72% of 0.0289 M is 0.0002 M. Subtract 0.0002 from the estimate to get an improved estimate of

$$[OH^-] = 0.0287\ M$$

Finally, we can compare this improved approximation with the quadratic value calculated from equation (VI-3). The results are

$$\text{approximation } (\sqrt{KC}) = 0.0289\ M$$

$$\text{improved approximation} = 0.0287\ M$$

$$\text{quadratic value} = 0.02867\ M$$

VI.2 THE EXACT SOLUTION FOR VERY WEAK OR VERY DILUTE ACIDS

In a solution in which both reactions

$$HA + H_2O \rightleftharpoons H_3O^+ + A^-$$

and

$$2H_2O \rightleftharpoons H_3O^+ + OH^-$$

produce $[H_3O^+]$ to a comparable extent, the laws of conservation of mass and conservation of charge form the basis for calculation of $[H_3O^+]$. Stoichiometry requires that

$$[HA] + [A^-] = C \tag{VI-8}$$

where C = initial molarity of acid. That is, there is no source of A^- other than the original HA added to the water. For each HA molecule that ionizes, one A^- ion appears in the solution; their total concentration is constant, equal to the initial value of acid concentration. Another way of saying this is that there is a certain amount of A present, regardless of whether it is protonated or unprotonated.

Since both water and nonionized acid are themselves uncharged, any mixture of them must also be electrically neutral: the total number of positively charged species must be the same as the total number of negatively charged species. In the case of an aqueous solution of the monoprotic acid HA, this means that

$$[H_3O^+] = [OH^-] + [A^-] \tag{VI-9}$$

Finally, we include the two equations describing the equilibria:

$$K_w = [H_3O^+][OH^-] \tag{VI-10}$$

$$K_a = \frac{[H_3O^+][A^-]}{[HA]} \tag{VI-11}$$

The result is a set of four equations, (VI-8), (VI-9), (VI-10), and (VI-11), containing the the four unknowns $[H_3O^+]$, $[OH^-]$, $[A^-]$, and $[HA]$ (C is a constant, not an unknown). The rules of algebra tell us that sets in which the number of equations is equal to the number of unknowns are solvable, although there are no guarantees that the solution will be easy or obvious.

We proceed by substituting (VI-11) into (VI-8) to eliminate $[HA]$ and get

$$C = \frac{[H_3O^+][A^-]}{K_a} + [A^-] \tag{VI-12}$$

Next, (VI-9) is used to eliminate $[A^-]$ from (VI-12):

$$C = \frac{[H_3O^+]([H_3O^+] - [OH^-])}{K_a} + ([H_3O^+] - [OH^-])$$

and the terms in parentheses are written out:

$$C = \frac{[H_3O^+]^2}{K_a} - \frac{[H_3O^+][OH^-]}{K_a} + [H_3O^+] - [OH^-] \tag{VI-13}$$

Equation (VI-10) is used to eliminate $[OH^-]$ from (VI-13):

$$C = \frac{[H_3O^+]^2}{K_a} - \frac{K_w}{K_a} + [H_3O^+] - \frac{K_w}{[H_3O^+]} \qquad \text{(VI-14)}$$

Both sides of (VI-14) are multiplied by $[H_3O^+]$, and all terms moved to the same side of of the equation:

$$\frac{1}{K_a}[H_3O^+]^3 + [H_3O^+]^2 - \left(C + \frac{K_w}{K_a}\right)[H_3O^+] - K_w = 0 \qquad \text{(VI-15)}$$

Equation (VI-15) is a cubic equation of the form

$$ax^3 + bx^2 + cx + d = 0$$

and although there exists a general method for its solution (see any college algebra text), it is easier to rely on the power of a computer to do the job for us. The mathematical details and a BASIC program for calculating $[H_3O^+]$ from equation (VI-15) are given in Appendix II.2. If K_b and $[OH^-]$ are used in place of K_a and $[H_3O^+]$, the same program may be used for weak bases.[1]

The results of these exact pH calculations, for a variety of acid concentrations and K_a values, are shown as the solid lines in Figure VI.1. Results obtained from the approximate equation (VI-4) and the quadratic equation (VI-3) are also shown. The dashed lines represent the quadratic form, and the dotted lines represent the approximate form. Deviations from the exact values can be considerable for both quadratic and approximate forms, since both ignore autoionization. The graphs confirm what intuition tells us, namely, that solvent autoionization becomes important at low concentration, regardless of acid strength. For very weak acids, autoionization dominates even at higher concentrations.

VI.3 GUIDELINES

The exact equation, (VI-15), *always* gives the correct value of $[H_3O^+]$. However, many problems can be solved with satisfactory accuracy without the use of a computer,

[1] If the acid is strong, K_a is very large and those terms in equation (VI-15) with K_a in the denominator become small and may be neglected. For strong acids, then, equation (VI-15) becomes

$$[H_3O^+]^2 - C[H_3O^+] - K_W = 0 \qquad \text{(VI-16)}$$

Unless the solution is very dilute, $C[H_3O^+] >> K_W$; K_W may be neglected in (VI-16), leaving

$$[H_3O^+]^2 - C[H_3O^+] = 0 \qquad \text{(VI-17)}$$

After division by $[H_3O^+]$, equation (VI-17) assumes the standard form for strong acids:

$$[H_3O^+] = C$$

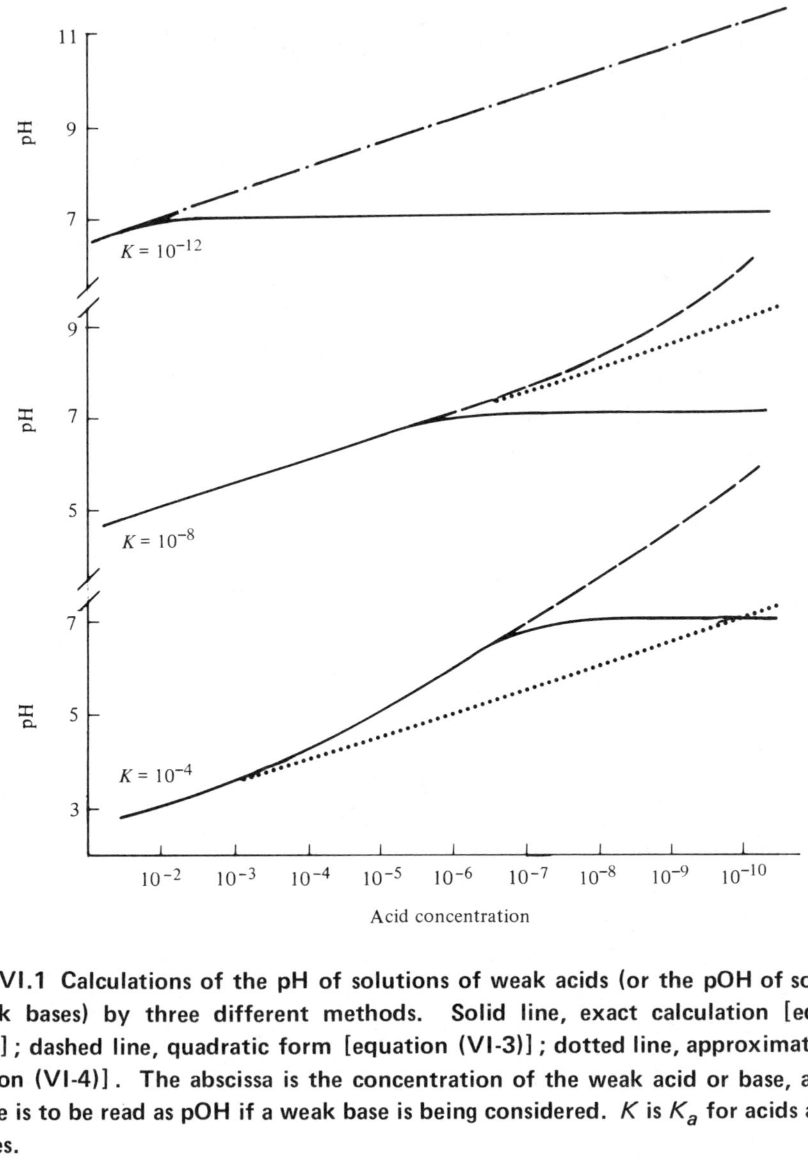

Figure VI.1 Calculations of the pH of solutions of weak acids (or the pOH of solutions of weak bases) by three different methods. Solid line, exact calculation [equation (VI-15)]; dashed line, quadratic form [equation (VI-3)]; dotted line, approximate form [equation (VI-4)]. The abscissa is the concentration of the weak acid or base, and the ordinate is to be read as pOH if a weak base is being considered. K is K_a for acids and K_b for bases.

using either the quadratic equation (VI-3) or the approximate equation (VI-4). Several rules of thumb can be extracted from the derivations of Sections VI.1 and VI.2, and from Figure VI.1, and these can be used to decide when the more convenient methods are appropriate.

(1) Only the exact solution specifically includes the effect of the autoionization of water. If the initial concentration C is less than about 10^{-6} M, or if the pH is greater than about 6, a significant portion of the H_3O^+ comes from autoionization. The rule is: Use the exact method only if $C < 10^{-6}$ or pH > 6.

(2) If autoionization is negligible, and if in addition the fractional extent of ionization is less than about 0.01, the approximate method of equation (VI-4) yields correct results. The fractional extent of ionization is given by $\sqrt{K/C}$ (see Section 3.8), so the rule is: When autoionization is negligible (pH < 6 and $C > 10^{-6}$), use the approximate equation (VI-4) if $K/C < 10^{-4}$.

(3) If autoionization is negligible, the quadratic equation gives the correct result regardless of the extent of ionization. The quadratic equation (VI-3) should be used whenever the approximate equation is inappropriate, that is when $K/C > 10^{-4}$.

If these rules of thumb are followed, calculations of $[H_3O^+]$ in solutions of weak acids will never have an error larger than 1%. (For weak base solutions, substitute pOH for pH and K_b for K_a in the rules.)

Index

Absorbance, 258, 259
Acetate, 77, 79
Acetic acid, 6, 11, 15, 39, 77, 79, 92
  acidity of substituted, 94
Acetic anhydride, 27
Acetylene, 91, 92
Acetylide ion, 91, 92
Acid anhydride, 22
Acid chloride, 66
Acid ionization, calculation of, 41-43
Acidifying principle, 7
Acidity function, 223, 224
  spectrometric determination, 219-221
Acids, 7, 8, 11, 12
  definition of, 11, 12
  dilute, 50
  diprotic, 33, 169
  monoprotic, 33
  polyprotic, 33, 169-171
  preparation of, 22
  properties of, 7, 8
  strength of, 8, 37, 45
  strong, 36, 222
  tetraprotic, 33
  tripotic, 33
  very weak, 50, 167, 225
  weak, 37, 50, 215, 216
Activation energy, 3
Activity, 215, 222, 223, 255, 256, 257
Activity coefficients, 222, 223, 255, 256, 257
Adduct, 59
Alcohols, 24, 112, 116
  butyl, 24
  dehydration of, 112
  isopropyl, 24
  methyl, 3, 24, 51
  preparation of, 108, 109
  primary, 108, 109
Aldehyde, 69, 109
Aldol condensation, 109-111
Alkali, 6
Alkali metals, 61
  reaction with water, 24
Alkali-metal hydrides, 26
Alkaline, 6
Alkaline earth metals, 61
  reaction with water, 24
Alkanes, 89, 90
Alkenes, 63, 89-91, 111
  basic character, 111, 113
  preparation of, 111, 112
Alkoxides, 18, 26
  preparation of, 27
Alkyl halides, 115
Alkynes, 89, 91, 92
Alpha hydrogens, acidity of, 109-111
Aluminum chloride, 60, 67, 115
Aluminum hydroxide, 33
Amide ion, 25
Amides, 66, 104, 105
Amines, 61, 99, 100, 105-107, 288
  definition of, 24
  methyl, 12, 43, 100

Amines (continued)
  primary, 99, 105-107
  propyl, 24
  secondary, 99, 105-107
  synthesis of, 105
  tertiary, 99, 105-107
Amino acid, 193-195
Ammonia, 8-10, 13, 15, 21, 24, 39, 61, 90 92, 99
  preparation of, 24, 25
  substituted, 99
Ammonium chloride, 25
Ammonium nitrite, hydrolysis of, 188-193
Amphiprotic, 13, 15, 45, 187-193
Amphoteric, 14
Anhydride, 66
Aniline, 50, 102, 103
  basicity of substituted, 100-103
Aqua regia, 6
Arrhenius, Svante, 7, 9
Arrhenius theory, 1, 7-9, 69
Association constant, 199
Autoionization, 17, 45, 46
Autoprotolysis, 46

Bare proton, 12
Barium hydroxide, 33, 172
Bases, 7, 8, 11, 12
  definition of, 11, 12
  dilute, 50
  ionization constant, 43
  monohydric, 33
  polyhydric, 172
  preparation of, 22
  properties of, 7, 8
  strong, 27
  trihydric, 33
  very weak, 50
  weak, 43
Basic functional groups, 99
Basicity, 45
Benzamidine, 161, 162
Benzene, 14, 89, 113-115
  electrophilic substitution, 114, 115
Benzenesulfonamide, 106
Benzenesulfonyl chloride, 106
Benzoic acid, 16, 34, 38, 40
  substituted, 94
Berzelius, Jon Jacob, 7
Bicarbonate, 14
Boron trihalides, 61, 62, 84-87
Boyle, Robert, 6
Brønsted, Johannes, 9
Brønsted-Lowry acid-base theory, 1, 9, 11 68, 69
Buffer capacity, 136-139
Buffer solutions, 130-144
  definition of, 130
  polyprotic, 183, 184
  preparation of, 138, 139
Butanoic acid, 83
Butanoic acids, chlorinated, 83
Butyl chloride (normal), 26
Butyllithium, 18, 26
Bz, 38

Carbocation, 115
Calcium carbonate, 6, 23, 25
Calcium cyanamide, 25
Calcium hydroxide, 23
Calcium oxide, 6, 23, 25
Carbanions, 25, 26, 225
  basic strength, 107
  methyl, 10, 90
Carbocations, 61, 113, 115
  allylic, 78
  stability of, 113
Carbon dioxide, 23
Carbonate, 14
Carbonic acid, 14, 15, 22, 169, 176, 177 180
  pH curve, 177
  titration curve, 185
Carboxylate anion, 78
Carboxylic acids, 92, 94, 95, 97, 116
Catalysis, 5
Chelation, 198
Chloramine, 24
Chlorobutane, 26
Chloromethane, 79

Citric acid, 176, 177, 179
  pH curve, 179
  titration curve, 187
Claisen condensation, 111
Common-ion effect, 131
Complex formation, 197-201
Computer programs, 228-245
  buffer system, 235-238
  errors of approximation, 230, 231
  polyprotic and amphoteric, 240-242
  polyprotic systems, 243-245
  very weak or dilute systems, 232-234
Concentration product, 203
Conjugate acid, 9, 10, 18
Conjugate base, 9, 10
Conjugate pairs, 9, 10
Contact process, 21
Coordinate covalent bond, 59, 64
Coordination number, 197
Coordination sites, 197
Covalent bond, 2
Cyanide ion, 10
Cyanohydrin, 62
Cycloalkanes, 89
Cyclohexane, 14
Cyclopentadienyl anion, 78

Davy, Humphry, 7
Dehydration, 112
Dehydrohalgenation, 111
Delocalization of charge, 72
Destructive distillation, 24
Dihydric bases, 33
Dihydrogen phosphate, 10
Dimethylamine, 24
Diprotic acids, 33
Displacement reactions, 64-67
Dissociation, 4, 8, 12
Donor-acceptor bond, 59
Double bonds, 63
  carbon-carbon, 63, 73
  polar, 62

EDTA, 198
Electrolysis, 23
Electron-poor, 1
Electron-rich, 1
Electronegativity, 71
Electrophile, 67, 114, 115
Electrophilic addition reactions, 112
Electrophilic substitution, 67, 114
End point, 148
Endothermic, 3
Enzyme, 6
Equilibrium, 11
  simultaneous, 204-207
Equilibrium constant, 11, 12, 170
Equivalence point, 148
Equivalent, definition of, 33
Errors of approximation, 260-266
  calculation of, 142-144
Esterification, mechanism of, 116
Esters, 66, 115, 116
  preparation of, 115, 116
Ethene (ethylene), 73, 90
  reaction with Ag^+, 64
Ethyl acetate, 16
Ethyl alcohol (ethanol), 6, 24, 77
Ethylamine, 24
Ethylenediamine, 172, 197
Ethylenediaminetetraacetic acid (EDTA), 198
Exothermic, 3

Fehling's solution, 69
Fluoride ion, 10
Fluoroacetic acid, 83
Formal charges, 59, 74-77
Formic acid, 34, 39
Fractional ionization, 41
Free radical(s), 1, 4, 5
Free-radical addition, 69
Friedel-Crafts reaction, 67, 115

Gabriel synthesis, 105
Gas-phase acidity, 226
Gas-phase basicity, 226
Gas-phase reaction, 4
  comparison with solution, 4
Glass, reaction with HF, 20

Glycine, 15
Grignard reagent, 107, 108
  basic strength, 108
  preparation of, 107, 108
  reaction with aldehydes, 108, 109
  reaction with ketones, 109

Haber process, 25
Hard and soft acids and bases, 68
Hartshorn, spirits of, 24
Henderson-Hasselbalch equation, 134, 141, 161, 184, 215, 223, 259
Heterolytic cleavage, 2, 5
Hinsberg test, 105-107
Homolytic cleavage, 2
Hybridization, 72, 89-92
  effect on acidity, 89-92
  effect on basicity, 89-92
Hydride ion, 26
Hydriodic acid, 19, 20
Hydrobromic acid, 19, 20, 84, 85
Hydrocarbons, 89-92
  aromatic, 89
  saturated, 89
  unsaturated, 89
Hydrochloric acid, 6, 7, 19, 20, 39, 84, 85
Hydrofluoric acid, 9, 10, 19, 20, 84, 85
Hydrogen bonding, 13
Hydrogen cyanide, 10
Hydrogen halides, 84, 85
Hydrogen ion concentration (pH), 35
Hydrogen sulfide, 170, 171
Hydrogen theory, 7
Hydroiodic acid, 84, 85
Hydrolysis, 124-127
Hydrolysis constant, 126-130
  of salts, 126, 127
Hydronium ion, 12
  concentration of, 48
Hydroxides, 10, 209, 210
  precipitation by pH control, 209, 210
Hydroxyquinoline, 198
Hypohalous acids, 83

Imides, 104, 105
Indicators, 106, 148, 163-166
Inductive effect, 78-83
Inner salts, 193-195
Inorganic acids, preparation of, 18
Inorganic bases, preparation of, 22
Ion cyclotron resonance, 226
Ion product, 46, 203
Ion size, 83
  effect on acidity, 83
  effect on basicity, 83
Ionic charge, 84
  effect on acidity, 84
  effect on basicity, 84
Ionization, 17, 38, 38
Ionization constants, 39, 127-129
  exact calculations, 172-176
  selected acids and bases, 246-254
Ionization equilibria, 11
  effect of temperature, 53
  extent of, 38, 40
Ions, 8
  positive, 61
Isophthalic acid, 171

Ketones, 109
  hydroxy, 110
  reaction with HCN, 62
K_{sp}, 202-204
K_w, 46

Lavoisier, Antoine, 7
Le Châtelier's principle, 52,131
Leveling effect, 17, 18
Lewis acids, 58-60, 115
  definition of, 58
  strength of, 60, 77
  types of 61-63
Lewis base, 58-60
  definition of, 58
  strength of, 60, 61
  types of, 63, 64
Liebig, Justus Von, 7
Ligands, 97, 172, 197
  bidentate, 198
  tetradentate, 198

tridentate, 198
Lime, 6
slaked, 6, 23
unslaked, 6
Lime-soda process, 23
Limestone, 23
Lithium aluminum hydride, 26
Litmus, 7
Lowry, Martin, 9

Magnesium hydride, 18
Magnesium oxide, 23
Malononitrile, 225
Meta isomers, 81
Metal alkyls, 18
Metal hydrides, 18, 25
preparation of, 26
Metal hydroxides, 23
Metal oxide, 23
Methane, 10, 89
Methane sulfonic acid, 92
Methyl alcohol (methanol), 3, 24, 51
Methyl bromide, 3
Methyl carbanion, 10, 90
Methyl red, 163, 165, 166
Methylamine, 12, 43, 100
Methylammonium ion, 12
Mineral acids, 17, 36
physical properties, 19
preparation of, 18
Molarity, 29, 38
calculation of, 29-32
definition of, 29
of water, 38
Muriatic acid, 20

Negative ions, 63
Neutral solutions, definition of, 47
Neutralization, 8
Neutralization reactions, 59
Nitrate ion, 10
Nitric acid, 6, 10, 19, 20, 22, 36
Nitric oxide, 21
Nitroaniline (*o, m,* and *p*), 100, 101
Nitrogen dioxide, 21
Nitrous acid, 19
Nonaqueous solvents, 11
Nonbonded electron pairs, 17
Nonmetal oxides, 22
Normality, 33
calculation of, 33, 34
definition of, 33
Nucleophile, 65

Octet deficiency, 61
Organic acids, 92-99
Ortho isomers, 81
Orthophosphoric acid, 10, 21
Ostwald process, 20, 25
Oxalic acid, 34, 35
Oxidation-reduction, 2, 5
Oxygen theory, 7

p character, 91
pH, 35, 47, 48, 53
approximate methods, 180-183
calculation of, 35-37
pOH, 45, 48
calculation of, 45
Para isomers, 81
Partial charges, 2, 59
Percent error, 50
Percent ionization, 41
Perchlorate ion, 10
Perchloric acid, 10, 19, 21, 22
Peroxide, 23
Phenanthroline, 198
Phenolphthalein, 166
Phenols, 10, 14, 16, 92, 95-99
acidity of, 95-99
substituted, 96-99
Phenoxide, 10, 167
Phosphoric acid, 10, 19, 21, 22, 33, 169-179
pH curve, 178
titration curve, 186
Photochlorination, 69
Phthalic acid, 171
Phthalic anhydride, 27
Phthalimide, 105
Pi bond, 73
Polyethylene, 69

Polymerizations, 69
Potash, 6
Potassium, 24
Potassium carbonate, 39
Potassium hydrogen phthalate, 15
Potassium hydroxide, 39
Potassium isopropoxide, 24
Potassium superoxide, 23
Potential energy, 3
Precipitations, pH dependence, 209-213
Propylamine, 24
Proton acceptor, 10
Proton donor, 10
Protonation, 11
Protons, aqueous, 29
Pyridine, 10, 14
Pyridinium, 10

Quadratic equation, 51
Quickline, 6

Radical, 1
Radical ion, 1
Reaction energy, 3
Redox reaction, 2
Resonance, 72-74, 77, 78, 86, 96, 97, 100 101, 104, 106, 110, 115
Reversibility, 9
Royal water, 6

S character, 91
Salts, 124, 125
  amphiprotic, 188
  definition of, 124, 125
Seashells, 6
Selenous acid, 22
Shorthand structural representation, 14, 16
Sidgwick, N. V., 58
Sigma bonds, 73
Soda, 23
Sodamide (sodium amide), 8, 18, 25
  preparation of, 26
Sodium, 24
Sodium bicarbonate, 25
Sodium borohydride, 26
Sodium carbonate, 8, 23, 25
  preparation of, 25
Sodium chloride, 22, 25
Sodium ethylamide, 24
Sodium hydride, 18
Sodium hydrogen carbonate, preparation of, 25
Sodium hydroxide, 23, 39
Sodium methoxide, 26
Sodium oxide, 23
Sodium peroxide, 23
Solubility product, 202-204
Solutions, saturated, 201
Solvation energy, 4
Solvay process, 25
Solvent interactions, 4
Solvent orientation, 13
Solvents, 222
  nonaqueous, 222
  nonpolar, 5
  polar, 4
Sorenson, S. P. L., 35
Spontaneous process, 3
Stability, 3, 4
Stabilization, 5
Substitution, 64-67
  electrophilic, 67, 114
  nucleophilic, 64-67
Successive approximations, 207-209
Sucrose, 21
Sulfate, 76
Sulfides, 211, 212
  precipitation by pH control, 211, 212
Sulfonamides, 105-107
Sulfonic acids, 92, 93, 95, 97
  acidity of, 93-95
Sulfur, 21
Sulfur trioxide, 73-75
Sulfuric acid, 6, 10, 15, 19, 21, 39
  fuming, 21
  protonated, 10
Superoxide, 23

TNT, 25
Terephthalic acid, 171
Tetrahedral geometry, 89, 90

Thermoneutral, 4
Titrations, 148-163, 167
  calculations, 152
  conductometric, 221
  polyprotic acids, 184-187
  potentiometric, 216-219
  strong acid with strong base, 150-153, 156, 157, 161
  weak acid, 159, 160
  weak acid with strong base, 154-156, 158, 166
Titration curves, 148-152
Transition state, 3
Tricyanomethane, 225

Usanovich acid-base theory, 68, 69

Valence shell expansion, 63
Vinegar, 6

Water, 10
  molarity in dilute solutions, 39
  pH of, 47
Wine, 6

Zwitterion (inner salt), 194, 195